300131402E

D0240039

Sodium–Calcium Exchange

Edited by

T. JEFF A. ALLEN
King's College, University of London

DENIS NOBLE
University of Oxford

and

HARALD REUTER
University of Bern

Oxford New York Tokyo
OXFORD UNIVERSITY PRESS
1989

Oxford University Press, Walton Street, Oxford OX2 6DP
Oxford New York Toronto
Delhi Bombay Calcutta Madras Karachi
Petaling Jaya Singapore Hong Kong Tokyo
Nairobi Dar es Salaam Cape Town
Melbourne Auckland
and associated companies in
Berlin Ibadan

Oxford is a trade mark of Oxford University Press

Published in the United States
by Oxford University Press, New York

British Library Cataloguing in Publication Data
Sodium–calcium exchange.
1. Organisms. Cells. Biopolymers
I. Allen, T. Jeff A. II. Noble, D. (Denis)
III. Reuter, Harald
574.87′6042
ISBN 0–19–854735–8

Library of Congress Cataloging in Publication Data
Sodium–calcium exchange / edited by T. Jeff A. Allen, Denis Noble, and Harald Reuter.
Includes bibliographies and index.
1. Calcium channels. 2. Sodium channels. I. Allen, T. Jeff
II. Noble, Denis. III. Reuter, Harald.
QP535.C2S63 1989 591.87′5—dc19 88–38243
ISBN 0–19–854735–8

Set by Latimer Trend & Company Ltd, Plymouth
Printed in Great Britain by
Biddles Ltd, Guildford and King's Lynn

Contents

Contributors

C. C. Aickin
University Department of Pharmacology, South Parks Road, Oxford, UK.

T. Jeff A. Allen
Department of Physiology, King's College London, Strand, London WC2R 2LS, UK.

Mordecai P. Blaustein
Department of Physiology, University of Maryland School of Medicine, 665 West Baltimore Street, Baltimore, Maryland 21201, USA.

Alison F. Brading
University Department of Pharmacology, South Parks Road, Oxford, OX1 3QT, UK.

Reg A. Chapman
Department of Physiology, School of Veterinary Sciences, Park Row, Bristol BS1 5LS, UK.

Reinaldo DiPolo
Centro Biofisica y Bioquimica IVIC, Apartado 21827, Caracas 1020–A, Venezuela.

David A. Eisner
Department of Physiology, University College, Gower Street, London, WC1E 6BT, UK

Maria L. Garcia
Department of Membrane Biochemistry and Biophysics, Merck Sharp and Dohme Research Laboratory, Rahway, New Jersey 07065–0900, USA.

Donald W. Hilgemann
Department of Physiology, University of Texas Health Science Center at Dallas, 5323 Harry Hines Boulevard, Dallas, Texas 75235–9040, USA.

Gregory J. Kaczorowski
Department of Membrane Biochemistry and Biophysics, Merck Sharp and Dohme Research Laboratory, Rahway, New Jersey 07065–0900, USA.

V. Frank King
Department of Membrane Biochemistry and Biophysics, Merck Sharp and Dohme Research Laboratory, Rahway, New Jersey 07065–0900, USA.

Leon Lagnado
Physiological Laboratory, Downing Street, Cambridge, CB2 3EG, UK.

W. J. Lederer
Department of Physiology, University of Maryland School of Medicine, 660 West Redwood Street, Baltimore, Maryland 21201, USA.

Peter A. McNaughton
Physiological Laboratory, Downing Street, Cambridge, CB2 3EG, UK.

Lorin J. Mullins
Department of Biophysics, University of Maryland School of Medicine, Baltimore, Maryland 21201, USA.

Denis Noble
University Laboratory of Physiology, Parks Road, Oxford, OX1 3PT, UK.

Kenneth D. Philipson
Departments of Medical Physiology, American Heart Association, Greater Los Angeles Affiliate, Cardiovascular Research Laboratories, University of California, Los Angeles, California 90024, USA.

Hannah Rahamimoff
Department of Biochemistry, The Hebrew University, Hadassah Medical School, Jerusalem, Israel.

John P. Reeves
Roche Institute of Molecular Biology, Roche Research Centre, Nutley, New Jersey 07110, USA.

Jaime Requena
Centro de Biociencias, Instituto Internacional de Estudios Avanzados (IDEA), Apartado 17606, Caracas 1015A, Venezuela.

Harald Reuter
Department of Pharmacology, University of Bern, 3010 Bern, Switzerland.

Torben Saermark
Laboratory for Protein Chemistry, University of Copenhagen, Sigurdsgade 34, DK 2200 Copenhagen N., Denmark.

Robert S. Slaughter
Department of Membrane Biochemistry and Biophysics, Merck Sharp and Dohme Research Laboratory, Rahway, New Jersey 07065–0900, USA.

Ann Taylor
University Laboratory of Physiology, Parks Road, Oxford, 0X1 3PT, UK.

Introduction

Harald Reuter

The historical roots of scientific developments are usually rather difficult to discover. Major advances and achievements often depend on new techniques, on new thoughts and questions pertinent to a scientific problem, and on ideas 'that have been around'. The discovery of the Na–Ca exchange system is no exception.

It has been known since Ringer's (1883) and Daly and Clark's (1921) classic work that contractility of the frog heart is directly related to the Ca concentration and inversely related to the Na concentration in the bathing solution. Wilbrandt and Koller (1948) investigated these relationships in greater detail and found that contractility of the frog heart depended on the ratio $[Ca^{2+}]:[Na^{+}]^2$ in the Ringer solution. They proposed a theory according to which the distribution of ions between fixed charges at cell surfaces and the surrounding solution follows a Donnan equilibrium. The theory included competition of $2Na^{+}$ and $1Ca^{2+}$ for fixed anionic charges, and application of the law of mass action predicted that occupancy of the anionic sites depends on the ratio $[Ca^{2+}]:[Na^{+}]^2$ in the external medium. However, it is amazing that Wilbrandt and Koller did not hypothesize *how* this interesting quantitative relationship might affect contractility of the heart. In the first paragraph of their paper the authors state explicitly that, 'according to several investigations, the membrane has to be considered as being impermeable for Ca ions'.

A decade later Lüttgau and Niedergerke (1958) addressed this question once more. They confirmed and extended Wilbrandt and Koller's basic experimental findings and suggested: 'Ca may compete with Na for membrane carriers which even in the combined state, say as CaR and Na_2R, bear negative charges and are, therefore, at the resting potential drawn mainly towards the external surface of the membrane. Reducing the membrane potential would lead to inward movement of CaR as well as Na_2R carriers.' A. L. Hodgkin reviewed the paper by Lüttgau and Niedergerke for the *Journal of Physiology* and suggested a quantitative theory which he generously offered for inclusion in the paper. Later, Niedergerke (1963) found that ^{45}Ca uptake was increased both after reduction of the external Na concentration and during depolarization. This agreed with the Lüttgau and Niedergerke hypothesis (see above). However, this scheme of Ca movement in the frog heart did not propose Na–Ca *exchange*, but rather Na–Ca *antagonism*. This *antagonism* could not account for *Na–Ca countertransport* which, several years later, was discovered simultaneously and independently in Plymouth (England) and Mainz (Germany). The finding of countertransport is the basis for the Na–Ca exchange system as we understand it today.

The story of the discovery of Na–Ca exchange in Plymouth began in the autumn of 1966 and has recently been told by M. P. Blaustein (1987). He testifies that Peter Baker was a key player in this story. Although work in Plymouth in the autumn of 1966 was aimed at further exploration of the Na pump in squid axon, it became quickly apparent that an ouabain-insensitive Na efflux from the axon was dependent on external Ca. This was a new finding, incompatible with the operation of the Na pump. Baker *et al.* (1969) pursued this question and not only found Ca uptake that was inversely related to the Na concentration in the medium, but also a component of Ca influx into squid axon that depended on the intracellular Na concentration. A Na-dependent Ca uptake was also found in crab nerve by Baker and Blaustein (1968). Moreover, Blaustein and Hodgkin (1969) in the squid seasons 1966–7 and 1967–8 discovered that Ca efflux from axons depended on the Na concentration in artificial sea-water. This Ca efflux component was particularly prominent in cyanide-poisoned axons. Within a year or so the framework of Na–Ca exchange was established. The Cambridge group had shown ouabain-insensitive Na_o–Na_i exchange, Na_i–Ca_o exchange, and Ca_i–Na_o exchange. Abstracts of their work were published in 1967–8 (Baker *et al.* 1967; Blaustein and Hodgkin 1968) and full papers in 1969 (Baker *et al.* 1969; Blaustein and Hodgkin 1969). Both Baker's and Blaustein's groups have continued to do fundamental work on Na–Ca exchange over the years.

I became interested in the question of Ca extrusion from cardiac cells following the finding that Ca ions contribute directly as charge carriers to the plateau phase of the cardiac action potential (Reuter 1966, 1967). In 1966, I participated in a course on 'Laboratory Techniques in Membrane Biophysics' organized by H. Passow and R. Stämpfli in Homburg/Saar. P. C. Caldwell and R. D. Keynes held a practical class on 'The exchange of ^{22}Na between frog sartorius muscle and the bathing medium'. Based on Ussing's (1947), Keynes and Swan's (1959), and Mullins and Frumento's (1963) results in frog muscle, they convinced us of the existence of an electroneutral Na–Na exchange that occurs without net flux by driving a Na-loaded 'carrier' in the membrane by means of the electrochemical Na gradient. Ussing had called this phenomenon 'exchange diffusion'. I suggested that not only could Na exchange for Na, but also Na for Ca. In contrast to Na–Na exchange, Na–Ca exchange could provide net fluxes of both ions driven by their respective electrochemical gradients. I rushed back to Mainz, where I was working in the Department of Pharmacology and together with a colleague, Norbert Seitz who is now in general practice, I started to do ^{45}Ca efflux experiments in guinea-pig auricles. Soon we discovered that ^{45}Ca efflux was greatly reduced when Ca and Na were eliminated from the bathing medium. As in Niedergerke's (1963) experiments there was a net uptake of Ca if the external Na concentration was reduced. Metabolic inhibitors increased rather than decreased Ca efflux. We divided the total ^{45}Ca efflux into Ca- and Na-activated fractions. There was competition between both fractions of ^{45}Ca efflux depending upon the ratio $[Ca^{2+}]:[Na^{+}]^2$ in the bathing medium. A similar competition was later found in cardiac membrane vesicles by Reeves and Sutko (1983). However, we were misled by this relationship in terms of the true stoichiometry of the exchanger. We proposed a 2Na for 1Ca exchange, while, for energetic reasons, Blaustein and Hodgkin (1969) suggested a higher exchange stoichiometry. Today a 3 to 1 exchange stoichiometry is fairly well-established. On the basis of our experiments we proposed a hetero-exchange diffusion process that controls Na and Ca influx and efflux in cardiac muscle. Our results were first published as abstracts in 1967–8 (Reuter and Seitz 1967, 1968a) and as a full paper in 1968 (Reuter and Seitz 1968b). Slightly later, together with H. G. Glitsch and H. Scholz, we found that ^{45}Ca influx into guinea-pig auricles depended on the intracellular Na concentration (1969, 1970). Thus, we had established Na_o–Ca_i exchange, Na_i–Ca_o exchange, and the competition between Na and Ca for the exchanger in heart muscle.

Since then, as this book shows, the field has greatly expanded. Peter Baker's scientific imagination and his experimental work were of paramount importance for this scientific development. His interest in the field continued until his untimely death on 10 March 1987. The first meeting on Na–Ca exchange, in Stowe in July 1987, was his idea and without his organizational contribution, it would not have taken place. We owe Peter Baker so much and are grateful to him.

References

Baker, P. F. and Blaustein, M. P. (1968). Sodium-dependent uptake of calcium by crab nerve. *Biochim. Biophys. Acta* **150**, 167–70.

Baker, P. F., Blaustein, M. P., Hodgkin, A. L., and Steinhardt, R. A. (1967). The effect of sodium concentration on calcium movements in giant axons. *J. Physiol.* **192**, 43–4P.

Baker, P. F., Blaustein, M. P., Hodgkin, A. L., and Steinhardt, R. A. (1969). The influence of calcium on sodium efflux in squid axons. *J. Physiol.* **200**, 431–58.

Blaustein, M. P. (1987). Citation classic. *Current Contents* **35** (Aug. 31), 14.

Blaustein, M. P. and Hodgkin, A. L. (1968). The effect of cyanide on calcium efflux in squid axons. *J. Physiol.* **198**, 46–8P.

Blaustein, M. P. and Hodgkin, A. L. (1969). The effect of cyanide on the efflux of calcium from squid axons. *J. Physiol.* **200**, 497–527.

Daly, I. de Burgh and Clark, A. Y. (1921). The action of ions upon the frog's heart. *J. Physiol.* **54**, 367–83.

Glitsch, H. G., Reuter, H., and Scholz, H. (1969). The influence of intracellular sodium concentration on calcium influx in isolated guinea pig auricles. *Nauny-Schmiedeberg's Arch. Pharmak.* **264**, 236–7.

Glitsch, H. G., Reuter, H., and Scholz, H. (1970). The effect of the internal sodium concentration on calcium fluxes in isolated guinea-pig auricles. *J. Physiol.* **209**, 25–43.

Keynes, R. D. and Swan, R. C. (1959). The effect of external sodium concentration on the sodium fluxes in frog skeletal muscle. *J. Physiol.* **147**, 591–625.

Lüttgau, H. C. and Niedergerke, R. (1958). The antagonism between Ca and Na ions on the frog's heart. *J. Physiol.* **143**, 486–505.

Mullins, L. Y. and Frumento, A. S. (1963). The concentration dependence of sodium efflux from muscle. *J. Gen. Physiol.* **46**, 629–54.

Niedergerke, R. (1963). Movements of Ca in frog heart ventricles at rest and during contractures. *J. Physiol.* **167**, 515–50.

Reeves, J. P. and Sutko, J. L. (1983). Competitive interactions of sodium and calcium with the sodium-calcium exchange system of cardiac sarcolemmal vesicles. *J. Biol. Chem.* **258**, 3178–82.

Reuter, H. (1966). Strom-Spannungsbeziehungen von Purkinje-Fasern bei verschiedenen extrazellulären Calcium-Konzentrationen und unter Adrenalineinwirkung. *Pflügers Arch.* **287**, 357–67.

Reuter, H. (1967). The dependence of slow inward current in Purkinje fibres on the extracellular calcium concentration. *J. Physiol.* **192**, 479–92.

Reuter, H. and Seitz, N. (1967). Untersuchungen zur Kinetik des ^{45}Ca-Efflux von Meerschweinchenvorhöfen. *Naunyn-Schmiedeberg's Arch. Pharmak.* **257**, 324.

Reuter, H. and Seitz, N. (1968a). The ionic dependence of calcium efflux from guinea pig auricles. *Naunyn-Schmiedeberg's Arch. Pharmak.* **259**, 190.

Reuter, H. and Seitz, N. (1968b). The dependence of calcium efflux from cardiac muscle on temperature and external ion composition. *J. Physiol.* **195**, 451–70.

Ringer, S. (1883). A further contribution regarding the influence of the different constituents of the blood on the contraction of the heart. *J. Physiol.* **4**, 29–47.

Ussing, H. H. (1947). Interpretation of the exchange of radiosodium in isolated muscle. *Nature* **160**, 262–3.

Wilbrandt, W. and Koller, H. (1948). Die Calciumwirkung am Forschherzen als Funktion des Ionengleichgewichts zwischen Zellmembran und Umgebung. *Helv. Physiol. Pharmacol. Acta.* **6**, 208–21.

1 The sodium–calcium exchange in intact cells

Reinaldo DiPolo

1.1 Introduction and general aspects

It seems to be a general characteristic of most living cells that they contain less ionized calcium (Ca_i) than the extracellular medium. Usual figures for invertebrates and mammalian cells indicate that the resting $[Ca^{2+}]_i$ is close to or lower than 10^{-7} M, while the extracellular [Ca] can be up to 10^{-2}–10^{-3} M. Cells use this enormous Ca gradient for the generation of an intracellular Ca signal which is the link between stimuli occurring at the cell surface and intracellular Ca-mediated processes (contraction, secretion, etc.). Calcium signalling (an increase in Ca_i^{2+}) must come from: (i) an increase in Ca influx through voltage-gated channels (Ca channels) and/or Ca entry through the Na–Ca exchange; (ii) Ca release from intracellular stores (sarcoplasmic and endoplasmic reticulum and mitochondria); or (iii) a combination of these. To keep a low Ca_i as well as a constant overall Ca content during resting periods, cells must extrude not only the Ca that enters during activity, but also that which leaks continuously into the cell. This job is accomplished by powerful Ca extrusion mechanisms located at the plasma membrane.

In excitable cells the two widely acceptable mechanisms that extrude Ca are the Na–Ca exchange and the ATP-driven Ca pump (DiPolo 1978; Caroni and Carafoli 1980). Although their relative contributions to Ca homeostasis depend on the conditions in which the cell is at a particular stage, the overview is that Na–Ca exchanger is a low-affinity, high-capacity Ca pumping system designed to extrude Ca effectively whenever its concentration reaches the micromolar range. Under these conditions, its low affinity will not be rate limiting and its high rate of pumping could be of special importance as in the case of cardiac relaxation. On the other hand, the ATP driven Ca pump can be defined as a high-affinity low-capacity Ca pumping system able to maintain the Ca_i at rest. Since the subject matter of this book is Na–Ca exchange, the reader is referred to several reviews on the parallel operation and physiological relevance of these two transport mechanisms (DiPolo and Beaugé 1980; Schatzmann 1985; Reeves 1985; Carafoli 1987).

Experimental evidence suggestive of the presence of a Na–Ca carrier-mediated mechanism came from cardiac muscle cells as early as 1957 (Lüttgau and Niedergerke 1958). The actual discovery of the Na–Ca exchange emerged in 1967 from the work of Baker *et al.* (1967) and Reuter and Seitz (1967). These two independent groups, working on squid axon and cardiac muscle, respectively, found that the exit of Ca was highly dependent on the presence of Na_o. Furthermore, the influx of Ca was enhanced by removal of Na_o or by increasing the Na_i, thus indicating the presence of a reversible carrier system able to couple the movement of Na for Ca across the membrane in either direction. These early pivotal experiments, as well as those of Blaustein and Hodgkin (1969) led to the concept of an ion-transport mechanism thought to be important in the transfer of Ca and Na ions into or out of excitable cells. The participation of this mechanism in a variety of physiological processes, including cardiac excitability (Mullins 1981) and photoreception (McNaughton, Nunn, and Hodgkin, 1986), has led to several studies on this transport system including: kinetics of the transported ions, electrogenicity, regulation, and biochemical characterization. The Na–Ca exchange has been studied in a wide variety of preparations including cardiac muscles, squid axons, barnacle muscle, brain synaptosomes, and plasma membrane vesicles from different sources. Although the exchanger seems to be most active in heart and nerve cells (Carafoli 1987; Schatzmann 1985), it is also found in non-excitable tissues like epithelial cells (Lee, Taylor, and Windhager 1980), endocrine tissues (Herchuelz, Sener, and Malaise 1980), bone cells (Krieger and Tashigian 1980), and dog and ferret erythrocytes (Parker 1978; Altamirano and Beaugé 1985; Milanick and Hoffman 1986). Giant nerve and muscle cells have been extensively used to study transport phenomena because they can give information that may be difficult to obtain from multicellular systems. This is based on the fact that ionic control of both intra- and extracellular medium as well as of the membrane potential, can be properly achieved by either internal perfusion (Baker, Hodgkin, and Shaw 1962) or intracellular dialysis (Brinley and Mullins 1967). In this article I will deal mainly with the results obtained from squid axons and barnacle muscle fibres assuming that Na–Ca exchange in these giant cells are homologous to those of other tissues and thus could provide information to understand the functioning of the exchanger in complex multicellular tissues. I will focus primarily on the kinetics, regulation, and interactions of the exchanger with different physiological ligands.

Quantitative kinetic studies of Ca transport are based on the assumption that the concentration of solutes, and specially of Ca ions, are known at both surfaces of cell membrane. In multicellular preparations it is difficult to obtain reliable kinetic data on Ca fluxes due to intra- and extracellular Ca-binding sites and Ca sequestration and release by internal organelles. Although some of the technical difficulties can be overcome in large single cells (squid, barnacle), there are some problems that should be considered. Cytoplasmic Ca uptake is a particularly acute problem when measuring Ca fluxes by the intracellular dialysis technique. This is specially important for Ca influx experiments, since all the Ca that enters the cell must be collected by the porous dialysis capillary. Although the energy-dependent Ca-sequestering systems can be turned off either by eliminating their natural substrates or by adding suitable inhibitors to the dialysis medium (cyanide, FCCP), the energy-independent systems must be neutralized with an exogenous Ca-sequestering agent (DiPolo 1979; DiPolo and Beaugé 1980; Rasgado-Flores and Blaustein 1981). This exogenous Ca buffer should have affinity and permeability characteristics (high permeability through the porous capillary) that enables a cytoplasmic concentration sufficient to overcome the endogenous Ca buffers (Baker and DiPolo 1984). Furthermore, it should be non-toxic to the cell. EGTA meets all these requirements for the measurement of steady-state unidirectional Ca fluxes in squid axons. Thus: (i) under a continuous dialysis experiment (1–3 mM total EGTA, and about 300 μM free EGTA), the influx of Ca measured at rest or following a trend of action potentials is not different from that obtained in a non-treated axon (DiPolo and Beaugé 1980); (ii) in the absence of EGTA in the dialysis fluid, the measured Ca influx is 5–10 times smaller than the anticipated one. However, addition of sufficient free $EGTA_i$ increases the influx to the expected value since further increases in free EGTA (at a constant Ca_i concentration) causes no variations in the level of the Ca influx (DiPolo 1979; DiPolo and Beaugé 1980). The use of exogenous buffer brings about an additional problem since if Ca–EGTA can escape across the plasma membrane this 'leak' could contribute to the total Ca efflux. Fortunately, in squid axons this leak is quite small (5 $fmol\ cm^{-2}\ s^{-1}$ per 1 mM EGTA) and insensitive to changes in Na_i, Na_o, Ca_o, and ATP (DiPolo and Beaugé 1984).

The binding of Ca ions is not restricted to the cytoplasm. Experiments by Baker and McNaughton (1978) in squid axons revealed the existence of an extracellular Ca binding matrix which can distort the measurement of Ca efflux especially under conditions of low external Ca. However, once recognized, this potential source of error can be overcome either by avoiding large changes in the external Ca, or by including adequate $[EGTA]_o$ in the experimental solutions. In the case of the barnacle muscles, the situation becomes more complex since their large diameter and the presence of profound membrane invaginations makes the control of Ca^{2+} difficult (Russell and Blaustein 1975). Nevertheless, in the presence of high concentrations of $EGTA_i$ and using the perfusion instead of the dialysis technique it has been possible to overcome this difficulty (Rasgado-Flores and Blaustein 1981; Lederer and Nelson 1983).

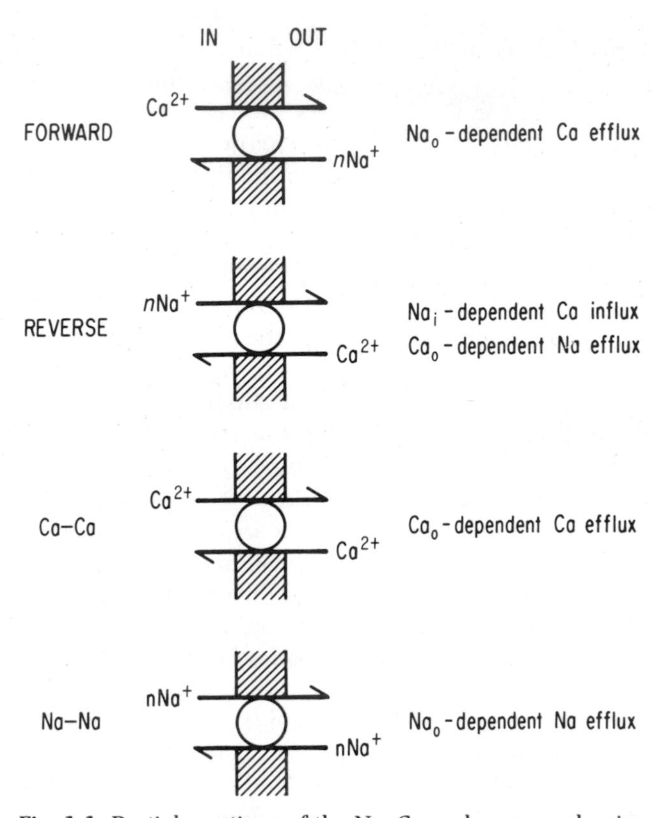

Fig. 1.1. Partial reactions of the Na–Ca exchange mechanism.

1.2 Modes of operation—kinetics

As illustrated in Fig. 1.1 the Na–Ca exchange can be operationally divided into four different modes: (i) forward, in which Ca efflux is coupled to Na influx; (ii) reverse, in which Ca influx is coupled to Na efflux; (iii) calcium–calcium, in which there is an homologous exchange of Ca ions; and (iv) sodium–sodium, in which there is an homologous exchange of Na ions.

1.2.1 Forward exchange (Na_o–Ca_i)

Na_o-dependent Ca efflux from unpoisoned injected squid axons or from axons dialysed with ATP and containing resting levels of Ca_i^{2+} amounts to about 10–20 per cent of the total Ca efflux. The rest occurs via an ATP-dependent Na_o-independent component (uncoupled Ca efflux; Ca pump) (DiPolo and Beaugé 1979; Baker and Singh 1982). In barnacle muscle the same seems to be true since changing extracellular Na has little effect on Ca efflux when

Ca_i^{2+} is in the low physiological range (Rasgado-Flores and Blaustein 1981; Lederer and Nelson 1983). In both preparations as Ca_i^{2+} is raised, the Na_o-dependent Ca efflux is largely increased to a value more than ten times that of the Ca pump flux at saturating Ca_i concentrations (see Fig. 1.2). In squid axons, the $K_{\frac{1}{2}}$ for the activation of the Na_o-dependent Ca efflux by Ca_i is 1–3 μM in the presence of ATP, and 10–15 μM in its absence, in contrast to a value of 0.2 μM for the ATP-dependent uncoupled Ca efflux through the Ca pump. The apparent low affinity of the exchanger for Ca_i seems to be also a general feature of several other preparations (Reeves 1985; Lüttgau and Niedergerke 1958). The apparent external K_{Na} (in squid axons) is between 50–80 mM in the presence of ATP and 110–140 mM in ATP-depleted axons (see Fig. 1.3). The curve relating the extent of activation to Na_o (Baker and Glitsch 1973; DiPolo 1974; Blaustein 1977) approximates a section of a rectangular hyperbola in ATP-containing axons although close inspection

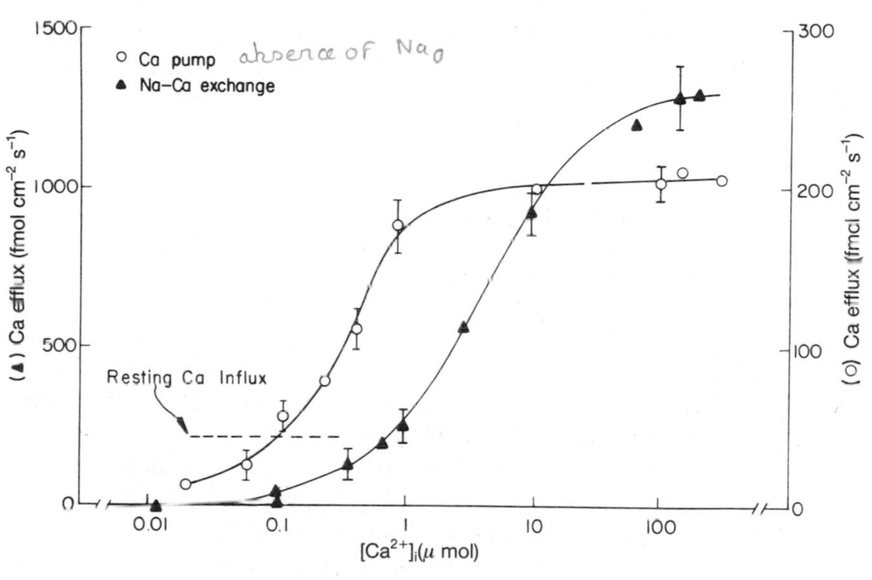

Fig. 1.2. The activation of the ATP-dependent uncoupled (Ca pump flux) and Na-dependent Ca efflux by Ca_i^{2+} in dialysed squid axons. Ordinate: Ca efflux in fmol $cm^{-2}s^{-1}$ (○) in the absence of external Na (Na_o substituted for $Tris_o$). (▲) in the presence of external Na. Notice the change in the ordinate scale for the Na–Ca exchange (▲) and the Ca pump (○) components. Axons were dialysed with internal solutions containing 310–350 mM K^+, 30 mM Na^+, 3–4 mM Mg^{2+}, 310–350 mM $aspartate^-$, and 2 mM ATP. Glycine was added to control the osmolarity, and EGTA (1–2 mM) to control the Ca_i. The horizontal dashed line corresponds to the value of the resting Ca influx. Its intersection at about 0.1 μM with the Ca efflux indicates that the passive Ca influx under resting conditions (external: 440 mM Na^+, 10 mM K^+, 3 mM Ca^{2+}, 50 mM Mg^{2+}, 556 mM Cl^-) is mostly balanced by the ATP-dependent uncoupled Ca efflux.

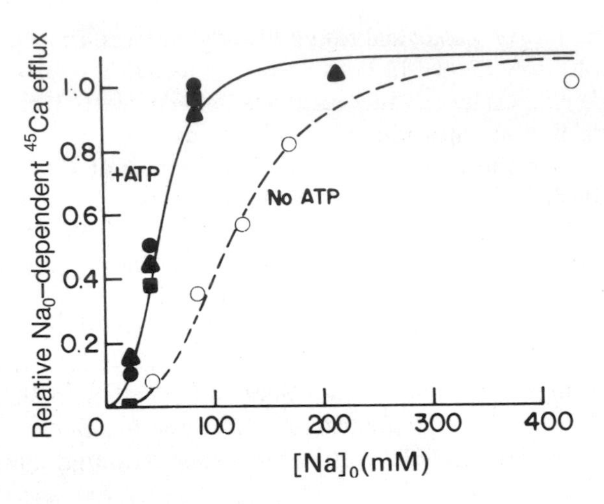

Fig. 1.3. The Na_0-dependent Ca efflux graphed as a function of the external Na concentration. Filled symbols, in the absence of ATP. Open symbols, in the presence of 4 mM ATP. Notice the increase in Na_o affinity induced by ATP. (Data from Blaustein 1977.)

reveals it to be slightly sigmoidal. This sigmoidicity becomes clearer in poisoned axons or in axons dialysed free of ATP (see Fig. 1.3). In internally perfused barnacle muscle fibres, the $K_{\frac{1}{2}}$ for Na_o is also in the millimolar range (about 60 mM). The activation curve is sigmoidal with a Hill coefficient of 2.6 (Lederer and Nelson 1983). One noticeable property of the Na_o-dependent Ca efflux is its high selectivity to Na_o ions, since in the absence of Ca_o (to eliminate Ca–Ca exchange) no other monovalent cation (Li, K, Rb, Cs) can replace Na_o (Blaustein 1977). The lack of effect of Ca_i on the affinity of the external Na site speaks against a mobile carrier, and favours a model in which the exchanger is simultaneously loaded with Na and Ca in order for translocation to occur (Blaustein 1977; Blaustein and Russell 1975).

Raising Na_i inhibits the Na_o-dependent Ca efflux in squid axons (Blaustein 1977; Requena 1978). The nature of this inhibition remains a controversial point. The experiments of Blaustein and Russell (1975) in dialysed axons are consistent with competition when the data is fitted to a square-law relationship between $[Na_i]$ and Na_o-dependent Ca efflux. Their results are compatible with competition only over a narrow range of Ca_i (6–10 μM). However, in the same preparation, Brinley (1975) and Requena (1978) have reported non-competitive inhibition between 0.03 and 300 μM Ca_i. It appears that the effects of Na_i on the Na–Ca exchange are far more complex than a simple competition between Na_i and Ca_i since this ion greatly affects other exchange parameters such as ATP activation, (see Section 1.3.2 and DiPolo and Beaugé 1984) and voltage sensitivity (DiPolo *et al.* 1985).

1.2.2 Reverse exchange (Na_i–Ca_o)

This operational mode has been explored by measuring either the Na_i-dependent Ca influx or the Ca_o-dependent Na efflux. Data obtained in different nerve cells (crab, Baker and Blaustein 1968; myxicola, Sjodin and Abercrombie 1978; and squid, DiPolo and Beaugé 1983) show that Ca_o activates the influx of Ca and the efflux of Na with low apparent affinity (millimolar range), and that activation by Ca_o is competitively inhibited by Na_o. In squid axons isosmotic replacement of Tris, choline, or glucose by Li or other external monovalent cations (K, Rb, Cs) not only increases the magnitude of Ca_o-dependent Na efflux (see Fig. 1.4), but also the affinity for Ca_o (see Fig. 1.4a and 1.4c). External Na at low concentrations activates the Ca_o-dependent Na efflux whereas at higher concentrations it inhibits the efflux (Fig. 1.4b). These findings have been interpreted on two bases: first, that Na_o and Ca_o competes for the external Ca binding site; and second, to the presence of an activatory external monovalent cation site (Baker and Blaustein 1968; Baker *et al.* 1967; Blaustein 1977; Allen and Baker 1986a). These results contrast with the absence of activation of the forward Na–Ca by other monovalent cations beside Na_o. In fact, the stimulation of the forward Na–Ca exchange by K^+_o (depolarization) is of an electrical nature since it can be removed by a similar electrical hyperpolarization (Allen and Baker 1986b). Therefore, the Na_o-dependent Ca efflux does not seem to display a comparable monovalent activatory external site as the reverse mode (Allen and Baker 1986a).

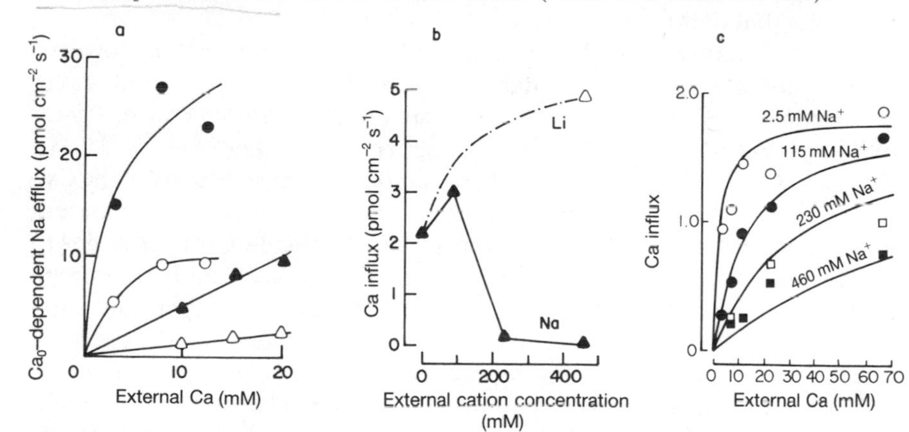

Fig. 1.4. (a) Ca_o-dependence of the Ca_o-activated Na efflux (reverse Na–Ca exchange) in choline (▲, △) and Li (●, △) sea-water. Open symbols, axon with low internal Na. Filled symbols, axon injected with Na_i. Notice the increase in affinity for Ca_o in the presence of an external monovalent cation. (Data from Allen and Baker, 1986a.) (b) Effect of external Na (▲) and Li (△) ions on Ca influx in squid giant axons. Observe the activation of the reversal at low Na_o and its inhibition at high Na_o (from Baker 1970.) (c) Ca influx as a function of Ca_o in intact crab nerve fibres. Na_o was replaced isosmotically by Li_o. (From Baker and Blaustein 1968).

The dependence of Ca influx and Ca_o-stimulated Na efflux on Na_i has been determined in squid (Baker 1978; DiPolo 1979) and barnacle muscle fibres (DiPolo 1973) in which activation seems to be proportional to Na_i. In dialysed squid axons Na_i stimulates Ca influx along a saturating curve with a $K_{\frac{1}{2}}$ between 40–60 mM (DiPolo 1979).

1.2.3 Calcium–calcium exchange (Ca_o–Ca_i)

Although there is no conclusive evidence that the same carrier system is involved in both Na_o–Ca_i and Ca_o–Ca_i exchanges, there is strong experimental evidence that they share the same carrier system. One important argument is that the Na_o-dependent and Ca_o-dependent Ca effluxes are not equal under conditions optimal for both fluxes to occur, thus implying that the two fluxes are not independent (Blaustein 1977). Moreover, the affinity of this component for Ca_i is remarkably similar for both modes of operation (Blaustein 1977). In squid, as well as in barnacle, the replacement of Na_o for Li_o greatly increases Ca efflux in a Ca-dependent manner (Blaustein 1977, Nelson and Blaustein 1981). Under low ionized Ca_i the magnitude of this Ca–Ca exchange component is negligible. When the Ca–Ca exchange is maximally activated (high Ca_i and presence of an external monovalent cation), Ca_o stimulates in a hyperbolic fashion with a K_m of about 3 mM (Baker 1978). The activation of Ca–Ca exchange by external monovalent cations closely resembles that of the reverse Na–Ca exchange.

It is still a matter of debate whether the stimulating external monovalent cations are also transported during the exchange process, and direct measurements of Rb uptake in cardiac sarcolemma vesicles during Ca–Ca exchange indicate that it is not (Slaughter, Sutko, and Reeves 1983). Recently, Allen and Baker (1986a), based on the voltage sensitivity of the Ca–Ca exchange in squid axons, suggested that the chemically activating monovalent cation could indeed be transported. The binding of Li or K ions to the external activatory site could provide the carrier with an extra charge that makes it sensitive to membrane potential. Against this hypothesis is the finding that in dialysed squid axons a similar Ca_o-dependent Ca efflux voltage sensitivity is found in the complete absence of monovalent cations on either side of the membrane (R. DiPolo and L. Beaugé, unpublished experiments). Although more information is needed, it is feasible that the carrier molecule might be directly affected by the electric field across the membrane.

1.2.4 Sodium–sodium exchange (Na_o–Na_i)

An homologous exchange of Na ions through Na–Ca countertransport system was first demonstrated in cardiac membrane vesicles (Reeves and Sutko 1979). Evidence for a similar process in intact cells came from recent experiments in dialysed squid axons based on the finding that Ca_i^{2+} is required for the operation of all of the partial reactions of the exchanger (see Section 1.3.1, and DiPolo 1979; DiPolo and Beaugé 1987a,b). The results can be summarized as follows:

1. Internal Ca^{2+} activates a Na_o-dependent Na efflux (Na–Na exchange) with a $K_{\frac{1}{2}}$ of about 10 µM in the absence of ATP, a value which is almost identical to the Ca_i activation of the forward, reverse, and Ca–Ca exchange (see Fig. 1.5c).
2. The Ca_i-activated Na–Na exchange is highly stimulated by internal alkalinity and inhibited by Mg_i, a feature common to the forward Na–Ca exchange (Baker and DiPolo 1984; DiPolo and Beaugé 1984; DiPolo and Beaugé 1987b).
3. External Na activates a Na_o-dependent Na efflux with a similar K_{Na} to that of the Na_o-dependent Ca efflux. In addition, the selectivity to Na_o is absolute since Li or K are not substitutes.

The fact that this mode of exchange will occur whenever Na_o and Na_i ions are present at both sides of the membrane and Ca_i is in the micromolar range should be considered when measuring the stoichiometry of the Na–Ca exchange through unidirectional flux measurements; otherwise the ratio of Na ions transported per Ca ions moved would be overestimated.

1.3 Regulation of the Na–Ca exchange

1.3.1 Ca_i, an essential activator of the Na–Ca exchange

In its simple conception the Na–Ca exchange can be thought of as a carrier-mediated process in which the transport of Ca is coupled to the reciprocal movement of Na ions, and that, depending on the magnitude and direction of the Na electrochemical gradient, the system can induce net movements of Ca ions into or out of the cell. Models for the operation of the exchange system are based on the assumption that a high degree of coupling and symmetry exist with respect to the ions being transported (Na and Ca). At the *trans* membrane side Na ions are required to bring about Ca translocation and vice versa. At the *cis* side Na and Ca act as competitive inhibitors (Mullins 1977; Wong and Bassingthwaighte 1981).

One interesting finding in dialysed squid axons, is that even in the presence of saturating concentrations of Ca_o and Na_i no reverse Na–Ca exchange (Ca_o-dependent Na efflux or Na_i-dependent Ca influx) can be detected unless micromolar amounts of $Ca^{2+}{}_i$ are present (see Fig. 1.5a and 1.5b; DiPolo 1979; DiPolo and Beaugé 1986). This Ca_i requirement, has also been recently extended to the Na_o-dependent Na efflux component of the exchanger (Dipolo and Beaugé 1987a). The results of Fig 1.5b and 1.5c show that Ca_i^{2+} activates the reverse and Na–Na exchange with a rather similar K_{Ca} (10 µM)

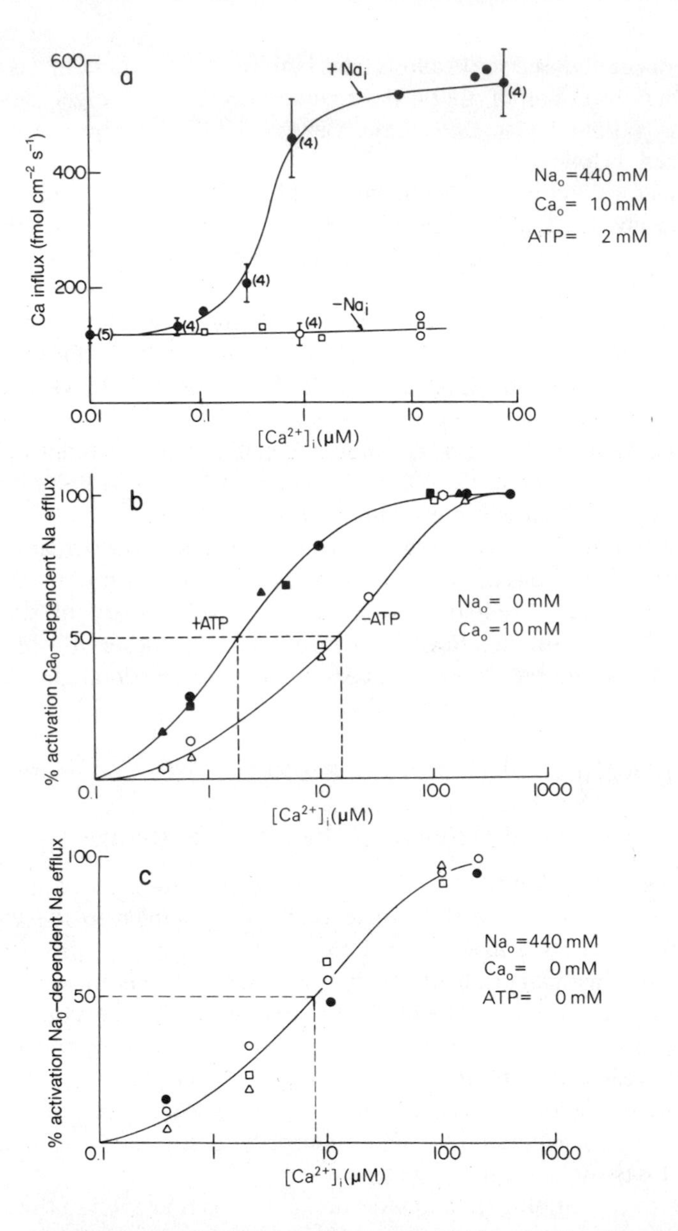

Fig. 1.5. (a) The effect of Ca_i^{2+} on the Ca influx in the absence and in the present of Na_i. In this experiment the concentration, Na_o was 440 mM and Ca_o 10 mM. The dialysis medium contained 2 mM ATP. (From DiPolo, Rojas, and Beaugé 1982.) (b) The effect of Ca_i^{2+} on the Ca_o-dependent Na efflux in the presence and in the absence of ATP in the dialysis fluid. External Na was substituted by Tris. (From DiPolo and Beaugé 1987a.) (c) The effect of Ca_i^{2+} on the Na_o-dependent Na efflux in the absence of internal ATP. External Ca was replaced by Mg ions. (From DiPolo and Beaugé 1987a.)

in the absence of ATP. This is in agreement with the activation of the forward Na–Ca and Ca–Ca exchange by Ca_i. Recently, experiments in perfused barnacle muscle fibres under conditions of minimal Ca–Ca and Na–Na exchange have confirmed the observation that intracellular Ca^{2+} promotes both Ca_o-dependent Na efflux and Na_i-dependent Ca influx (Rasgado-Flores and Blaustein 1981).

It is unknown why Ca_i^{2+} is needed for the activation of Ca influx and Na efflux mediated by the Na/Ca exchange. Nevertheless, it could certainly explain the inhibition of the Ca_o-dependent Na efflux observed by Baker (1970) upon injecting EGTA into squid axons as well as the inhibition of the reverse exchange caused by internal quin-2, a Ca chelating agent (Allen and Baker 1985). Of particular significance in this respect is the observation in Na loaded myocytes and squid axons of an outward current, caused by the Na–Ca exchange working in its reverse mode, that requires the presence of Ca_i^{2+} (Kimura, Miyamae, and Noma 1987; DiPolo, Caputo, and Bezanilla 1987). If the dependence of the reverse Na–Ca exchange on Ca_i^{2+} is a general property of the exchanger, then its concentration at the membrane level might regulate the magnitude of Ca entry. Supporting this possibility is the observation that for a given K_o depolarization the increase in Ca influx through the exchanger is much greater at high than at low Ca_i^{2+} concentration (DiPolo, Rojas, and Beaugé 1982).

1.3.2 The effect of ATP

One important problem that has become apparent from Ca transport studies in intact cells is the role of ATP in the modulation of the activity of the Na–Ca exchange. Although not essential for its operation, ATP induces marked modifications in the functioning of the exchanger. The most significant findings derived from experiments in injected and dialysed squid axons can be summarized as follows:

1. ATP stimulates the Na–Ca exchanger in all its operational modes (forward, reverse, Ca–Ca and Na–Na) (Baker and Glitsch 1973; DiPolo 1974, Blaustein 1977; DiPolo and Beaugé 1987a). Activation is larger at Ca_i^{2+} ranging from 1–10 μM, being considerably reduced at exceedingly high Ca_i^{2+} concentrations (>200 μM) (Blaustein 1977; DiPolo 1977).
2. ATP stimulation is exerted with low affinity. The half-saturating concentration is approximately 250 μM. This value is ten time higher than that required to activate the ATP-dependent Ca pump component, (DiPolo 1974; DiPolo and Beaugé 1979).
3. The MgATP complex is strictly required for activation (DiPolo and Beaugé 1984).
4. The nucleotide site is highly selective to ATP. ADP, GTP, UTP, ITP, AMP, c-AMP, acetylphosphate, and phosphoarginine are all ineffective (DiPolo

1976). The hydrolysable ATP analogues AMp-Cpp and 2-deoxy-ATP can mimic the effect of ATP although with lower affinity. The non-hydrolysable ATP analogue AMp-pCp competes with ATP being ineffective in activating the exchanger (DiPolo 1977).

5. ATP modified the kinetics of activation of the Na–Ca exchanger by increasing the affinity of the transport and 'regulatory' sites towards Na and Ca ions (Blaustein 1977; Baker and DiPolo 1984; DiPolo and Beaugé 1987b) (see Fig. 1.3 and 1.5b). In unpoisoned axons or in axons dialysed in the presence of ATP, half-maximal activation of the Na_o-dependent Ca efflux requires 40–60 mM Na_o. In poisoned axons or in axons dialysed without ATP the curve relating the extent of activation becomes very sigmoidal, and half-maximal activation may increase to 120–150 mM (see Fig. 1.3). The Ca_i^{2+} concentration required for half-maximal activation of the Na_o-dependent Ca efflux is also modified by ATP, being 1–3 μM in the presence of the nucleotide and about 10–15 μM in its absence (Blaustein 1977; DiPolo and Beaugé 1980). Recent experiments in squid axons on the activation of the reverse Na–Ca exchange by Ca_i show that ATP substantially modifies the curve relating Ca_o-dependent Na efflux to Ca_i^{2+}. The half-maximal activation increases from about 1.5 μM in the presence of ATP to 15 μM in the absence of ATP (see Fig. 1.5b).

A puzzling observation concerning the effect of ATP is the lack of stimulation of the Na_o-dependent Ca efflux in the absence of Na_i. This effect occurs even at unsaturating values of Ca efflux (DiPolo and Beaugé 1984). A possible explanation is that since Na_i inhibits Na_o-dependent Ca efflux, ATP could reduce Na_i inhibition (Requena 1978). A problem with this interpretation is that, as expected, Na_i does not seem to change the $K_{\frac{1}{2}}$ for ATP (DiPolo and Beaugé 1984). Further experiments are required to settle this point.

A stimulation of the Na–Ca exchange by ATP has also been shown in different experimental preparations including barnacle muscle (DiPolo and Caputo 1977; Nelson and Blaustein, 1981), guinea pig atria (Jundt and Reuter 1977), and isolated adult rat heart cells (Haworth *et al.* 1987). Some considerations should be devoted to the effect of ATP in barnacle muscle. Although ATP activates the Na_o-dependent Ca efflux at low Ca_i^{2+}, Nelson and Blaustein (1981) reported that at high Ca_i ATP depletion markedly decreases Ca_o-dependent Ca efflux into Li sea-water. This has been explained as if ATP inhibits the turnover of the carrier in the Ca–Ca exchange modality thus shifting the carrier from Ca–Ca to Na–Ca exchange. Such an effect has not been found in squid axons. On the contrary, ATP as well as the ATP anologue adenosine 5′-*O*-(3-thiotriphosphate) ([γ-S]ATP) causes activation of the Ca–Ca exchange (DiPolo and Beaugé, unpublished information). Squid axons and barnacle muscles also differ in their response to internal vanadate. In the former, orthovanadate, a potent inhibitor of the Ca pump has no effect on Na–Ca exchange in the absence of ATP but it activates the exchanger in its presence (DiPolo and Beaugé 1981). In barnacle fibres, internal vanadate inhibits Na_o-dependent Ca efflux at low Ca_i but not at high Ca_i (Nelson and Blaustein 1981).

Since the discovery of the ATP effect on the Na–Ca exchange, the mechanism by which ATP regulates this process has been a subject of debate. The early finding that ATP activates the forward exchange with low apparent affinity (app $K_{ATP} = 200–250\ \mu M$) suggests that ATP may activate allosterically, by binding to a low-affinity site on the exchanger (Post, Hegyvary, and Kume 1972; Simons 1975; Beaugé and Glynn 1980). Nevertheless, the observations that only hydrolysable ATP analogues are active and that Mg_i is a requirement for activation strongly suggest a phosphorylation step during ATP stimulation. To examine further this possibility we have explored in squid axons the effect of the ATP analogue, [γ-S]ATP on the modes of operations of the Na–Ca exchange (DiPolo and Beaugé 1987b). This analogue of ATP can act as substrate for kinases but not for ATPases. The thiophosphoester induced by this analogue is slowly hydrolysed by protein phosphatases (Sherry *et al.* 1978; Cassidy, Hoar, and Kerrick 1979). The data can be summarized as follows:

1. All the modes of operation of the Na–Ca exchanger are activated by this ATP analogue. Activation is substantially greater than that of ATP alone, in agreement with a slower dephosphorylation of the exchanger in the presence of the analogue.
2. Activation requires the presence of Ca_i^{2+}.
3. Activation is on the Na–Ca exchange mechanism since the Na–K pump, Ca pump and ATP-dependent-Ca_i-independent Na–Na exchange are unaffected.

The above results, obtained in whole nerve cells, constitute evidence that the effects of ATP on the exchanger are most likely to be mediated by a Ca_i^{2+}-dependent protein kinase responsible either for a direct phosphorylation of the exchanger, or to another protein which could then regulate the carrier. Caroni and Carafoli (1983) have studied the mechanism of ATP activation of the Na–Ca exchange in heart sarcolemma. Their experiments show that ATP activation (phosphorylation process) is mediated by a Ca^{2+} and calmodulin-dependent protein kinase. Deactivation of the activated exchanger (dephosphorylation) is catalysed by a membrane bound phosphatase also dependent on Ca_i and calmodulin (Caroni and Carafoli 1983). Although the evidence that the Na–Ca exchanger is regulated by a phosphorylation–dephosphorylation mechanism are quite firm, the possible physiological meaning of this regulation is still poorly understood. A fundamental question which has not been answered yet, concerns the way in which the concentrations of Ca_i^{2+} attained during a physiological response, control the degree of phosphorylation and hence the activity of the exchanger.

1.4 Asymmetric ligand interactions of the Na–Ca exchange

In recent years, a vast amount of literature has evolved on the characterization of the Na–Ca exchanger in intact cells and in isolated membrane preparations. The picture that emerged is that the Na–Ca exchange is a complex carrier-mediated system far from a simple symmetrical transport process. The purpose of this section is to provide information on the asymmetric properties of the Na–Ca exchange in a particular cell, the squid axon. The term asymmetry is used to denote: (i) the existence of specific sites present only on one side of the exchanger (examples of this are the already described nucleotide (ATP) and Ca_i effects); or (ii) to denote differences in the affinity of a given site (Ca site) on either side of the membrane.

1.4.1 Monovalent cations

If Li ions are used as external Na-substitute instead of Tris, or choline, both the Ca_o-dependent Na efflux (also Na_i-dependent Ca influx) and Ca_o-dependent Ca efflux are markedly activated (Blaustein 1977). As mentioned above, the presence of an external monovalent cation that activates the 'Ca arm' of the Na–Ca exchange, contrasts with the lack of effect of monovalent cations on the Na_o-dependent Ca efflux and Na_o-dependent Na efflux ('Na arm').

Is there a similar monovalent cation site on the inner side of the exchanger? Blaustein (1977) found that replacement of K_i with tetramethylammonium ions inhibit Na_o-dependent Ca efflux by only 21 per cent as compared to 85 per cent inhibition of the Ca_o-dependent Ca efflux. This feature of the Ca–Ca exchange (i.e., being activated by external and internal alkali metal ions) has led to the conclusion that *both* external cation sites need to be loaded simultaneously in order for the exchange to proceed (Blaustein 1977). In recent experiments in dialysed squid axons, we have found that under voltage-clamp conditions replacement of K_i with Tris causes a much larger inhibition (about 70 per cent) of the Na_o-dependent Ca efflux. Furthermore, internal K ions *per se* (chemical effect), activate ($K_{\frac{1}{2}} = 90$ mM) all the modes of operation of the Na–Ca exchange (DiPolo and Rojas 1984). This differs from the fact that activation by external monovalent cations is only on the Ca_o-dependent fluxes. We have further examined the possible asymmetry of the monovalent cation site by studying the effect of Li ions on the activation of the Ca_o-dependent Ca efflux under conditions in which both intra- and extracellular solutions resemble each other. This approach minimizes possible changes in the carrier molecule which might be induced by the normally different intra- and extracellular environments. The result of these types of experiments (see Fig. 1.6) indicates:

1. In the complete absence of internal monovalent cations (Tris substitution), external Li promotes a large Ca_o-dependent Ca efflux.
2. In the complete absence of external monovalent cations (Tris substitution), internal Li has very little effect on the Ca_o-dependent Ca efflux.
3. In the presence of Li ions (similar concentrations) on both sides of the membrane the magnitude of the Ca_o-dependent Ca efflux is greater than in any of the above conditions.
4. Even if the absence of monovalent cations on either side of the membrane, a Ca_o-dependent Ca efflux, although much reduced in size, can be demonstrated. The data suggest an asymmetry in the activation of the Na–Ca exchange by monovalent cations. Although internal Li (as well as K_i ions) exert a stimulating effect on Ca–Ca exchange, this effect only occurs in the presence of Li_o thus indicating that the monovalent activated site faces the external medium.

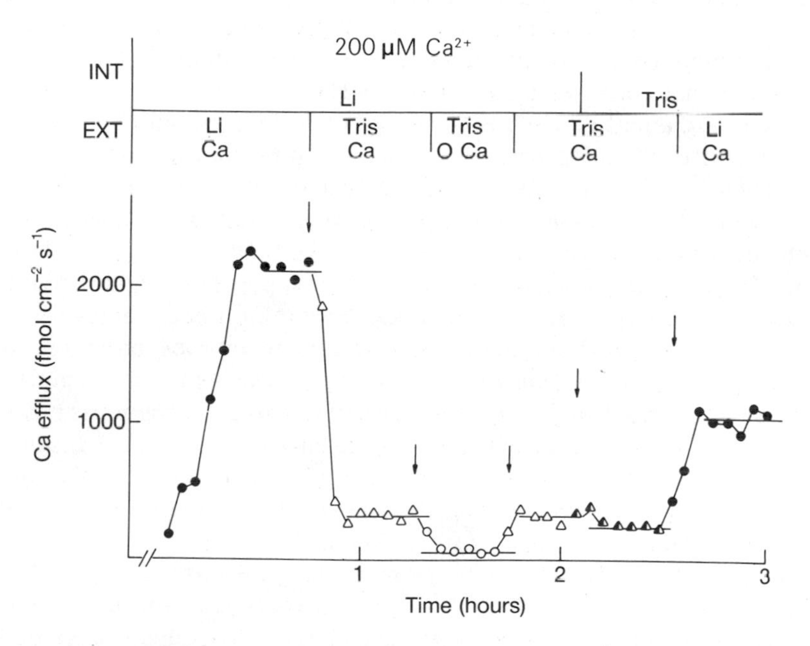

Fig. 1.6. The effect of Li_o and Li_i on the Ca_o-dependent Ca efflux. Ordinate: Ca efflux in fmol $cm^{-2}s^{-1}$. Abscisa: time in hours. The internal and external solutions contained (mM): Tris-Mops, 200; Tris-Cl or LiCl, 200; $MgCl_2$, 20; EGTA, 1. Glycine was used to control the osmolarity. $[Ca^{2+}]_i$ and $[Ca^{2+}]_o$ were 0.2 and 20 mM respectively. The axon was voltage clamped at zero membrane potential. Notice the small effect of Li_i on Ca_o-dependent Ca efflux in the absence of Li_o as compared with that in its presence (from R. DiPolo and L. Beaugé, unpublished.)

1.4.2 Divalent cations

One of the noticeable asymmetric features of the Na–Ca exchange found in intact cells is that the apparent affinity for Ca^{2+} appears to be different on the two sides of the membrane. The high-affinity Ca binding site is located at the inner side of the carrier and the external site exhibits a low affinity (Blaustein 1977; Baker 1978).

In contrast to these findings, work done in membrane vesicles and reconstituted preparations suggests that the Na–Ca exchange system is symmetric with respect to its affinity for Ca ions. The K_m for Ca uptake in vesicles selectively loaded with the Na–K pump (inside-out subpopulation) can not be distinguished from the K_m of the entire population, thus suggesting that the Ca binding sites on the two faces of the exchanger are symmetrical (Philipson and Nishimoto 1982). It has been argued that the existence of unstirred layers in intact cells could be responsible for the observed differences in K_m for Ca_i and Ca_o (Reeves 1985). Although this is a factor to be considered when studying ionic fluxes in multicellular preparations, this does not seem to be the case in squid axons in which Ca fluxes are measured under well-controlled ionic conditions (using adequate concentrations of Ca chelating buffers and fluxes determined under steady-state).

Another argument which has been used to explain this affinity difference between intact cell and membrane vesicles is that *in vivo* the inner and outer surface of the bilayer are exposed to different ionic environments. This factor could induce different properties of the Na–Ca exchange at both sides of the membrane (Philipson and Nishimoto 1982). We have recently (DiPolo and Beaugé, manuscript in preparation) explored this possibility by determining the apparent affinity of the exchanger for Ca_i and Ca_o under identical intra- and extracellular ionic conditions and at zero membrane potential. The results of several experiments are shown in Fig. 1.7A. Even under symmetric ionic conditions, the Na–Ca exchange still behaves as an asymmetric system. The half-maximal activation of the Ca_o-dependent Ca efflux by Ca_i^{2+} is still ten times less than that for Ca_o. At the moment, there is no clear answer as to why membrane vesicle preparations do not exhibit some of the asymmetrical properties found in intact cells (but see Section 1.5).

Interaction of the Na–Ca exchange with Mg^{2+} ions are of interest since millimolar concentrations of this ion exist in the intra- and extracellular medium. In squid axons internal Mg^{2+} is an important inhibitor of all of the modes of operation of the Na–Ca exchange, in particular the Na_o-dependent Ca efflux and the Na_o-dependent Na efflux components (DiPolo and Beaugé 1984, 1987). At physiological levels of Mg_i (2–3 mM) the Na–Ca exchange in squid axons is inhibited by 50 per cent. From the experiment of Fig. 1.7B, it is evident that the removal of Mg_i causes a large increase in the Na_o-dependent Ca efflux. The $K_{\frac{1}{2}}$ for the Mg_i inhibition is around 1.4 mM. The experimental

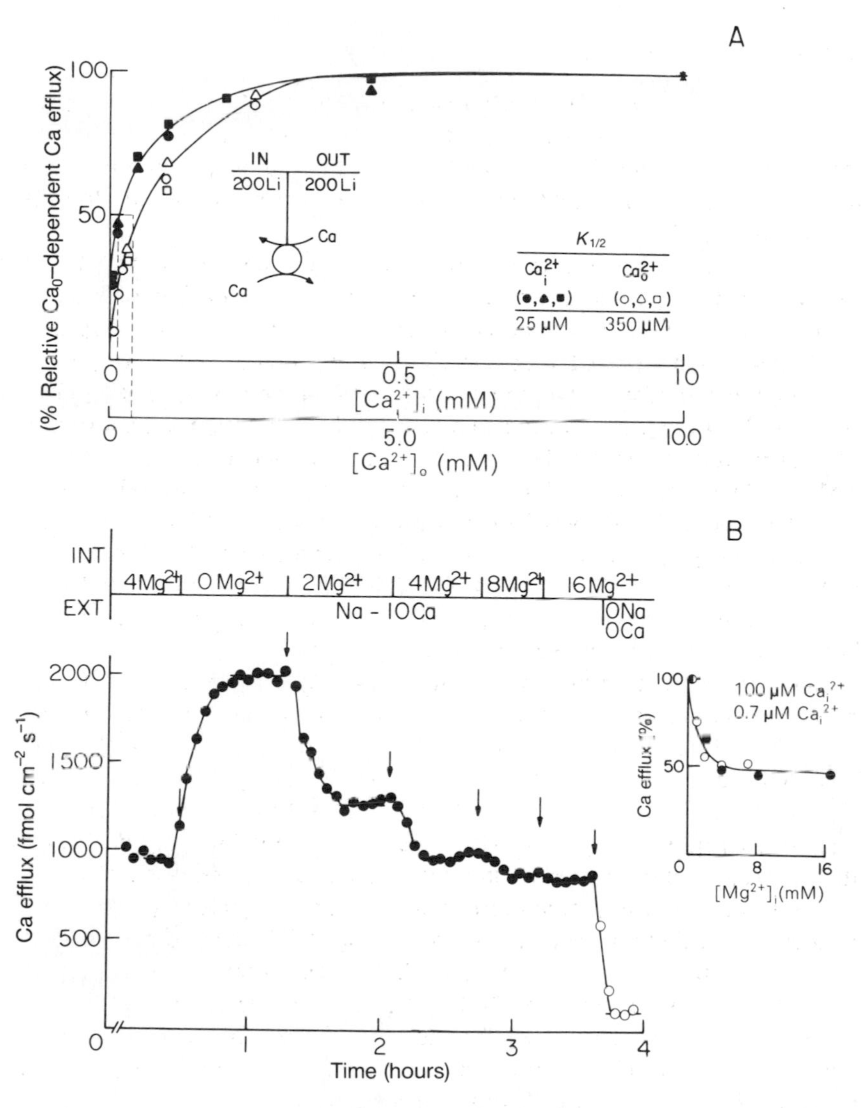

Fig. 1.7. (A) Ca_o and Ca_i dependence of the Na–Ca exchange in axons internally dialysed and externally perfused with identical solutions. The internal and external solutions contained (mM): Tris-Mops, 200; LiCl, 200; $MgCl_2$, 20, EGTA, 2. Osmolarity was adjusted with glycine. Axons were voltage clamped at zero membrane potential. The $K_{1/2}$ for the activation of the Ca_o-dependent Ca efflux by Ca_i was 25 μM as compared with a value of 300 μM for its activation by Ca_o. Observe that under identical symmetrical conditions ($Ca_i = Ca_o = 800$ μM) the internal Ca site is completely saturated while the external site is not. (From DiPolo and Beaugé unpublished.) (B) The inhibitory effect of Mg_i on Na-dependent Ca efflux. The insert shows the effect at two different Ca_i^{2+} concentrations (from DiPolo and Beaugé 1984).

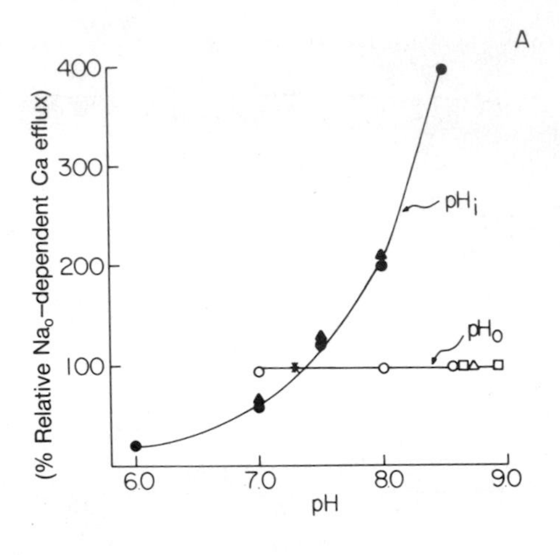

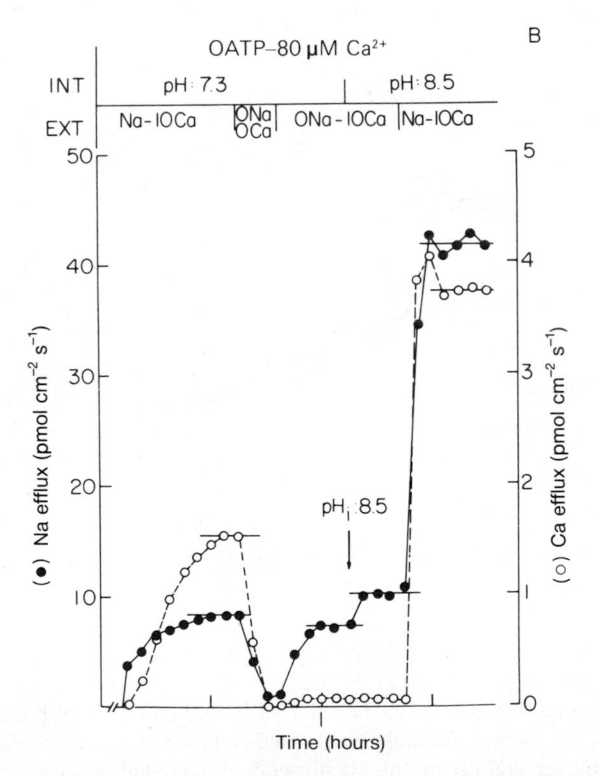

Fig. 1.8. (A) The effect of internal and external pH on the Na$_o$-dependent Ca efflux (from DiPolo, Rojas and Beaugé 1982.) (B) The effect of internal alkalinity on the Na$_o$ and Ca$_o$ components of the Ca and Na effluxes. Notice that internal alkalinity induces a much larger activation of the Na$_o$-dependent Ca efflux compared to that of the Ca$_o$-dependent Na efflux component (from DiPolo and Beaugé 1987a).

data also show that the Mg_i inhibition is only partial (50 per cent maximal inhibition), and apparently not a simple consequence of competition with Ca_i since a similar inhibition is found with two extreme Ca_i concentrations (see insert to Fig. 1.7). Magnesium ions interact asymmetrically with the exchanger since an Mg_o of up to 50 mM has no effect on any of the operational modes of the Na–Ca exchange (DiPolo and Beaugé, unpublished experiments).

1.4.3 Hydrogen ions

In intact cells, alteration in Ca^{2+} as a consequence of changes in the pH_i could be the result of alterations in the intracellular Ca^{2+} binding and/or to modifications in the membrane Ca transport systems. In experiments on intact axons, it has been reported that decreasing the pH_i either by exposing the axons to CO_2, or by injecting acid reduces the Na_o-dependent Ca efflux, the Na_i-dependent Ca influx and the associated Ca_o-dependent Na efflux. In contrast, intracellular alkalinity increases the Ca_o-dependent Na efflux (Baker and DiPolo 1984). Dialysed axons are ideal to study the differential effects of external and internal pH on the Na–Ca exchange since the use of high concentrations of pH buffers allows the independent modification of H^+ concentration on either side of the membrane. The data of Fig. 1.8A show an important inhibition of the forward Na–Ca exchange at acid pH_i, and a marked activation of the flux at alkaline pH_i (up to 400 per cent between pH 7.3 and 8.5). Interestingly, this activation seems to occur mostly on the Na-dependent fluxes with little effect on the Ca-dependent ones (see Fig. 1.8B). Since no effect of external pH is found over a wide range (see Fig. 1.8A), the ionizable groups responsible for the alterations in the rate of Na–Ca exchange must be located on the cytoplasmic side. Philipson and Nishimoto (1982) working on Na–Ca exchange in sarcolemma vesicles found identical responses of the inside-out and total population of vesicles to pH, thus suggesting that the chemical nature of the Ca binding site is similar on the opposite sides of the membrane. This constitutes more evidence of the difference in the properties of the Na–Ca exchanger present in intact cells and isolated membrane preparations.

1.5 Concluding remarks

In the previous sections I have reviewed some of the most relevant aspects of the modes of operation of the Na–Ca exchange as they are found in intact or whole cell preparations. The main source of information has been from work on squid axons and barnacle muscle fibres. Special emphasis has been devoted to the analysis of the kinetics, ligand interactions, and regulation of the exchanger. The first general conclusion is that the Na–Ca exchange is far from being a simple symmetrical carrier system. On the contrary, our present view of the Na–Ca antiporter, as found in intact cells, is that it is a highly asymmetric mechanism subject to ionic and metabolic regulation.

Much work remains to be done in these particular areas, however, especially concerning the effects of ATP on the kinetics of the Na–Ca exchange and the participation of Ca_i^{2+} in the process of phosphorylation. The role of Ca_i^{2+} as an 'essential' activator of the Na–Ca exchange and its involvement in a positive feedback mechanism, should be carefully explored in cells in which the activity of the Na–Ca exchange can be unequivocally related to a physiological function.

A special effort should be made to relate findings in intact cells with those obtained in isolated membrane preparations. There is a wide impression among different workers in the field that the membrane vesicle preparations presently used to study Na–Ca exchange may not exhibit several properties known to exist in intact cells (Reeves 1985; Baker and Allen 1986). First, a symmetrical exchange such as that found in membrane vesicles does not seem to be a property of intact cells. Second, contrary to data in intact cells, most workers have found little evidence for a role of ATP on Na–Ca exchange in vesicular preparations (however, see Caroni and Carafoli 1983). A possible explanation for these discrepancies, although not proved is that during the membrane isolation procedure a component of the exchanger is either lost or modified (proteolytic modification of the exchanger has been suggested by Baker and Allen 1986; Allen *et al.* 1988). Therefore, experiments on isolated preparations should be related to transport parameters obtained in intact cells.

Finally, other areas of study of the Na–Ca exchange should receive major attention including (see other contributions in this book): (i) voltage-dependence properties; (ii) isolation, biochemical characterization, and reconstitution studies; and (iii) a search for a specific inhibitor of the Na–Ca exchange a valuable tool in the purification and structural characterization of the exchange molecule.

Acknowledgements

I wish to thank Madalina Condrescu and Guillermo Whittembury for reading and criticizing the manuscript. The author would also like to thank the IVIC squid supply team at Mochima Edo. Sucre-IVIV Caracas, and the Director and staff of the Marine Biological Laboratory for the facilities and services put at his disposal. This work was supported by CONICIT (S1-1144, S1-1934) Venezuela, National Science Foundation RNS 8500595 USA, NIH-1RO1HL39243, USA, Fundacion Polar, Venezuela, and CONICET, Argentina.

References

Allen, T. J. A. and Baker, P. F. (1985). *Nature* **316**, 755–6.
Allen, T. J. A. and Baker, P. F. (1986a). *J. Physiol.* **378**, 53–76.
Allen, T. J. A. and Baker, P. F. (1986b). *J. Physiol.* **378**, 77–96.
Allen, T. J. A., DiPolo, R., and Rojas, H. (1988). *J. Physiol.* **407**, 136.
Altamirano, A. A., and Beaugé, L. A. (1985). *Cell Calcium* **6**, 503–25.
Baker, P. F. (1970). In: *Calcium and cellular function* (ed. A. W. Cuthbert) Macmillan, New York.
Baker, P. F. (1978). *Ann. N.Y. Acad. Sci.* **307**, 250–68.
Baker, P. F. and Allen, T. J. A. (1986). In: *Intracellular calcium regulation* (eds. H. Bader, K. Gietzen, J. Rosenthal, R. Rudel, and H. H. Wolf) pp. 35–47. Manchester University Press.
Baker, P. F. and Blaustein, M. P. (1968) *Biochim. Biophys. Acta* **150**, 167–70.
Baker, P. F. and DiPolo, R. (1984). *Curr. Top. Membr. Trans.* **22**, 195–242.
Baker, P. F. and Glitsch, H. G. (1973). *J. Physiol.* **233**, 44–6.
Baker, P. F. and McNaughton, P. A. (1978). *J. Physiol.* **276**, 127–150.
Baker, P. F. and Singh, R. (1982). *J. Physiol.* **330**, 373–92.
Baker, P. F., Hodgkin, A. L., and Shaw, T. I. (1962). *J. Physiol.* **164**, 330–54.
Baker, P. F., Blaustein, M. P., Hodgkin, A. L., and Steinhadt, R. A. (1967). *J. Physiol.* **192**, 43–44.
Beaugé, L. and Glynn, I. M. (1980). *J. Physiol.* **299**, 367–83.
Blaustein, M. P. (1977). *Biophys. J.* **20**, 79–110.
Blaustein, M. P. and Hodgkin, A. L. (1969). *J. Physiol.* **200**, 497–527.
Blaustein, M. P. and Russell, J. M. (1975). *J. Membr. Biol.* **22**, 285–312.
Brinley, F. J. and Mullins, L. J. (1967). *J. Gen. Physiol.* **50**, 2203–331.
Brinley, F. J., Spangler, S. G., and Mullins, L. J. (1975). *J. Gen. Physiol.* **66**, 223–50.
Carafoli, E. (1987). *Ann. Rev. Biochem.* **56**, 395–33.
Caroni, P. and Carafoli, E. (1980). *Nature* **283**, 765–67.
Caroni, P. and Carafoli, E. (1983). *Eur. J. Biochem.* **132**, 451–60.
Cassidy, P., Hoar, P., and Kerrick, G. (1979). *J. Biol. Chem.* **21**, 1148–53.
DiPolo, R. (1973). *Biochim. Biophys. Acta* **298**, 279–83.
DiPolo, R. (1974). *J. Gen. Physiol.* **64**, 503–17.
DiPolo, R. (1976). *Fed. Proc.* **35**, 2579–82.
DiPolo, R. (1977). *J. Gen. Physiol.* **69**, 795–814.
DiPolo, R. (1978). *Nature* **274**, 390–2.
DiPolo, R. (1979). *J. Gen. Physiol.* **73**, 91–113.
DiPolo, R. and Beaugé, L. (1979). *Nature* **278**, 271–3.
DiPolo, R. and Beaugé, L. (1980). *Cell Calcium* **1**, 147–69.
DiPolo, R. and Beaugé, L. (1981). *Biochim. Biophys. Acta* **645**, 229–36.
DiPolo, R. and Beaugé, L. (1983). *Ann. Rev. Physiol.* **45**, 3313–24.
DiPolo, R. and Beaugé, L. (1984). *J. Gen. Physiol.* **84**, 895–914.
DiPolo, R. and Beaugé, L. (1986). *Biochim. Biophys. Acta* **854**, 298–306.
DiPolo, R. and Beaugé, L. (1987a). *J. Gen. Physiol.* **90**, 505–25.
DiPolo, R. and Beaugé, L. (1987b). *Biochim. Biophys. Acta* **897**, 347–53.
DiPolo, R. and Caputo, C. (1977). *Biochim. Biophys Acta* **470**, 389–94.

DiPolo, R. and Rojas, H. (1984). *Biochim. Biophys. Acta* **776**, 313–16.
DiPolo, R., Rojas, H., and Beaugé, L. (1982). *Cell Calcium* **3**, 19–41.
DiPolo, R., Caputo, C., and Bezanilla, F. (1987). *Biophys. J.* **51**, 686a.
DiPolo, R., Bezanilla, F., Caputo, C., and Rojas, H. (1985). *J. Gen. Physiol.* **86**, 457–78.
Haworth, R. A., Goknur, A. B., Hunter, D. R., Hegge, O. J., and Berkoff, H. A. (1987). *Circ. Res.* **60**, 586–94.
Herchuelz, A., Sener, A., and Malaise, W. J. (1980). *J. Membr. Biol.* **57**, 1–12.
Jundt, H. and Reuter, H. (1977). *J. Physiol.* **266**, 78P–79P.
Kimura, J., Miyamae, S., and Noma, A. (1987). *J. Physiol.* **384**, 199–222.
Krieger, N. S. and Tashigian, A. H. Jr (1980). *Nature* **287**, 843–5.
Lederer, W. J. and Nelson, M. T. (1983). *J. Physiol.* **341**, 325–39.
Lee, C. O., Taylor, A. and Windhager, E. C. (1980). *Nature* **287**, 859–61.
Lüttgau, H. C. and Niedergerke, R. (1958). *J. Physiol.* **143**, 486–505.
McNaughton., Nunn, B. J., and Hodgkin, A. L. (1986). In: *The molecular mechanism of photoreception* (ed. H. Stieve). Springer-Verlag, Berlin.
Milanick, M. A. and Hoffman, J. (1986). *Ann. N.Y. Acad. Sci.* **488**, 174–86.
Mullins, L. J. (1977). *J. Gen. Physiol.* **70**, 681–96.
Mullins, L. J. (1981). *Ion transport in heart.* Raven Press, New York.
Nelson, T. M. and Blaustein, M. P. (1981). *Nature* **289**, 314–316.
Parker, J. C. (1978). *J. Gen. Physiol.* **71**, 1–17.
Philipson, K. D. and Nishimoto, A. Y. (1982). *J. Biol. Chem.* **257**, 5111–17.
Post, R. L., Hegyvary, C., and Kume, S. (1972). *J. Biol. Chem.* **247**, 6530–40.
Rasgado-Flores, H. and Blaustein, M. P. (1981). *Am. J. Physiol.* **252**, C499-C504.
Reeves, J. P. (1985). *Curr. Top. Membr. Trans.* **25**, 77–119.
Reeves, J. and Sutko, J. L. (1979). *Fed. Proc.* **38**, 1199 Abs.
Requena, F. (1978). *J. Gen. Physiol.* **72**, 443–70.
Reuter, H. and Seitz, N. (1967). *Arch. Exp. Pathol. Pharmacol.* **257**, 324–5.
Russell, J. M. and Blaustein, M. P. (1975). *J. Membr. Biol.* **23**, 157–79.
Schatzmann, H. J. (1985). In: *Calcium and cell physiology* (ed. D. Marme). Springer-Verlag, Berlin.
Sherry, J., Goreka, A., Aksoy, M., Debrowska, R., and Heartborne, D. (1978). *Am. Chem. Soc.* **17**, 4411–18.
Simons, T. J. B. (1975). *J. Physiol.* **244**, 731–9.
Sjodin, R. A. and Abercrombie, R. F. (1978). *J. Gen. Physiol.* **71**, 453–66.
Slaughter, R. S., Sutko, J. L., and Reeves, J. P. (1983). *J. Biol. Chem.* **258**, 3183–90.
Wong, A. Y. K. and Bassingthwaighte, J. B. (1981). *Math. Biosci.* **53**, 275–310.

2 Sodium–calcium exchange activity in plasma membrane vesicles

John P. Reeves and Kenneth D. Philipson

2.1 Introduction

In the first decade following the experimental demonstratration of Na–Ca exchange activity in guinea-pig atria (Reuter and Seitz 1968) and squid axons (Baker *et al.* 1969), a great deal of experimental effort was devoted to examining the characteristics of this transport process in a variety of cellular and subcellular systems. These included squid axons, cardiac muscle, barnacle muscle, and synaptosomes (for reference, see reviews by Reeves 1985; Philipson 1985a; Sheu and Blaustein 1986). The most versatile systems were the squid axon and barnacle muscle where the cytoplasmic environment could be reasonably well-controlled by internal perfusion or dialysis. A great deal of valuable information resulted from these studies and the essential characteristics of the Na–Ca exchange system, such as its bidirectionality, transmembrane kinetic asymmetry, the effects of ATP, and its operation in a Ca–Ca exchange mode, were defined during this period.

The stoichiometry of the exchange process was still a matter of debate, however; studies by Mullins and Brinley (1975) and Blaustein, Russell and de Weer (1974) indicated that in squid axons, Na_o^+-dependent $^{45}Ca^{2+}$ efflux increased upon hyperpolarization and decreased upon depolarization, suggesting that the system was electrogenic and that its stoichiometry was greater than 2 Na^+ per Ca^{2+}. On the other hand, in cardiac tissue the available evidence on flux measurements favoured an electroneutral system having a stoichiometry of 2 Na^+ per Ca^{2+} (Jundt *et al.* 1975), although measurements of mechanical activity suggested a higher stoichiometry (Chapman and Tunstall 1980).

Evidence was also beginning to emerge that the exchange system was a much more complex transport process than had been realized; La^{3+}, for example, appeared to inhibit Na_i^+-dependent Ca^{2+} influx, but not Na_o^+-dependent Ca^{2+} efflux (Barry and Smith 1982; Lederer and Nelson 1983).

These and other observations would lead to suggestions that two different transport systems might mediate Na^+-dependent Ca^{2+} movements in opposite directions (cf. Reeves 1985). By the late 1970s, it was clear that the development of a simple subcellular system for the study of exchange activity would be necessary before a full understanding of the exchange system could be attained.

The use of membrane vesicles to study Na–Ca exchange activity was introduced in 1979, with the demonstration by Reeves and Sutko (1979) that Ca^{2+} movements in a crude preparation of cardiac membranes could be elicited by generating appropriate concentration gradients of Na^+ across the vesicle membrane; this approach was quickly extended to other experimental systems such as synaptosomal plasma membranes (Rahamimoff and Spanier 1979). The vesicle methodology is based on the techniques developed by Kaback (1974) to study ion gradients and active transport activity in membrane vesicles from bacteria. Plasma membranes are prepared under conditions where they form osmotically sealed vesicles; under most isolation conditions, a mixed population of inside-out and right-side-out orientations is obtained. The intravesicular compartment is essentially free of the normal cytoplasmic constituents of the cell and its ionic composition can be manipulated by simple equilibration procedures.

To study Na_i^+-dependent Ca^{2+} uptake, the vesicles are equilibrated in a medium containing the desired concentration of NaCl and then diluted 20-to 50-fold into an iso-osmotic Na^+-free medium containing $^{45}Ca^{2+}$; the dilution step creates an outward-directed concentration gradient of Na^+ which energizes the accumulation of $^{45}Ca^{2+}$ by the vesicles via the Na–Ca exchange system. The data in Fig. 2.1A, taken from the paper of Reeves and Sutko (1979), show the results of an experiment of this type; Ca^{2+} uptake by the vesicles is rapid and achieves a steady-state in less than a minute. When the Na^+-loaded vesicles are diluted into a medium containing a high concentration of Na^+, or when Na^+ is absent from the vesicle interior, very little Ca^{2+} accumulation occurs (Fig. 2.1A). Na^+-dependent Ca^{2+} movements in the opposite direction are studied by preloading the vesicles with Ca^{2+}, either by passive equilibration or by allowing the vesicles to accumulate Ca^{2+} by Na–Ca exchange or the ATP-dependent Ca^{2+} transport system; the latter approach selectively loads the inside-out subpopulation of vesicles with Ca^{2+} because only this orientation will offer proper access of the ATPase to extravesicular ATP. As shown in Fig. 2.1B, when Ca^{2+}-loaded vesicles are diluted into a medium containing no Na^+, efflux of Ca^{2+} from the vesicles is quite slow, exhibiting a half-time of approximately 10 minutes in cardiac sarcolemmal vesicles. The presence of Na^+ in the dilution medium dramatically accelerates the rate of Ca^{2+} efflux (Fig. 2.1B) and this effect is blocked by inhibitors of Na–Ca exchange, such as La^{3+} or amiloride analogues (data not shown).

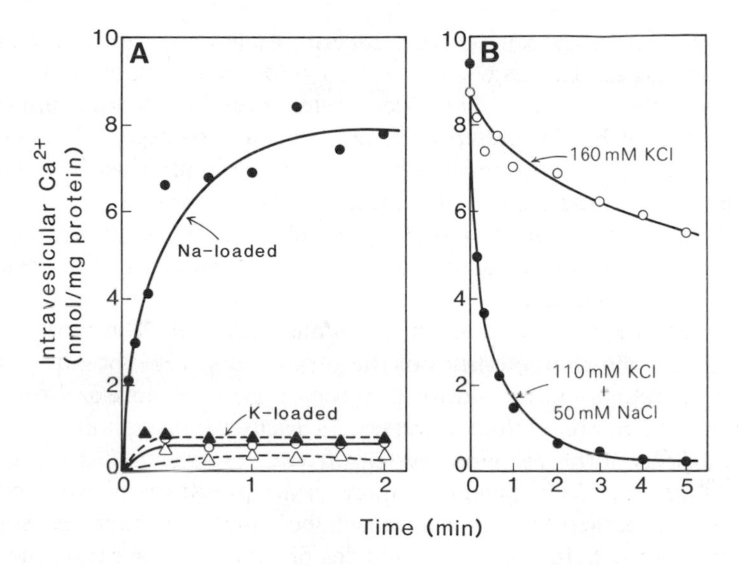

Fig. 2.1. Na–Ca exchange in cardiac sarcolemmal vesicles. (A) Ca^{2+} uptake by vesicles loaded internally with either 160 mM NaCl (●, ○) or 160 mM KCl (▲,△). Ca^{2+} uptake was assayed in either 160 mM LiCl (closed symbols) or 160 mM NaCl (open symbols); the $[Ca^{2+}]$ was 40 μM. (B) Efflux of Ca^{2+} from preloaded vesicles. Vesicles were preloaded with $^{45}Ca^{2+}$ by Na–Ca exchange and then diluted 67-fold with either 160 mM KCl (0) or 50 mM NaCl + 110 mM KCl (●); the solutions were buffered to pH 7.4 with 20 mM Mops/Tris and contained 0.1 mM EGTA. (Data taken from Reeves and Sutko (1979) with permission.) The sarcolemmal preparation used in these early studies was a relatively crude microsomal preparation; current sarcolemmal preparations are more highly purified and exhibit levels of Ca^{2+} accumulation nearly 10-fold greater than shown here.

The vesicles can also be used to study Ca^{2+}-dependent Na^+ movements (cf. Pitts 1979); this aspect of exchange activity has not been extensively studied in vesicles, however, because it is difficult to measure accurately the exchange-mediated Na^+ fluxes due to high background permeability of the vesicles to Na^+ and the low specific activity of $^{22}Na^+$.

The vesicles offer several advantages over intact cells or tissues as an experimental system. The small size of the vesicles and their spherical geometry minimizes the effects of unstirred water layers, which may exist at both the cytoplasmic and extracellular surfaces of large cells such as squid giant axons or barnacle muscles; diffusion of ions through these regions may constitute an appreciable kinetic barrier for rapidly transporting systems like the Na–Ca exchanger and could produce significant distortions of transport kinetics (Dietschy 1978). Intracellular compartmentation and buffering of Ca^{2+} ions can lead to uncertainty as to the concentration of free Ca^{2+} at the cytoplasmic membrane surface; this uncertainty is minimized in the case of inside-out vesicles (cf. below) where the cytoplasmic membrane surface is accessible to the external medium. Measurements of Ca^{2+} fluxes in intact cells are complicated by the presence of multiple Ca^{2+} transport pathways; in vesicles, Ca^{2+} movements attributable to Na–Ca exchange can be unequivocally determined. The vesicles are also an extremely versatile system experimentally, and the effects of various biochemical modifications and putative regulators or inhibitors can be assessed and evaluated in a relatively straightforward manner.

The vesicles are not without their pitfalls, however. The most obvious worry is that the characteristics of the exchange system may be modified through alterations induced during the preparation procedure or through the loss of intracellular regulatory processes. As discussed throughout this article, certain aspects of the exchange system in vesicles appear to differ from its generally understood behaviour in more intact preparations; some possible explanations for these discrepancies will be suggested where possible. In addition, the size heterogeneity of vesicles precludes the accurate measurement of true initial rates of transport (Hopfer 1981). In cardiac vesicles, there is the further complication that the accumulation of Ca^{2+} within the vesicle interior during the transport assay alters the apparent K_m for external Ca^{2+} (cf. below). For these and other reasons (cf. Reeves 1985), the values for the kinetic parameters obtained in vesicles may not provide a true index of the kinetic behaviour of the exchange system in intact cells. On the other hand, because of the potential difficulties discussed in the previous paragraph, it is also uncertain whether the apparent behaviour of exchange activity in preparations like the squid giant axon provides a true picture of the biochemical properties of the Na–Ca exchange system.

It would appear that a complete understanding of Na–Ca exchange is still a distant goal and that both the physiological and biochemical approaches will be important in achieving that understanding. The rather limited goals of the present article are to summarize the behaviour of the exchange system in plasma membrane vesicles, with special emphasis on cardiac sarcolemma, to point out areas where the vesicle data and the physiological results are in sharpest disagreement, and, wherever possible, to suggest possible approaches to reconcile the two sets of results.

2.2 Stoichiometry

In one of the initial studies carried out with membrane vesicles, Pitts (1979) measured the ratio of the exchange-mediated Na^+ and Ca^{2+} fluxes and obtained a stoichiometry of 3 Na^+ per Ca^{2+} (cf. Pitts, Okhuysen, and Entman 1980). While the data appeared to be clear-cut, there were several technical reasons why this value for the stoichiometry was greeted with skepticism. First, there was the general difficulty of accurately measuring $^{22}Na^+$ fluxes in the vesicles (cf. Section 2.1). In addition, the contribution of Na–Na exchange and the influence of intravesicular Ca^{2+} on passive Na^+ fluxes (e.g. through Ca^{2+}-activated channels) were not fully evaluated. Nevertheless, these results were the first direct measurement of the exchange stoichiometry in any system and have the additional virtue, in retrospect, of being correct.

Other, less direct, approaches soon provided evidence suggesting that the exchange system was electrogenic and that its stoichiometry was therefore greater than 2 Na^+ per Ca^{2+}. Bers, Philipson, and Nishimoto (1980) and, in a more detailed study, Philipson and Nishimoto (1980), showed that the establishment of positive (inside) membrane diffusion potentials stimulated Na_i^+-dependent Ca^{2+} accumulation by the vesicles; negative potentials were later shown to stimulate Na_o^+-dependent Ca^{2+} efflux (Kadoma *et al.* 1982). As pointed out by Eisner and Lederer (1985), however, the voltage sensitivity of exchange activity does not necessarily indicate electrogenicity, or a stoichiometry greater than 2, since the rate of an electroneutral process might exhibit rate-limiting reaction steps that are voltage-sensitive.

Another approach was adopted by Reeves and Sutko (1980), and subsequently by Caroni, Reinlib and Carafoli (1980), who demonstrated with lipophilic cations that Na_i^+-dependent Ca^{2+} uptake in cardiac sarcolemmal vesicles was associated with the development of a negative transmembrane potential, consistent with an electrogenic exchange process. The potential dissipated as Ca^{2+} uptake approached a steady-state (at which point net current generation by the exchanger should cease) and it correlated strongly in its dependence upon temperature and ionic concentrations with Na–Ca exchange activity. The possibility that this potential was not due to Na–Ca exchange but reflected the activity of a Ca_i^{2+}-dependent increase in cation conductance could not be completely eliminated, however (cf. Eisner and Lederer 1985).

Compelling evidence for the electrogenicity of the exchange system was provided by studies in which exchange-mediated Ca^{2+} fluxes were elicited in cardiac sarcolemmal vesicles by diffusion potentials in the absence of concentration gradients for Na^+ or Ca^{2+} (Reeves and Hale 1984). Briefly, vesicles were equilibrated with 100 μM $^{45}Ca^{2+}$ and 30 mM Na^+, and positive or negative K^+-diffusion potentials were then generated (using the K^+-ionophore valinomycin) without changing the concentrations of either Na^+ or $^{45}Ca^{2+}$. Positive (inside) membrane potentials caused a transient accumulation of $^{45}Ca^{2+}$ while negative potentials caused a transient efflux. These Ca movements were completely dependent upon the presence of valinomycin, indicating that they were not driven by K^+ gradients *per se*. (The high K^+-conductance of cardiac cells is not retained in the isolated vesicles (Schilling *et al.* 1984) and a K^+-ionophore like valinomycin is required to generate significant K^+-diffusion potentials in this system.) The results demonstrate that a membrane potential alone can act as a driving force for Na–Ca exchange in the absence of concentration gradients for either Na^+ or Ca^{2+}.

This general procedure was used to determine the stoichiometry of the exchange system by establishing a series of $[Na^+]$ gradients to oppose the effects on Ca^{2+} movements of a constant imposed membrane potential (either positive or negative). The stoichiometry was determined from the magnitude of the Na^+ gradient that exactly compensated for the membrane potential so that no net Ca^{2+} movement occurred. For both positive and negative values of the membrane potential, this balance point yielded a stoichiometry of 3; the mean stoichiometry ($\pm$SE) of nine such determinations (three at negative and six at positive membrane potentials) was 2.97 ± 0.03 (Reeves and Hale 1984).

Recently, Barzilai and Rahamimoff (1987) have proposed a stoichiometry of 5 Na^+ per Ca^{2+} based on their studies with synaptosomal membranes. Two different approaches were used: the direct measurement of exchange-mediated Na^+ and Ca^{2+} fluxes and a thermodynamic approach where the steady state accumulation of Ca^{2+} was determined as a function of the $[Na^+]$ gradient across the vesicle membrane. Both procedures yielded a value of 5, but both procedures are subject to a number of criticisms as well. As in Pitts' (1979) study, the possible effects of intravesicular Ca^{2+} on passive cation permeability were not evaluated; this could lead to an overestimate of the exchange-mediated Na^+ fluxes. The thermodynamic approach is undermined by the uncertainty as to the intravesicular concentrations of either Ca^{2+} or Na^+ in the steady-state. A stoichiometry of 5 would imply that the synaptosomal exchange carrier is a different entity from the exchanger in cardiac tissue, where a stoichiometry of 3 seems to be well-established, both by data in vesicles (cf. above) and by exchange current measurements in perfused cardiac cells (Kimura, Noma, and Irisawa 1986; Kimura, Miyamae, and Noma, 1987). A stoichiometry of 3, determined by direct flux measurements, has also been reported for barnacle muscles (Rasgado-Flores and Blaustein 1987). Continued investigations into the stoichiometry of Na–Ca exchange are necessary to determine whether this parameter varies with different tissues or experimental conditions.

2.3 Sodium–calcium exchange

2.3.1 $[Na^+]$ dependence of Ca^{2+} movements in vesicles

Ca^{2+} accumulation by membrane vesicles is highly dependent upon the presence of intravesicular Na^+; indeed, this was one of the initial criteria for the demonstration of Na–Ca exchange activity in the vesicle system (cf. Fig.

2.1A). No other monovalent cations have been found that will substitute for exchanger from the mitochondrial Na–Ca exchange system, where Li^+ is an effective substitute for Na^+ (Crompton, Künzi, and Carafoli 1977). The steady-state levels of Ca^{2+} accumulation tend to vary in a linear fashion with intravesicular $[Na^+]_i$; the initial rates of Ca^{2+} uptake, however, show a sigmoidal dependence upon $[Na^+]_i$ with a Hill coefficient of approximately 2 and a K_m of 13–30 mM (Reeves and Sutko 1983; Wakabayashi and Goshima 1982). $Na^+{}_i$ increases the V_{max} for Ca^{2+} uptake, but does not affect the K_m for Ca^{2+} (Philipson and Nishimoto 1982a). The latter result provides suggestive evidence in favour of a simultaneous translocation of Na^+ and Ca^{2+} ions by the exchange carrier as opposed to a sequential (ping-pong) mechanism (cf. Chapter 6).

The bidirectional nature of Na–Ca exchange was evident from the results of initial studies in guinea-pig atria (Glitsch, Reuter, and Scholz 1970) and squid axons (Baker *et al.* 1969). As mentioned previously (cf. Reeves 1985), the complex behaviour of the exchange system has led, on occasion, to suggestions that the Na_i^+-dependent Ca^{2+} influx and Na_o^+-dependent Ca^{2+} efflux in intact cells might be mediated by two different systems. This certainly is not the case in vesicles, however, because in reconstituted systems where exchange activity is present in only a small percentage of the proteoliposomes (Cheon and Reeves 1988a), the same population of vesicles displays both Na_i^+-dependent Ca^{2+} uptake and Na_o^+-dependent Ca^{2+} efflux. If two independent systems were involved in mediating Ca^{2+} uptake and efflux, one would not expect their activities to overlap to any significant degree in the reconstituted preparations.

In cardiac sarcolemmal vesicles, Ca^{2+} efflux exhibits a sigmoidal dependence upon $[Na^+]_o$, with a K_m of approximately 30 mM and a Hill coefficient of 2–3 (Kadoma *et al.* 1982; Philipson 1985b; de la Peña and Reeves 1987). The kinetics of Na_o^+-dependent Ca^{2+} efflux have not been extensively examined in other vesicle systems, although Michaelis and Michaelis (1981) reported a K_m (Na^+) of 6.6 mM for Ca^{2+} efflux in synaptic plasma membranes.

2.3.2 $[Ca^{2+}]$ dependence

Initial velocities of Ca^{2+} uptake by cardiac sarcolemmal vesicles exhibit Michaelis–Menten kinetics with respect to $[Ca^{2+}]_o$. For cardiac sarcolemmal vesicles, reported values for the K_m (Ca^{2+}) vary over a wide range, from a low value of 1.5 μM to values greater than 200 μM (cf. Reeves 1985). As shown in Table 2.1, however, most values fall within the 20–40 μM range; for the most part, similar values have been obtained with vesicles from other membrane systems, and these are also listed in Table 2.1. The variablity in K_m (Ca^{2+}) values for the cardiac sarcolemmal vesicles may reflect, in part, the kinetic malleability of this system; as discussed in detail below (cf. section

Table 2.1. Kinetics of Na–Ca exchange in membrane vesicles

Source	K_m (Ca^{2+}) (μM)	V_{max} (nmol per mg per second)	References (see legend)
Cardiac sarcolemma	27±15 (N=26)	25±18 (N=13)	(1)
Synaptosome membrane	34±21 (N=5)	0.17±0.17 (N=5)	(2)
Rabbit thymocytes	61	0.19	(3)
Squid optic nerve	7.5	0.45	(4)
Coated vesicles (brain)	0.8	0.006	(5)
Artemia	20	14	(6)
Kidney basolateral membranes	0.2–8.0	0.05–0.09	(7)
Cultured pituitary cells (GH_3)	20	0.27	(8)
Smooth muscle	24±18 (N=4)	0.33	(9)

Mean values±SD of *N* different determinations are reported where indicated. (1) References available from JPR upon request. The data exclude three reports of K_m values>100 μM (Philipson *et al.* 1982; Reeves and Poronnik 1987; de la Peña and Reeves 1987); if these are included, the mean K_m (Ca) becomes 44±52 μM. (2) Schellenberg and Swanson 1981, 1982; Michaelis and Michaelis 1981; Rahamimoff *et al.* 1980; Rahamimoff and Spanier 1984. (3) Ueda 1983. (4) Osses *et al.* 1986. (5) Saermark and Gratzl 1986. (6) Cheon and Reeves 1988b. (7) van Heeswijk, Geertsen, and van Os 1984, Jayakumar *et al.* 1984. (8) Kaczorowski *et al.* 1984. (9) Slaughter, Welton, and Morgan, 1987; Morel and Godfraind 1984; Grover *et al.* 1983. The V_{max} value is from Slaughter *et al.* 1987.

2.6), the K_m (Ca^{2+}) in cardiac vesicles can be modified over a broad range by a host of different experimental treatments, including proteolysis, phospholipase treatment, redox reagents, and intravesicular Ca^{2+}.

The K_m for Ca^{2+} in vesicles is similar in magnitude to that which has been measured in squid axons and barnacle muscles in the absence of intracellular ATP (Blaustein 1984). In the latter systems, the presence of ATP reduces the K_m at the cytoplasmic surface to the low micromolar range; in vesicles, most investigators have not observed significant effects of ATP on Na–Ca exchange activity (cf. Section 2.6). It seems likely that an ATP-dependent regulatory process is missing in most vesicle preparations and that the kinetic behaviour of the exchange system in vesicles is similar to that observed *in the absence of ATP* in more intact preparations.

2.3.3 V_{max} and turnover number

The data in Table 2.1 show that the V_{max} for Ca^{2+} uptake by cardiac sarcolemmal vesicles is more than 50 times greater than that of other membrane systems (with the exception of the brine shrimp *Artemia*). Ram sperm flagellar plasma membranes also exhibit high Na–Ca exchange activities, although V_{max} values were not reported (Bradley and Forrester 1980). Thus, cardiac sarcolemma is a particularly rich source of Na–Ca exchange activity, a conclusion which is in line with the large currents in cardiac cells that are thought to reflect exchange activity (Kimura *et al.* 1986, 1987; Hume and Uehara 1986; Mechmann and Pott 1986; Kenyon and Sutko 1987). Thus, a Ca^{+} flux of 25 nmol per mg protein per second in vesicles (Table 2.1) is equivalent to a current density of approximately 1 $\mu A\ \mu F^{-1}$, assuming that 1 mg protein = 2000 cm^2 membrane surface area (Chapman, Corey, and McGuigan 1983), that the membrane capacitance is 1 $\mu F\ cm^{-2}$ and that the stoichiometry is 3 Na^{+} per 1 Ca^{2+}. This is similar in magnitude to the currents measured in cardiac cells (e.g. Kimura, Miyamae, and Norma 1987), although it should be pointed out that these measurements were not necessarily measured under V_{max} conditions for Na–Ca exchange.

Recently, the density of the cardiac exchange carrier was estimated in reconstituted proteoliposomes by measuring the fraction of proteoliposomes which exhibited exchange activity (Cheon and Reeves 1988a). The details of these measurements will not be presented here; however, the results suggested that approximately 3 per cent of the proteoliposomes contained the exchange carrier, a figure which corresponds to 10–20 pmol of exchanger per mg of protein in the reconstituted preparations. The V_{max} for Na–Ca exchange in these preparations was 20 nmol per mg protein per second. Therefore, the turnover number of the exchanger (k_{cat}) is approximately 1000 per second, a value which is similar to the turnover number recently suggested for the Na^{+}–H^{+} exchanger (Vigne *et al.* 1985; Dixon *et al.* 1987). The K_m for Na–Ca exchange in the reconstituted preparations was 14 μM; this yields a value of $10^{8}\ M^{-1}\ s^{-1}$ for the second-order rate constant ($k = k_{cat}/K_m$) of the exchange reaction. This is similar in magnitude to the rate of association of Ca^{2+} with a variety of Ca^{2+}-binding proteins (e.g. parvalbumin, troponin) and suggests that the binding of Ca^{2+} to the exchange carrier may be the rate-limiting step of the exchange reaction under certain conditions. Moreover, if the K_m for Ca^{2+} in intact cells, in the presence of ATP, is reduced to the 1–2 μM range, as suggested by the data cited in Section 2.3.2., the second order rate constant would approach the diffusion controlled limit for enzyme substrate reactions

($10^9\ M^{-1}\ s^{-1}$; Eigen and Hammes 1963), assuming that k_{cat} remains the same in the presence of ATP.

While diffusion control of Na–Ca exchange cannot be ruled out at present, it is also possible that there are mechanisms for circumventing this limitation. One possibility is that the local concentration of Ca^{2+} at the cytoplasmic surface of the membrane is considerably higher than the bulk concentration because of the negative surface potential of the membrane. Thus, if expressed in terms of the local Ca^{2+} concentration, the K_m for Ca^{2+} may be considerably higher than the 1–2 μM suggested above; this would reduce the second-order rate constant safely below the diffusion control range. However, as discussed further below, the exchanger appears to be sensitive to Ca^{2+} in the bulk solution rather than Ca^{2+} in the diffuse double layer (cf. Section 2.6.2 and Bers, Philipson, and Peskoff 1985).

Another possibility is that Ca^{2+} bound to surface of the membrane, or the Ca^{2+}-buffering constituents in the cell cytoplasm, can serve as a direct source of Ca^{2+} for the exchange carrier. Such a source could prevent local Ca^{2+} depletion and prevent diffusion limitations. Support for this possibility is provided by the finding that the presence of EGTA or other Ca^{2+}-chelating agents lowers the apparent K_m for Ca^{2+} in both cardiac sarcolemma and synaptosomal membrane vesicles (cf. Section 2.4.4). Although the explanation for the effects of Ca^{2+} chelators remains uncertain, these observations certainly suggest that the kinetics of the exchange reaction in intact cells can be affected not only by the presence of ATP, but also by Ca^{2+}-binding constituents of the cell cytoplasm as well.

2.3.4 Na–Ca antagonism

Na_i^+-dependent Ca^{2+} uptake in vesicles is inhibited by high concentrations of Na_o^+ (cf. Fig. 2.1A). In vesicles from either cardiac sarcolemma or synaptosomal membranes, the inhibition by Na_o^+ exhibits a sigmoidal concentration dependence (cf. e.g. Miyamoto and Racker 1980; Reeves and Sutko 1983; Schellenberg and Swanson 1982). This feature was examined in detail by Reeves and Sutko (1983), who observed a biphasic concentration profile for Na_o^+, with a Hill coefficient of 1 at low $[Na^+]_o$ and values approaching 2 at higher $[Na^+]_o$. The inhibition by Na_o^+ was competitive with respect to Ca^{2+} and the K_i for Na^+ was 16 mM.

The question arises as to whether the apparent co-operativity at high $[Na^+]_o$ reflects the co-operative interaction of Na^+ with the exchange carrier or a less interesting result, e.g. a decrease in the driving force of Ca^{2+} uptake or the promotion of Ca^{2+} efflux by Na_o^+. The latter interpretation seems unlikely because Hill coefficients approaching 2 were also observed for inhibition of $^{45}Ca^{2+}$ uptake by Ca–Ca exchange (cf. Section 2.4) under two sets of conditions: (a) equilibrium exchange (i.e. in the absence of $[Na^+]$

gradients) and (b) when the vesicles were preloaded with a high concentration (1 mM) of unlabelled Ca^{2+} to minimize Na_o^+-dependent $^{45}Ca^{2+}$ backflux through the exchanger. Moreover, when $^{45}Ca^{2+}$ efflux is prevented by the presence of intravesicular oxalate (a Ca^{2+} precipitating anion) in a reconstituted preparation, Hill coefficients approaching 2 for Na^+ inhibition are still observed (K. D. Philipson, unpublished observations). Thus, it seems likely that Na^+ interacts in a co-operative manner with the exchange carrier to antagonize Ca^{2+} binding.

2.3.5 A model for the Na–Ca exchanger

The results on exchange activity in vesicles have been interpreted in terms of a model for the exchange carrier (Slaughter, Sutko, and Reeves 1983) which represents an elaboration of a model advanced by others (cf. Baker *et al.* 1969; Blaustein and Russell 1975). This model features two classes of cation binding sites: a divalent site (A) which binds either a single Ca^{2+} ion or one to two Na^+ ions, and a second, monovalent site (B) which binds the third Na^+ involved in Na–Ca exchange (cf. Fig. 2.2). The binding of either Na^+ or Ca^{2+} to site A was seen as the simplest explanation for the competitive inhibition of Ca^{2+} uptake by Na^+. It is obvious that more complicated models of the exchanger, with separate, but allosterically interacting, sites for Na^+ and Ca^{2+} would explain the inhibition data equally well; in the absence of

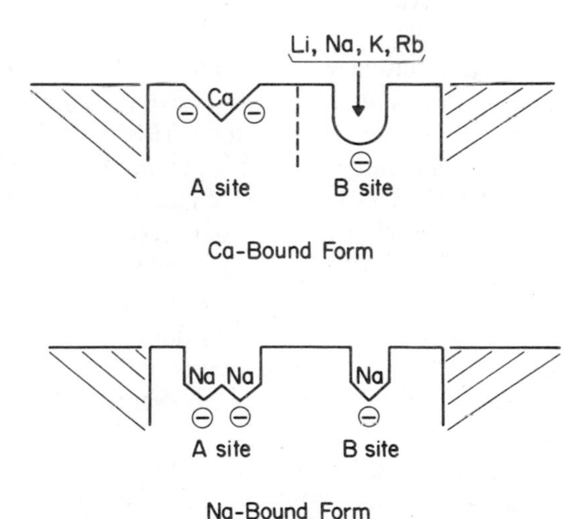

Fig. 2.2. Model for the Na–Ca exchange carrier. The diagram depicts the proposed cation binding sites of the carrier at one membrane surface. The dotted line in the Ca-bound form indicates that the cation bound to the B site is not translocated. See text for details. Reprinted from Reeves (1985) with permission.

additional data showing separate sites for Na^+ and Ca^{2+}, however, the more parsimonious model seems preferable.

Since Na^+ and Ca^{2+} have similar crystal ionic radii, the accomodation of two Na^+ ions at the A site would undoubtedly require a markedly different site geometry than the binding of a single Na^+ or Ca^{2+} ion. With this in mind, it has been postulated that the Na^+-loaded conformation of the exchange carrier differs from the Ca^{2+}-loaded conformation. Some evidence favouring this interpretation is that the stimulation of exchange activity by redox reagents (cf. Section 2.6.5, Reeves, Bailey, and Hale 1986) is only observed in the presence of Na^+. Indeed, even when Na^+ is present, Ca^{2+} antagonizes the effect of redox reagents on exchange activity. Moreover, the presence of Na^+ is required to retain exchange activity (after reconstitution) during the detergent solubilization of sarcolemmal membranes, suggesting that Na^+ stabilizes a detergent-resistant conformation of the exchange carrier; Ca^{2+} will not substitute for Na^+ in this regard (Cheon and Reeves 1988a).

The model has been useful in interpreting the effects of inhibitors of Na–Ca exchange (cf. Chapter 4) and for explaining the stimulatory effects of monovalent cations on Ca–Ca exchange (cf. Section 2.4). Whether or not these interpretations prove to be correct is perhaps less important, at this stage, than providing a discrete focus for the design of new experiments.

2.4 Ca–Ca exchange

Ca–Ca exchange is an alternative transport mode of the Na–Ca exchanger. In sarcolemmal vesicles, Ca–Ca exchange is conveniently studied as Ca_o^{2+}-dependent Ca^{2+} efflux (Philipson and Nishimoto 1981), Ca_i^{2+}-dependent Ca^{2+} uptake (Bartschat and Lindenmayer 1980), and equilibrium Ca–Ca exchange (Slaughter, Sutko, and Reeves 1983). The exchange is electroneutral and has a stoichiometry of 1:1 (Bartschat and Lindenmayer 1980). Ba^{2+}, Sr^{2+}, and even Cd^{2+} can substitute for Ca^{2+} in inducing Ca^{2+} efflux from cardiac membrane vesicles suggesting that these divalent cations can participate in exchange reactions with Ca^{2+} (Trosper and Philipson 1983). Recently, it has been found that Na–Ca and Ca–Ca exchange activities copurify, confirming that these activities are alternative modes of the same transporter (Philipson and Ward 1988).

Interestingly, vesicular Ca–Ca exchange activity is stimulated by certain monovalent cations. Either K^+, Rb^+, or Li^+ can fully activate Ca–Ca exchange (Philipson and Nishimoto 1981; Slaughter, Sutko, and Reeves 1983). The monovalent cations increase the V_{max} for the Ca–Ca exchange reaction without affecting the K_m (Ca^{2+}). The monovalent cation is not cotransported with Ca^{2+}, and optimal activity occurs when the monovalent cation is present on both sides of the vesicle membrane. Low concentrations of Na^+ can also activate Ca–Ca exchange; at higher Na^+ concentrations, however, the Na^+ will compete with Ca^{2+} for the transport sites on the exchanger and inhibit Ca–Ca exchange (Philipson and Nishimoto 1981; Slaughter *et al.* 1983).

It has been suggested that the monovalent cations that stimulate Ca–Ca exchange interact with the B site of the Na–Ca exchange carrier (cf. Section 2.3.5, Slaughter, Sutko, and Reeves 1983). According to this proposal, when Ca^{2+} is bound to the A site of the exchanger (as in Ca–Ca exchange) the B site exhibits a rather broad specificity and can accommodate K^+, Rb^+, and Li^+, as well as Na^+. In this mode of operation, the filling of the B site aids the translocation of Ca^{2+} bound to the A site, although the ion occupying the B site is not translocated during the exchange process.

Vesicular Na^+-dependent Ca^{2+} fluxes are generally unaffected by the presence of other monovalent cations (Philipson and Nishimoto 1981; Slaughter, Sutko, and Reeves 1983). For example, Slaughter, Sutko, and Reeves (1983) found that external K^+ had no effect on Na–Ca exchange when, under similar conditions, Ca–Ca exchange was stimulated 2-fold. The results with vesicles differ from results from squid axons; although Blaustein (1977) found little effect of K^+ on Na–Ca exchange, at least three other studies (Baker *et al.* 1969; Sjodin and Abercrombie 1978; DiPolo and Rojas 1984) report substantial effects of alkali metal ions on Na^+–Ca^{2+} exchange in contrast to the vesicle results.

2.5 Symmetry

The Na–Ca exchanger can exchange Na^+ on either side of the membrane for Ca^{2+} on the opposite side. Therefore, the Na–Ca exchange protein must have binding sites for both Na^+ and Ca^{2+} on both intra- and extracellular surfaces of the protein. It would be of much interest to characterize both the intra- and extracellular binding sites of the exchanger for Na^+ and Ca^{2+}. Plasma membrane vesicle preparations, however, contain mixtures of inside-out and right-side-out vesicles, and this introduces experimental difficulties. Vesicular Na–Ca exchange is most conventionally studied as Na_i^+-dependent Ca^{2+} uptake. To prepare vesicles for the Ca^{2+} uptake reaction, Na^+ loading is usually first accomplished by passive diffusion. Since both inside-out and right-side-out vesicles will become Na^+ loaded during this step, both types of vesicles will then participate in a subsequent Na_i^+-dependent Ca^{2+} uptake reaction. If the two types of vesicles have different properties, the interpretation of data becomes complicated. The ideal approach to solve this problem would be to physically separate the inside-out and right-side-out vesicles, but, despite much effort, this has not proven feasible.

Methods to circumvent this problem, however, have been developed for use with cardiac sarcolemmal vesicles. These methods take advantage of ion pumps which are also present in the isolated vesicles. Thus, the inside-out vesicles in a mixed population of vesicles can be selectively preloaded with either Na^+ or Ca^{2+} through the actions of either the ATP-dependent Na^+ or Ca^{2+} pumps. These inside-out vesicles can then be used for Na_i^+-dependent Ca^{2+} uptake or Na_o^+-dependent Ca^{2+} efflux reactions, respectively. Experiments are performed in parallel using the total population of vesicles (preloaded with Na^+ or Ca^{2+} by passive diffusion) and the results are compared.

Philipson and Nishimoto (1982a) have used this approach to study the symmetry of Na_i^+-dependent Ca^{2+} uptake reactions. They concluded that the Ca^{2+} binding sites on the two surfaces of the exchanger had similar K_m values. Likewise, the Na–Ca exchange of both the inside-out and right-side-out vesicles responded in an identical manner to changes in pH and membrane potential.

The nature of the Na^+ binding sites has been addressed in two studies. Kadoma *et al.* (1982) reported a modest difference in the K_m values for Na^+ on the intra- and extracellular surfaces of the exchanger of 22 *vs.* 31 mM respectively), while Philipson (1985b) found no difference in the affinities (30 mM) of the Na^+ binding sites on the opposing surfaces of the membrane. Thus, the data from experiments with sarcolemmal vesicles suggest that the exchanger can behave as a functionally symmetric transporter. *In vivo*, however, the exchanger will be exposed to different environments at its two surfaces. Thus, the exchanger will experience very different ionic, metabolic and regulatory environments at the intra- and extracellular surfaces. This environmental asymmetry could then induce a functional asymmetry in the Na^+–Ca^{2+} exchange process.

In contrast to the vesicle results, experiments with more intact preparations have demonstrated apparent asymmetries in the affinity of the exchanger for Ca^{2+}. For example, in heart (Wakabayashi and Goshima 1981; Kimura, Miyamae, and Noma 1987), the apparent affinity for Ca^{2+} at the external cell surface is lower than that found in vesicles. Also, in squid axon (Baker *et al.* 1969; Blaustein, Russell, and de Weer 1974), the apparent affinity of the exchanger for Ca^{2+} is much higher at the internal surface than at the external surface of the membrane. The reason for the discrepancy between results with vesicles and cells is not clear but could have several causes.

1. During the preparation of vesicles an important regulatory factor may be lost.
2. The experiments with intact preparations may not always have been performed under symmetrical conditions. For example, Na^+ will raise the K_m for Ca^{2+} due to competitive interactions. If the determinations of

K_m (Ca^{2+}) are not done with equal concentrations of Na^+ present, apparent asymmetries could arise. Likewise, since EGTA lowers the K_m (Ca^{2+}), measurements of the external K_m (Ca^{2+}) in the absence of EGTA and measurements of the internal K_m (Ca^{2+}) in the presence of EGTA could also produce apparent asymmetries.

3. Perhaps the exchangers of heart and squid axon have different properties, and it is improper to assume that experiments with axon and sarcolemmal vesicles should agree.
4. Both measurements with vesicles and intact preparations have limitations (cf. Section 2.1), and one should not assume that either technique produces 'correct' results.

2.6 Regulation

Many interesting and novel ways to modulate Na–Ca exchange in plasma membrane vesicles have been described. Studies on the regulation of Na–Ca exchange in vesicles, however, have in some ways been disappointing. No regulatory mechanism of unequivocal physiological significance has been described. For example, although activity can be modulated by Ca^{2+}-dependent phosphorylation reactions in the squid giant axon (DiPolo and Beaugé 1987), analogous regulatory pathways have not been described in vesicular preparations. The one exception is the study of Caroni and Carafoli (1983) who found that a calmodulin-dependent kinase and phosphatase system in cardiac sarcolemmal vesicles could modulate exchange activity. This observation, however, has been difficult for others to reproduce. It is unclear whether important regulatory factors are lost from plasma membranes during isolation procedures or whether, perhaps, the exchanger of squid axons has some unique properties. Nevertheless, the regulatory pathways which have been described are of potential physiological significance and have provided much information on the exchanger. We will describe the properties and speculate on the possible roles of each regulatory mechanism.

2.6.1 Proteinase activation

Mild treatment of cardiac sarcolemmal vesicles with a variety of proteinases stimulates Na–Ca exchange activity (Philipson and Nishimoto 1982b). The stimulation can be as much as 5-fold and is primarily due to an increase in the affinity of the exchanger for Ca^{2+}. The exchanger is remarkably resistant to inactivation by proteolysis and activity survives even harsh treatment (Wakabayashi and Goshima 1982; Hale *et al.* 1984). The stimulation of the exchanger by proteinase treatment may be due to a direct modulation of the exchanger protein, and it has recently been suggested that mild chrymotrypsin treatment converts the exchanger from a 120 000 M_r protein to a 70 000 M_r protein (Philipson and Ward 1988). Activation of the exchanger by intracellular proteases during pathological situations may be a mechanism for maintaining a normal intracellular ionic environment.

2.6.2 Sensitivity to lipid environment

Na–Ca exchange activity is readily modulated by changes in the membrane lipid environment. The effects have been studied using phospholipases (Philipson, Frank, and Nishimoto 1983; Philipson and Nishimoto 1984), a variety of amphiphiles (Philipson 1984; Philipson and Ward 1985, 1987), and by solubilization–reconstitution techniques (Soldati, Longoni, and Carafoli 1985; Vemuri and Philipson 1988a). The most notable effect is that Na–Ca exchange is stimulated by particular anionic lipid components. For example, treatment with phospholipase D, which converts neutral phospholipids to anionic phosphatidic acid, stimulates sarcolemmal exchange activity by 300 per cent (Philipson and Nishimoto 1984). Philipson and collaborators have developed a model (Philipson and Ward 1987) for the action of amphiphiles on Na–Ca exchange. This model has the following elements:

1. To stimulate exchange, an amphiphile must have an anionic head group. The degree of stimulation depends on the nature of the head group, implying some specificity in the interaction with the exchanger. This information comes from experiments in which the potency of a variety of dodecyl derivatives with different head group substituents was examined (Philipson 1984). For example, dodecylsulphate was a more potent stimulator then lauric acid.
2. The stimulation of Na–Ca exchange by an anionic amphiphile is potentiated if the amphiphile also causes some disorder within the lipid bilayer. This is demonstrated by the greater potency of unsaturated *vs.* saturated fatty acids (Philipson and Ward 1985) or by doxylstearic acids *vs.* stearic acids (Philipson and Ward 1987) to stimulate exchange activity.
3. Lipid disorder is most effective in stimulating Na–Ca exchange activity if it is localized toward the centre of the lipid bilayer. Thus, for example, doxylstearates stimulate Na–Ca exchange activity more potently if the lipid-perturbing doxyl group is in the 16-position (near the bilayer centre) rather than in the 5-position (toward the lipid–water interface) (Philipson and Ward 1987).

The stimulation by anionic amphiphiles is primarily due to an increase in

Ca^{2+} affinity and implies the anionic charge modifies the exchanger Ca^{2+} binding site. In contrast, many cationic amphiphiles (e.g. dodecylamine) are moderately potent inhibitors of Na–Ca exchange, and uncharged amphiphiles (e.g. laurylacetate or methyl esters of fatty acids) have little effect on exchange. The effects of charged amphiphiles on Na–Ca exchange are not due to a modification of the membrane surface charge. In fact, Na–Ca exchange activity is impervious to the Ca^{2+} concentration in the diffuse double layer but responds to the Ca^2 concentration of the bulk solution. This was demonstrated in experiments using dimethonium, an organic divalent cation, to screen membrane surface charge and thus modulate double layer Ca^{2+} (Bers, Philipson and Peskoff 1985).

Solubilization–reconstitution experiments have helped define the interactions between phospholipids and the Na–Ca exchanger (Soldati, Longoni, and Carafoli 1985; Vemuri and Philipson 1988a). Detergent-solubilized sarcolemma was reconstituted into defined lipid mixtures. The presence of specific anionic phospholipids was required for optimal exchange activity in the proteoliposomes. For example, high exchange activity was observed after reconstitution into vesicles containing phosphatidic acid, cardiolipin, or phosphatidylserine. In contrast, phosphatidylinositol and phosphatidylglycerol were unable to support high exchange activity. Phosphatidic acid, cardiolipin, and phosphatidylserine have well exposed anionic charges whereas for phosphatidylinositol and phosphatidylglycerol the negative charge is somewhat shielded. This may explain the differences in the ability of the different anionic phospholipids to interact with the exchanger. When sarcolemma was reconstituted into asolectin rather than a defined lipid mixture, exchange activity was also quite high. The high activity is surprising since asolectin does not have a high content of anionic phospholipids. An additional finding was that, in addition to an anionic phospholipid, cholesterol was required for recovery of activity (at least in the case of proteoliposomes containing a mixture of phosphatidylcoline and phosphatidylserine).

In general, the exchanger was found to be more sensitive to the lipid environment than other plasma membrane transporters such as the Na^+, K-ATPase, or the ATP-dependent Ca^{2+} pump. Nevertheless, the lipid composition of plasma membranes does not usually undergo large variation. Thus, although a proper lipid environment is essential for the exchanger, it is not obvious whether exchanger–lipid interactions are part of physiologically relevant regulation pathways. A possible situation where lipid alteration might affect the exchanger is myocardial ischaemia. After relatively short periods of ischaemia, free fatty acid concentrations rise to such an extent that they could alter Na–Ca exchange activity (Chien *et al.* 1984; Philipson and Ward 1985). Also, anionic lipid components of the phosphatidylinositol cycle (e.g. phosphatidic acid or phosphatidylinositol 4,5-bisphosphate) could rapidly modulate exchange activity in response to a variety of agonists.

2.6.3 Modulation by pH

The Na_i^+-dependent Ca^{2+} uptake of cardiac sarcolemmal vesicles is affected by extravesicular pH (Philipson *et al.* 1982; Wakabayashi and Goshima 1982; Slaughter, Sutko, and Reeves 1983). Activity is inhibited at low pH and stimulated at high pH. For example, at a Ca^{2+} concentration of 15 μM, the Ca^{2+} uptake is inhibited by about 50 per cent at pH 6.7 and more than doubled at pH 8.5 in comparison with activity at pH 7.4. The effects of protons appear to be competitive with Ca^{2+} and are therefore more marked at lower Ca^{2+} concentrations. The dependence of Na–Ca exchange on pH suggests that a histidine residue may be involved in the exchange process. Reconstitution experiments (Vemuri and Philipson 1988a) indicate that the effects of pH are due to changes in the ionization state of the exchanger itself rather than to effects on the ionization of membrane lipids. Analogous effects of pH on Na–Ca exchange have also been observed in the squid axon (DiPolo and Beaugé 1982). During myocardial ischaemia and other pathological conditions intracellular pH can fall. Under these conditions, protons would effectively compete with Ca^{2+} for binding sites on the exchanger and compromise the ability of cells to extrude Ca^{2+} via Na–Ca exchange.

2.6.4 Interaction with Ca–EGTA complexes

Extravesicular EGTA can directly affect Na_i^+-dependent Ca^{2+} uptake. For example, Trosper and Philipson (1984) measured the Na–Ca exchange activity of sarcolemmal vesicles at equal free Ca^{2+} concentrations in the presence and absence of EGTA. The presence of EGTA increased the exchanger Ca^{2+} affinity and stimulated activity. The reason for the stimulation is unclear, but the exchanger behaves as if some of the Ca^{2+} bound to the EGTA were directly accessible to the exchanger binding site. An analogous phenomenon has also been observed for the plasma membrane ATP-dependent Ca^{2+} pump (Schatzmann 1973; Sarkadi, Shubert, and Gardos 1979).

Modulation of Ca^{2+} transport pathways by Ca^{2+} chelators presents an experimental difficulty. Use of low Ca^{2+} solutions (< 2 μM) requires the presence of Ca^{2+} buffers such as EGTA. If the chelator itself can affect results, however, the interpretation of measurements may be questionable. Measurements of the Ca^{2+} dependency of the plasma membrane Ca^{2+} pump are usually done in the presence of EGTA. This results in the determination of a low K_m value. In contrast, vesicular exchange measurements are most commonly made in the absence of Ca^{2+} buffering and relatively high K_m values are obtained. When both transport systems are measured under similar conditions (Gill, Chueh, and Whitlow 1984), however, much of the large apparent difference in the K_m (Ca^{2+}) values disappears. The Ca^{2+}

concentration ranges where the two cellular Ca^{2+} extrusion mechanisms operate may not be as different as is sometimes thought. In this regard, it should be pointed out that virtually all measurements of exchange activity in internally perfused cell systems (e.g. squid axons, barnacle muscles, myocardial cells) have relied on high concentrations of EGTA to buffer Ca_i^{2+}. Given the effects of EGTA and other Ca^{2+} chelators on the kinetics of Na–Ca exchange in vesicles one must wonder whether the use of these agents in internally perfused cells distorts measurements of the exchange system.

In vivo, intracellular Ca^{2+} buffers may act to lower the apparent K_m (Ca^{2+}). Cheon and Reeves (1988a) have recently proposed that the Na–Ca exchanger has a relatively high turnover and under some conditions may be limited by Ca^{2+} diffusion. Intracellular Ca^{2+} buffers might then prevent local Ca^{2+} depletion and help maintain exchange activity (cf. Section 2.3.3).

2.6.5 Redox modification

Reeves, Bailey, and Hale (1986) have found that redox modification can markedly stimulate the Na–Ca exchange activity of cardiac sarcolemmal vesicles. The stimulation could be observed under a variety of conditions. For example, pretreatment of vesicles with either $FeSO_4$ plus DTT, or with a mixture of oxidized and reduced glutathione, gave stimulation. The critical requirement appeared to be the simultaneous presence of both an oxidizing and reducing agent, and a transition metal ion (e.g. Fe^{2+}). A possible explanation is that the redox modulation involves a thiol–disulphide interchange within the exchanger protein. That is, the reducing agent would break an existing disulphide linkage, and the oxidizing agent would cause a different disulphide bond to form.

It was speculated that the redox stimulation of exchange activity may have physiological significance as a protective response of myocytes to pathological situations that involve oxidative stress. Alternatively, redox modification of Na–Ca exchange activity may be part of more general regulatory pathways involving the cellular redox state.

2.6.6 Ca^{2+} regulatory sites

In cardiac sarcolemmal vesicles, Na_i^+-dependent Ca^{2+} uptake is stimulated by intravesicular Ca^{2+}, an effect which is due to a decrease in the K_m for external Ca^{2+} (Reeves and Poronnik 1987). The effect requires Ca^{2+} concentrations in the range of 0.1–0.5 mM and can be reversed by treating the vesicles with EGTA. It was shown that these effects represented a true modulation of Na–Ca exchange activity and were not due to Ca–Ca exchange.

De la Peña and Reeves (1987) have suggested that intravesicular Ca^{2+} stimulates Ca^{2+} uptake by a local electrostatic effect of internally bound Ca^{2+}

on the Na–Ca exchanger. Consistent with this interpretation, they found that certain organic cations which bind to membranes (i.e. quinacrine, tetraphenylphosphonium) could replace internal Ca^{2+} as a stimulator of Na–Ca exchange. There was some specificity to the effect in that several other lipophilic cations could not stimulate exchange under any conditions. Exchange activity may be regulated by the distribution of charged groups between the inner and outer surfaces of the vesicle membrane. This interpretation could provide a unitary explanation for the activating effects of extravesicular anionic amphiphiles (described above) and intravesicular bound cations.

The effect of intravesicular Ca^{2+} on Na–Ca exchange activity complicates the analysis of the kinetic properties of the vesicular exchange system. When the Na_i^+-dependent Ca^{2+} uptake reaction is initiated, the intravesicular Ca^{2+} concentration will immediately rise. The rise in Ca_i^{2+} lowers the K_m for external Ca^{2+} and stimulates further uptake. Since the K_m (Ca^{2+}) changes during the measurement, true initial rate measurements are difficult to obtain. This may partially account for the wide range of apparent K_m (Ca^{2+}) values for the exchanger in the literature.

In squid giant axons (DiPolo 1979; DiPolo and Beaugé 1987), barnacle muscle fibres (Rasgado-Flores and Blaustein 1987), and internally perfused heart cells (Kimura *et al.* 1986, 1987), Na_i^+-dependent Ca^{2+} influx appears to require the presence of cytoplasmic Ca^{2+}. While it is tempting to attribute the vesicular and cellular effects of internal Ca^{2+} to the same phenomenon, the intact systems require only submicromolar concentrations of Ca^{2+} (at least in the presence of ATP) while much higher concentrations are required in the vesicles. Thus, it seems unlikely that the two effects are related.

Moreover, in vesicles, the operation of the Na–Ca exchange system does not *require* the presence of *trans* Ca^{2+}. This was clear from the earliest studies of Na_o^+-dependent Ca^{2+} efflux in vesicles, which were carried out in the presence of 0.1 mM EGTA (cf. Fig. 2.1B). More recent studies indicate that the presence of 10 mM EDTA in the external medium does not reduce the ability of either low or high concentration of Na^+ to stimulate Ca^{2+} efflux from the vesicles. Moreover, reconstituted proteoliposomes containing 80 mM EDTA appear to take up Ca^{2+} via Na–Ca exchange at the same rate as control proteoliposomes containing NaCl (J. Durkin and J. P. Reeves, unpublished observations). This, then, represents a rather striking disparity between the properties of the exchange system in internally perfused cells and vesicles; an explanation for this difference is not apparent at the present time.

2.6.7 Methylation of the exchanger

Methylation of cardiac sarcolemmal vesicles using S-adenosyl-L-methionine as methyl donor inhibits Na–Ca exchange activity by up to 50 per cent. The data strongly suggest that the inhibition is due to a direct methylation of the exchanger protein (Vemuri and Philipson 1988b), although an alternative, and less likely explanation, involving methylation of phosphatidylethanolamine has been published (Panagia *et al.* 1987). In general, the regulatory role of membrane methylation reactions had always been speculative and controversial, and it is unclear whether the effect on the exchanger has any physiological significance.

2.6.8 Concluding comments on regulation

All of the methods just discussed for stimulating Na_i^+-dependent Ca^{2+} uptake lower the K_m of the exchanger for Ca^{2+}, although in some cases V_{max} may also increase. This raises the possibility that these various stimulatory effects may involve a common mechanism. However, studies with the sulphhydryl reagent *p*-chloromercuribenzene sulphonic acid (*p*CMBS) suggest that there are at least two different modes of regulation of the exchanger (Bailey and Reeves 1986; Bailey, Poronnik, and Reeves 1988). Briefly, the findings that lead to this conclusion are that vesicles stimulated with chymotrypsin or redox reagents showed an increased sensitivity to inhibition by 100 μM *p*CMBS compared to control vesicles. In contrast, vesicles in which exchange activity had been stimulated by intravesicular Ca^{2+}, sodium dodecylsulphate, or phospholipase D treatment showed the same sensitivity to inhibition by *p*CMBS as control vesicles. It is interesting to note that the former treatments involve probable changes in the structure of the exchange carrier, either through proteolytic cleavage or (perhaps) thiol–disulphide interchange; the latter treatments, however, appear to involve ionic interactions with the exchange carrier and, in the case of intravesicular Ca^{2+} and dodecylsulphate, the stimulatory effects are reversible. The results suggest that activity of the exchange carrier can be regulated by two different pathways, one involving reversible ionic interactions which do not affect the sensitivity of the carrier's sulphydryl groups to *p*CMBS, and the others involving irreversible structural changes in the exchange carrier.

It is interesting that the plasma membrane ATP-dependent Ca^{2+} pump can also be stimulated by some of the same interventions as the Na–Ca exchanger. Both plasma membrane Ca^{2+} transport pathways can be stimulated by proteinase treatment, EGTA, and by anionic amphiphiles. Perhaps the two Ca^{2+} transporters have some common structural features. A noticeable difference in the modulation of Ca^{2+} pump and the exchanger is that Ca^{2+} pump activity can be enhanced by a direct interaction with calmodulin. Such an interaction has not been reported for the exchanger.

Finally, the regulatory effects of the various treatments discussed above have been described only in cardiac sarcolemmal vesicles. In contrast, membrane vesicle preparations from bovine adrenal chromaffin cells (Bailey and Reeves 1987), tracheal or bovine aorta smooth muscle (Slaughter, Welton, and Morgan 1987; R. S. Slaughter and G. Kaczorowski, personal communication), and the brine shrimp *Artemia* (Cheon and Reeves 1988b), do not exhibit stimulation of Na–Ca exchange activity when treated with chymotrypsin, redox reagents, or intravesicular Ca^{2+}. Thus, it seems likely that the plasticity of the cardiac system may be a tissue specific property of the cardiac exchanger; perhaps this is a manifestation *in vitro* of a specific cardiac regulatory process which is required to meet the special demands imposed by continual fluctuations of $[Ca^{2+}]_i$ in heart cells, and the need to regulate total cellular Ca^{2+} content in a highly adaptive manner.

References

Bailey, C. A. and Reeves, J. P. (1986). Vicinal dithiol groups and activation of the cardiac sodium–calcium exchange system. *Biophys. J.* **48**, 345a.

Bailey, C. A. and Reeves, J. P. (1987). Sodium–calcium exchange in plasma membrane vesicles prepared from bovine adrenal chromaffin cells. *Fed. Proc.* **46**, 367.

Bailey, C. A., Poronnik, P., and Reeves, J. P. (1988). Na–Ca exchange in cardiac sarcolemmal vesicles. In: *Cellular calcium regulation* (eds. D. R. Pfeiffer, J. B. McMillan, and S. Little) pp. 97–104. Plenum Press, New York.

Baker, P. F., Blaustein, M. P., Hodgkin, A. L., and Steinhardt, R. A. (1969). The influence of calcium on sodium efflux in squid axons. *J. Physiol.* **200**, 431–58.

Barry, W. H. and Smith, T. W. (1982). Mechanisms of transmembrane calcium movement in cultured chick embryonic ventricular cells. *J. Physiol.* **325**, 243–60.

Bartschat, D. K. and Lindenmayer, G. E. (1980). Calcium movements promoted by vesicles in a highly enriched sarcolemma preparation from canine ventricle. *J. Biol. Chem.* **255**, 9626–34.

Barzilai, A. and Rahamimoff, H. (1987). Stoichiometry of the sodium–calcium exchanger in nerve terminals. *Biochemistry* **26**, 6113–18.

Bers, D. M., Philipson, K. D., and Nishimoto, A. Y. (1980). Sodium–calcium exchange and sidedness of isolated cardiac sarcolemmal vesicles. *Biochim. Biophys. Acta* **601**, 358–371.

Bers, D. M., Philipson, K. D., and Peskoff, A. (1985). Calcium at the surface of cardiac plasma membrane vesicles: Cation binding, surface charge screening and Na–Ca exchange. *J. Membr. Biol.* **85**, 251–61.

Blaustein, M. P. (1977). Effects of internal and external cations and of ATP on sodium–calcium and calcium–calcium exchange in squid axons. *Biophys. J.* **20**, 79–111.

Blaustein, M. P. (1984). The energetics and kinetics of sodium–calcium exchange in barnacle muscles, squid axons and mammalian heart: The role of ATP. In: *Electrogenic transport: fundamental principles and physiological implications* (eds. M. P. Blaustein and M. Lieberman) pp. 129–47. Raven Press, New York.

Blaustein, M. P. and Russell, J. M. (1975). Sodium–calcium exchange and calcium–calcium exchange in internally dialyzed squid giant axons. *J. Membr. Biol.* **22**, 285–312.

Blaustein, M. P., Russell, J. M., and deWeer, P. (1974). Calcium efflux from internally dialyzed squid axons: The influence of external and internal cations. *J. Supramol. Struct.* **2**, 558–81.

Bradley, M. P. and Forrester, I. T. (1980). A sodium–calcium exchange mechanism in plasma membrane vesicles isolated from ram sperm flagella. *FEBS Lett.* **121**, 15–18.

Caroni, P. and Carafoli, E. (1983). The regulation of the Na–Ca exchanger of heart sarcolemma. *Eur. J. Biochem.* **132**, 451–60.

Caroni, P., Reinlib, L., and Carafoli, E. (1980). Charge movements during the Na–Ca exchange in heart sarcolemmal vesicles. *Proc. Natl. Acad. Sci. USA*, **77**, 6354–8.

Chapman, R. A. and Tunstall, J. (1980). The interaction of sodium and calcium ions at the cell membrane and the control of contractile strength in frog atrial muscle. *J. Physiol.* **305**, 109–23.

Chapman, R. A., Corey, A., and McGuigan, J. A. S. (1983). Sodium–calcium exchange in mammalian heart: The maintenance of low intracellular calcium concentration. In: *Cardiac metabolism* (A. J. Drake-Holland and M. I. M. Noble), pp. 117–49. Wiley, New York.

Cheon, J. and Reeves, J. P. (1988a). Site density of the sodium–calcium exchange carrier in reconstituted vesicles from bovine cardiac sarcolemma. *J. Biol. Chem.* **263**, 2309–15.

Cheon, J. and Reeves, J. P. (1988b). Sodium–calcium exchange in membrane vesicles from brine shrimp. *Biophys. J.* **53**, 141a.

Chien, K. R., Han, A., Sen, A., Buja, M., and Willerson, J. T. (1984). Accumulation of unesterified arachidonic acid in ischemic canine myocardium. *Circ. Res.* **54**, 313–22.

Crompton, M., Künzi, M., and Carofoli, E. (1977). The calcium-induced and sodium-induced effluxes of calcium from heart mitochondria. *Eur. J. Biochem.* **79**, 549–58.

de la Peña, P. and Reeves, J. P. (1987). Inhibition and activation of Na–Ca exchange activity by quinacrine. *Am. J. Physiol.* **252**, C24–C29.

Dietschy, J. (1978). The uptake of lipids in the intestinal mucosa. In: *Physiology of membrane disorders* (eds T. E. Andreoli, J. F. Hoffman and D. P. Fanestil) pp. 577–92. Plenum Press, New York.

DiPolo, R. (1979). Ca influx in internally dialyzed squid axons. *J. Gen. Physiol.* **73**, 91–113.

DiPolo, R. and Beaugé, L. (1982). The effect of pH on Ca^{2+} extrusion mechanisms in dialyzed squid axons. *Biochim. Biophys. Acta* **688**, 227–45.

DiPolo, R. and Beaugé, L. (1987). In squid axons, ATP modulates Na^+–Ca^{2+} exchange by a Ca_i-dependent phosphorylation. *Biochim. Biophys. Acta* **897**, 347–54.

DiPolo, R. and Rojas, H. (1984). Effect of internal and external K^+ on Na^+–Ca^{2+} exchange in dialyzed squid axons under voltage clamp conditions. *Biochim. Biophys. Acta* **776**, 313–16.

Dixon, S. J., Cohen, S., Cragoe, E. J., Jr., and Grinstein, S. (1987). Estimation of the number and turnover rate of Na^+–H^+ exchangers in lymphocytes. *J. Biol. Chem.* **262**, 3626–32.

Eigen, M. and Hammes, G. G. (1963). Elementary steps in enzyme reactions. *Adv. Enzymol.* **25**, 1–38.

Eisner, D. A. and Lederer, W. J. (1985). Na–Ca exchange: Stoichiometry and electrogenicity. *Am. J. Physiol.* **248**, C189–202.

Gill, D. L., Chueh, S-H., and Whitlow, C. L. (1984). Functional importance of the synaptic plasma membrane calcium pump and sodium–calcium exchanger. *J. Biol. Chem.* **259**, 10807–13.

Glitsch, H. G., Reuter, H., and Scholz, H. (1970). The effect of internal sodium concentration on calcium fluxes in isolated guinea pig auricles. *J. Physiol.* **209**, 25–43.

Grover, A. K., Kwan, C-Y., Rangachari, P. K., and Daniel, E. E. (1983). Na–Ca exchange in a smooth muscle plasma membrane-enriched fraction. *Am. J. Physiol.* **244**, C158–65.

Hale, C. C., Slaughter, R. S., Ahrens, D., and Reeves, J. P. (1984). Identification and partial purification of the cardiac sodium–calcium exchange protein. *Proc. Natl. Acad. Sci. USA*, **81**, 6569–73.

Hopfer, U. (1981). Kinetic criteria for carrier-mediated transport mechanisms in membrane vesicles. *Fed. Proc.* **40**, 2480–5.

Hume, J. R. and Uehara, A. (1986). 'Creep Currents' in single frog atrial cells may be generated by electrogenic Na–Ca exchange. *J. Gen. Physiol.* **87**, 857–84.

Jayakumar, A., Cheng, L., Liang, C. T., and Sacktor, B. (1984). Sodium gradient-dependent calcium uptake in renal basolateral membrane vesicles. *J. Biol. Chem.* **259**, 10827–33.

Jundt, H., Porzig, H., Reuter, H., and Stucki, J. W. (1975). The effect of substances releasing intracellular calcium ions on sodium-dependent calcium efflux from guinea pig auricles. *J. Physiol.* **246**, 229–53.

Kaback, H. R. (1974). Transport studies in bacterial membrane vesicles. *Science* **186**, 882–92.

Kaczorowski, G. J., Costello, L., Dethmers, J., Trumble, M. J., and Vandlen, R. L. (1984). Mechanisms of Ca^{2+} transport in plasma membrane vesicles prepared from cultured pituitary cells. I. Characterization of Na–Ca exchange activity. *J. Biol. Chem.* **259**, 9395–403.

Kadoma, M., Froehlich, J., Reeves, J., and Sutko, J. (1982). Kinetics of sodium-ion-induced calcium ion release in calcium-ion-loaded cardiac sarcolemmal vesicles: Determination of initial velocities by stopped flow spectrophotometry. *Biochemistry* **21**, 1914–18.

Kenyon, J. and Sutko, J. L. (1987). Calcium- and voltage-activated plateau currents of cardiac Purkinje fibers. *J. Gen. Physiol.* **89**, 921–58.

Kimura, J., Noma, A., and Irisawa, H. (1986). Na–Ca exchange current in mammalian heart cells. *Nature* **319**, 596–7.

Kimura, J., Miyamae, S., and Noma, A. (1987). Identification of sodium–calcium exchange current in single ventricular cells in guinea pig. *J. Physiol.* **384**, 199–222.

Lederer, W. J. and Nelson, M. T. (1983). Effects of extracellular sodium on calcium efflux and membrane current in single muscle cells from the barnacle. *J. Physiol.* **341**, 325–39.

Mechmann, S. and Pott, L. (1986). Identification of Na–Ca exchange current in single cardiac myocytes. *Nature* **319**, 597–9.

Michaelis, M. L. and Michaelis, E. K. (1981). Ca^{++} fluxes in resealed synaptic plasma membrane vesicles. *Life Sci.* **28**, 37–45.

Miyamoto, H. and Racker, E. (1980). Solubilization and partial purification of the

Ca^{2+}–Na^{+} antiporter from the plasma membrane of bovine heart. *J. Biol. Chem.* **255**, 2656–8.

Morel, N. and Godfraind, T. (1984). Sodium–calcium exchange in smooth muscle microsomal fractions. *Biochem. J.* **218**, 421–7.

Mullins, L. J. and Brinley, F. J., Jr. (1975). Sensitivity of calcium efflux from squid axons to changes in membrane potential. *J. Gen. Physiol.* **65**, 135–52.

Osses, L., Condrescu, M., and DiPolo, R. (1986). ATP-dependent calcium pump and Na–Ca exchange in plasma membrane vesicles from squid optic nerve. *Biochim. Biophys. Acta* **860**, 583–91.

Panagia, V., Makino, N., Ganguly, P. K., and Dhalla, N. S. (1987). Inhibition of Na^{+}–Ca^{2+} exchange in heart sarcolemmal vesicles by phosphatidylethanolamine N-methylation. *Eur. J. Biochem.* **166**, 597–603.

Philipson, K. D. (1984). Interaction of charged amphiphiles with Na^{+}–Ca^{2+} exchange in cardiac sarcolemmal vesicles. *J. Biol. Chem.* **259**, 13999–14002.

Philipson, K. D. (1985a). Sodium–calcium exchange in plasma membrane vesicles. *Ann. Rev. Physiol.* **47**, 561–71.

Philipson, K. D. (1985b). Symmetry properties of the Na^{+}–Ca^{2+} exchange mechanism in cardiac sarcolemmal vesicles. *Biochim. Biophys. Acta* **821**, 367–76.

Philipson, K. D. and Nishimoto, A. Y. (1980). Na–Ca exchange is affected by membrane potential in cardiac sarcolemmal vesicles. *J. Biol Chem.* **255**, 6880–2.

Philipson, K. D. and Nishimoto, A. Y. (1981). Efflux of Ca^{2+} from cardiac sarcolemmal vesicles. Influence of external Ca^{2+} and Na^{+}. *J. Biol. Chem.* **256**, 3698–3702.

Philipson, K. D. and Nishimoto, A. Y. (1982a). Na–Ca exchange in inside-out cardiac sarcolemmal vesicles. *J. Biol. Chem.* **257**, 5111–17.

Philipson, K. D. and Nishimoto, A. Y. (1982b) Stimulation of Na–Ca exchange in cardiac sarcolemmal vesicles by proteinase pretreatment. *Am. J. Physiol.* **243**, C191–C195.

Philipson, K. D. and Nishimoto, A. Y. (1984). Stimulation of Na–Ca exchange in cardiac sarcolemmal vesicles by phospholipase D. *J. Biol. Chem.* **259**, 16–19.

Philipson, K. D. and Ward, R. (1985). Effects of fatty acids on Na^{+}–Ca^{2+} exchange and Ca^{2+} permeability of cardiac sarcolemmal vesicles. *J. Biol. Chem.* **260**, 9666–71.

Philipson, K. D. and Ward, R. (1987). Modulation of Na^{+}–Ca^{2+} exchange and Ca^{2+} permeability in cardiac sarcolemmal vesicles by doxylstearic acids. *Biochim. Biophys. Acta* **897**, 152–8.

Philipson, K. D. and Ward, R. (1988), Purification of the cardiac Na^{+}–Ca^{2+} exchange protein. *Biochim. Biophys. Acta* (in press).

Philipson, K. D., Bersohn, M. M., and Nishimoto, A. Y. (1982). Effects of pH on Na^{+}–Ca^{2+} exchange in canine cardiac sarcolemmal vesicles. *Circ. Res.* **50**, 287–93.

Philipson, K. D., Frank, J. S., and Nishimoto, A. Y. (1983). Effects of phospholipase C on the Na^{+}–Ca^{2+} exchange and Ca^{2+} permeability of cardiac sarcolemmal vesicles. *J. Biol. Chem.* **258**, 5905–10.

Pitts, B. J. R. (1979). Stoichiometry of sodium–calcium exchange in cardiac sarcolemmal vesicles *J. Biol. Chem.* **254**, 6232–5.

Pitts, B. J. R., Okhuysen, C. H., and Entman, M. L. (1980). Stoichiometry and kinetics of sodium–calcium exchange. In: *Calcium Binding Proteins: structure and function*, (eds F. L. Siegel, E. Carafoli, R. M. Kretsinger, D. H. MacLennan, and R. H. Wasserman) pp. 39–46. Elsevier Publishing, Amsterdam.

Rahamimoff, H. and Spanier, R. (1979). Sodium-dependent calcium uptake in membrane vesicles derived from rat brain synaptosomes. *FEBS Lett.* **104**, 111–14.

Rahamimoff, H. and Spanier, R. (1984). The asymmetric effect of lanthanides on Na^+-gradient-dependent Ca^{2+} transport in synaptic plasma membrane vesicles. *Biochim. Biophys. Acta* **773**, 297–89.

Rahamimoff, H., Abramovitz, E., Papazian, D., Goldin, S. M., and Spanier, R. (1980). Calcium transport systems in nerve terminals. Studies on membrane vesicles. *J. Physiol. (Paris)* **76**, 487–95.

Rasgado-Flores, H. and Blaustein, M. P. (1987). Na–Ca exchange in barnacle muscle cells has a stoichiometry of 3 Na: 1 Ca. *Am. J. Physiol.* **252**, C499–C504.

Reeves, J. P. (1985). The sarcolemmal sodium–calcium exchange system. *Curr. Top. Membr. Trans.* **25**, 77–127.

Reeves, J. P. and Hale, C. C. (1984). The stoichiometry of the cardiac sodium–calcium exchange system. *J. Biol. Chem.* **259**, 7733–9.

Reeves, J. P. and Poronnik, P. (1987). Modulation of Na^+–Ca^{2+} exchange in sarcolemmal vesicles by intravesicular Ca^{2+}. *Am. J. Physiol.* **252**, C17–C23.

Reeves, J. P. and Sutko, J. L. (1979). Sodium–calcium ion exchange in cardiac membrane vesicles. *Proc. Natl. Acad. Sci. USA* **76**, 590–4.

Reeves, J. P. and Sutko, J. L. (1980). Sodium–calcium exchange activity generates a current in cardiac membrane vesicles. *Science* **208**, 1461–4.

Reeves, J. P. and Sutko, J. L. (1983). Competitive interactions of sodium and calcium with the sodium–calcium exchange system of cardiac sarcolemmal vesicles. *J. Biol. Chem.* **258**, 3178–82.

Reeves, J. P., Bailey, C. A., and Hale, C. C. (1986). Redox modification of sodium–calcium exchange activity in cardiac sarcolemmal vesicles. *J. Biol. Chem.* **261**, 4948–55.

Reuter, H. and Seitz, N. (1968). The dependence of calcium efflux from cardiac muscle on temperature and external ion composition. *J. Physiol.* **195**, 451–70.

Saermark, T. and Gratzl, M. (1986). Na–Ca exchange in coated microvesicles. *Biochem. J.* **233**, 643–8.

Sarkadi, B., Shubert, A., and Gardos, G. (1979). Effect of Ca-EGTA buffers on active calcium transport in inside-out red cell membrane vesicles. *Experientia* **35**, 1045–7.

Schatzman, H. J. (1973). Dependence on calcium concentrations and stoichiometry of the calcium pump in human red cells. *J. Physiol.* **235**, 551–69.

Schellenberg, G. D. and Swanson, P. D. (1981). Sodium-dependent and calcium-dependent calcium transport by rat brain microsomes. *Biochim. Biophys. Acta* **648**, 13–27.

Schellenberg, G. D. and Swanson, P. D. (1982). Solubilization and reconstitution of membranes containing the Na^+–Ca^{2+} exchange carrier from rat brain. *Biochim. Biophys. Acta* **690**, 133–44.

Schilling, W. P., Schuil, D. W., Bagwell, E. E., and Lindenmayer, G. E. (1984). Sodium and potassium permeability of membrane vesicles in a sarcolemma-enriched preparation from canine ventricle. *J. Membr. Biol.* **77**, 101–14.

Sheu, S-S. and Blaustein, M. P. (1986). In: *The heart and cardiovascular system*, (eds H. A. Fozzard, E. Haber, R. B. Jennings, A. M. Katz, and H. E. Morgan) pp. 509–35. Raven Press, New York.

Sjodin, R. A. and Abercrombie, R. F. (1987). The influence of external cations and membrane potential on Ca-activated Na efflux in *Myxicola* giant axons. *J. Gen. Physiol.* **71**, 453–66.

Slaughter, R. S., Sutko, J. L., and Reeves, J. P. (1983). Equilibrium calcium-calcium exchange in cardiac sarcolemmal vesicles. *J. Biol. Chem.* **258**, 3183–90.

Slaughter, R. S., Welton, A. F., and Morgan, D. W. (1987). Sodium-calcium exchange in sarcolemmal vesicles from tracheal smooth muscle. *Biochim. Biophys. Acta* **904**, 92–104.

Soldati, L., Longoni, S., and Carafoli, E. (1985). Solubilization and reconstitution of the Na^+–Ca^{2+} exchanger of cardiac sarcolemma. *J. Biol. Chem.* **260**, 13321–7.

Trosper, T. L. and Philipson, K. D. (1983). Effects of divalent and trivalent cations on Na^+–Ca^{2+} exchange in cardiac sarcolemmal vesicles. *Biochim. Biophys. Acta* **731**, 63–8.

Trosper, T. L. and Philipson, K. D. (1984). Stimulatory effect of calcium chelators on Na^+–Ca^{2+} exchange in cardiac sarcolemmal vesicles. *Cell Calcium*, **5**, 211–22.

Ueda, T. (1983). Na^+–Ca^{2+} exchange activity in rabbit lymphocyte plasma membranes. *Biochim. Biophys. Acta* **734**, 342–46.

van Heeswijk, M. P. E., Geertsen, J. A. M., and van Os, C. N. (1984). Kinetic properties of the ATP-dependent Ca^{2+} pump and the Na^+–Ca^{2+} exchange system in basolateral membranes from rat kidney cortex. *J. Membr. Biol.* **79**, 19–31.

Vemuri, R. and Philipson, K. D. (1988a). Phospholipid composition modulates the Na^+–Ca^{2+} exchange activity of cardiac sarcolemma in reconstituted vesicles. *Biochim. Biophys. Acta* **937**, 258–68.

Vemuri, R. and Philipson, K. D. (1988b), Protein methylation inhibits Na^+–Ca^{2+} exchange activity in cardiac sarcolemmal vesicles. *Biochim. Biophys. Acta* **939**, 503–8.

Vigne, P., Jean, T., Barbry, P., Frelin, C., Fine, L. G., and Lazdunski, M. (1985). [^{3}H]ethyl-propylamiloride, a ligand to analyze the properties of the Na^+–H^+ exchange system in the membranes of normal and hypertrophied kidneys. *J. Biol. Chem.* **260** (14) 120–125.

Wakabayashi, S. and Goshima, K. (1981). Comparison of kinetic characteristics of Na^+–Ca^{2+} exchange in sarcolemma vesicles and cultured cells from chick heart. *Biochim. Biophys. Acta* **645**, 311–17.

Wakabayashi, S. and Goshima, K. (1982). Partial purification of Na–Ca antiporter from plasma membrane of chick heart. *Biochim. Biophys. Acta* **693**, 125–33.

3 Sodium–calcium exchange in secretory vesicles

T. Saermark

3.1 Introduction

The secretion of hormones and neurotransmitters involves the fusion of a secretory vesicle containing the secretory product with the plasma membrane, a process called exocytosis. The secretory vesicles contain several other substances in addition to the hormones and transmitter substances depending on the tissue, but certain generalizations can be made. They seem to contain binding proteins for their hormonal content and cations such as Ca^{2+}. Since the vesicles eventually release their contents to the cell surface they are conceptually very attractive as an internal Ca^{2+} buffering system with a role in the regulation of intracellular free Ca^{2+} in secretory cells. Such a role is further supported by their relative abundance in secretory cells. An Na–Ca exchange mechanism has recently been demonstrated in the membrane of secretory vesicles from chromaffin cells and from the posterior pituitary. This chapter will focus on the Na–Ca exchange process in secretory vesicles and its possible role in Ca^{2+} regulation in secretory cells.

3.2 The evidence for Na–Ca exchange across the secretory vesicle membrane

Intracellular Ca^{2+}-accumulating systems have been expected to involve an ATP-dependent mechanism in analogy to the sarcoplasmic reticulum, although mitochondria accumulate Ca^{2+} by other mechanisms (see elsewhere in this volume). Ca^{2+}-transporting ATPases were thought to be present in secretory vesicles from several tissues. However, their presence in secretory vesicle preparations seems to be due to contamination by other types of membrane. Chromaffin secretory vesicles were thought to have ATPases (Grafenstein and Neuman 1983; Hausler, Burger, and Niedermaier 1981; Niedermaier and Burger 1981), but they were found to be sensitive to inhibitors of mitochondrial Ca^{2+} uptake and may therefore have been contaminated. An ATP-dependent Ca^{2+}-accumulating system in secretory vesicles from the posterior pituitary was suggested by Douglas (1974), but the use of improved isolation methods failed to confirm this (Russell and Thorn 1975). Insulin-containing granules were supposed to have a Ca^{2+} ATPase (Howell 1984a), but its presence was probably due to contamination or a Mg^{2+}-dependent proton pump also activated by Ca^{2+} (Howell 1984b; Hutton 1984; Jones, Saermark, and Howell 1987). It is, however, quite clear that secretory vesicles in most tissues contain Ca^{2+}. Secretory vesicles from pancreatic beta-cells contain Ca^{2+} (Howell, Montague, and Tyhurst 1975). In chromaffin cells, secretory vesicles contain 60 nmol of Ca^{2+} per mg protein (Borowitz, Fuwa, and Weiner 1965; Bulenda and Gratzl 1985; Ravazolla 1976), but ATP does not support Ca^{2+} accumulation (Kostron *et al.* 1977; Krieger-Brauer and Gratzl 1982). Posterior pituitary secretory vesicles contain 25 nmol of Ca^{2+} per mg protein (Saermark *et al.* 1983), but ATP stimulates only microsomal Ca^{2+} accumulation (Conigrave *et al.* 1981; Russell and Thorn 1975), a type of Ca^{2+} accumulation also found in brain nerve endings (Blaustein *et al.* 1978). Cytochemical evidence also indicates that the secretory vesicles from most tissues contain a high concentration of Ca^{2+} (Wick and Heppler 1982). ^{31}P-NMR has been used to demonstrate high concentrations of Ca^{2+} in isolated cholinergic secretory vesicles (Whittaker 1982). But how are the secretory vesicles accumulating this calcium? The answer seems to be that it is catalysed by an Na–Ca exchange mechanism.

The first indication that an Na–Ca exchange mechanism operated across the secretory vesicle membrane came from work on secretory vesicles from chromaffin cells (Krieger-Brauer and Gratzl 1981; Phillips 1981) and pituitary nerve endings (Saermark *et al.* 1983). The basis for this discovery lay in the use of KCl, instead of NaCl, in buffers used for the isolation of secretory vesicles on Percoll gradients. Secretory vesicles purified by this procedure showed pronounced Ca^{2+} uptake and it was found that this was inhibited by Na^+ ions at concentrations above 30 mM (Krieger-Brauer and Gratzl 1982; Saermark *et al.* 1983). The Ca^{2+} accumulated by both posterior pituitary secretory vesicles and chromaffin secretory vesicles could be released by Na^+ but not K^+ (Krieger-Brauer and Gratzl 1982; Saermark *et al.* 1983). It was shown that $^{22}Na^+$ accumulated in intact secretory vesicles when Ca^{2+} was released (Saermark, Thorn, and Gratzl 1983). The uptake of Na^+ was reflected by an increase in the osmotic content of secretory vesicles from the posterior pituitary indicating an energy-dependent process (Saermark, Thorn, and Gratzl 1983). The conclusion from these studies was that an Na–Ca exchange mechanism was present in the secretory vesicle membrane (Krieger-Brauer and Gratzl 1982; Phillips 1981; Saermark *et al.* 1983; Saermark, Thorn, and Gratzl 1983). The properties of this Na^+–Ca^{2+} exchange system are summarized in Table 3.1.

Further studies were carried out on secretory vesicle ghosts from the

Table 3.1 Properties of the Na^+–Ca^{2+} exchange system in secretory vesicles

	Secretory vesicles from	
	Chromaffin cells	Posterior pituitary
Ca^{2+} content per mg	60–80 nmol	25 nmol
Na^+ content per mg	90 nmol	30 nmol
K_m for Ca^{2+}	0.2 μM	0.7 μM
V_{max} per mg*	14 nmol min^{-1}	10 nmol min^{-1}
Half-maximal Na^+ inhibition	30 mM	30 mM
Half-maximal Na^+ uptake (a: vesicles; b: ghosts)	15 mM (b)	15 mM (a)
Ca^{2+}–Ca^{2+} exchange inhibited by Mg^{2+}	yes	yes
Effect of ATP	no	no
Inhibition by ruthenium red	no	no
Intravesicular free Ca^{2+}	4–8 μM	?

This summarizes the more important features of the secretory vesicle Na^+–Ca^{2+} exchange system found in chromaffin cells and in the nerve endings from the posterior pituitary. All data refer to isolated vesicles and are taken from Krieger-Brauer and Gratzl (1982), Saermark *et al.* (1983), Saermark, Thorn, and Gratzl (1983), and from Bulenda and Gratzl (1985). Values are given per mg of protein except for * where mg of membrane-bound protein is used.

adrenal medulla (Krieger-Brauer and Gratzl 1983). These studies supported the concept of an Na–Ca exchanger operating in the secretory vesicle membrane. Only Na^+-loaded ghosts could accumulate Ca^{2+} and Ca^{2+} accumulation ceased when the outside Na^+ concentration was identical to the internal concentration (Krieger-Brauer and Gratzl 1983).

3.3. Characteristics of the Na–Ca exchange mechanism in secretory vesicles

The Na^+–Ca^{2+} exchange is not the only ion exchange mechanism across the membrane of secretory vesicles. Ca^{2+} can also exchange for Ca^{2+} (Saermark *et al.* 1983). This Ca–Ca exchange made the quantification of Ca^{2+} uptake difficult. The initial experiments showed a Ca^{2+} uptake that increased linearly with log(Ca^{2+}) (Krieger-Brauer and Gratzl 1981; Saermark *et al.* 1983b). However, when Mg^{2+} was added (1 mM) the Ca^{2+} accumulation showed saturation kinetics (Krieger-Brauer and Gratzl 1983; Saermark, Thorn, and Gratzl 1983). The affinity for Ca^{2+} is high and indicated that the Na–Ca exchange in secretory vesicles may operate at cytoplasmic concentrations of free Ca^{2+}. The apparent K_m value is about 7×10^{-7} mM free Ca^{2+} for posterior pituitary secretory vesicles (Saermark, Thorn, and Gratzl 1983). Also secretory vesicles from the chromaffin cells take up Ca^{2+} below 10^{-6} mM with a K_m value of 2×10^{-7} mM (Krieger-Brauer and Gratzl 1983). Under these conditions (1 mM Mg^{2+}) Na^+ is inhibited half-maximally at about 15 mM, somewhat lower than the 30 mM found in the absence of Mg^{2+} (Krieger-Brauer and Gratzl 1982; Saermark *et al.* 1983). This lower value is similar to the K_m value of 20 mM for $^{22}Na^{2+}$ uptake found for posterior pituitary secretory vesicles (Saermark, Thorn, and Gratzl 1983).

The data obtained from both the posterior pituitary and adrenal medulla indicated that the affinity for Ca^{2+} was within the cytoplasmic concentrations found in most cells during stimulation (between 10^{-7} and 5×10^{-6} mM, Baker, Hodgkin, and Ridgway 1971; Murphy *et al.* 1980; Simons 1982). The stoichiometry for Na^+–Ca^{2+} exchange appeared to be close to 2 both in the adrenal medulla (Krieger-Brauer and Gratzl 1982) and posterior pituitary (Saermark *et al.* 1983). However, these data are not very reliable since the contribution of H^+ exchange was not controlled. No data exists on the contribution of H^+ gradients to the exchange process in intact vesicles. In ghosts it is clear that Ca^{2+} uptake is strictly dependent on Na^+ exchange (Krieger-Brauer and Gratzl 1983). The anion is apparently without importance (Krieger-Brauer and Gratzl 1983).

The use of inhibitors showed that the secretory vesicle Na–Ca exchange mechanism was different from other Ca^{2+}-sequestering systems known to be present in secretory cells. First, plasma membrane contamination of the secretory vesicle preparations was not measurable using the Na–K ATPase as marker (Gratzl *et al.* 1980; Saermark, Thorn, and Gratzl 1983). This was further supported by the lack of effect by ATP and ouabain on secretory vesicle Ca^{2+} accumulation (Krieger-Brauer and Gratzl 1982; Saermark *et al.* 1983). Oligomycin and ruthenium red were also without effects (Krieger-Brauer and Gratzl 1982; Saermark *et al.* 1983), indicating the purity of the preparations with respect to other Ca^{2+}-accumulating systems such as mitochondria.

3.4 The driving forces

Na^+ uptake in exchange for Ca^{2+} is accompanied by osmotic changes in the secretory vesicle content (Saermark, Thorn, and Gratzl 1983). This indicates that the Na^+ ion is driven against a gradient. A determination of the internal Ca^{2+} concentration has been achieved by the use of ionophores such as A23187 to carry out null-point titrations (Bulenda and Gratzl 1985). Under conditions excluding Mg^{2+} or H^+ transport by the ionophore an internal free Ca^{2+} concentration of 4×10^{-6} mM was found for isolated secretory vesicles from the adrenal medulla (Bulenda and Gratzl 1985). This is much lower than the total concentration of about 40 mM (Bulenda and Gratzl 1985). This discrepancy between the total and free Ca^{2+} concentration indicates that the secretory vesicles must have a high internal buffer capacity. This is also indicated by the observation that Na^+ releases 60 per cent of the total Ca^{2+} from intact vesicles (Krieger-Brauer and Gratzl 1982; Saermark, Thorn, and Gratzl 1983), but almost 100 per cent from loaded ghosts (Krieger-Brauer and Gratzl 1983).

Figure 3.1 shows the results of an experiment revealing the ability of secretory vesicles from the adrenal medulla to sequester large amounts of Ca^{2+}. In this experiment a calcium-sensitive electrode was used to follow uptake of Ca^{2+} by a suspension of secretory vesicles before and after the addition of the ionophore A23187. During the whole experiment the

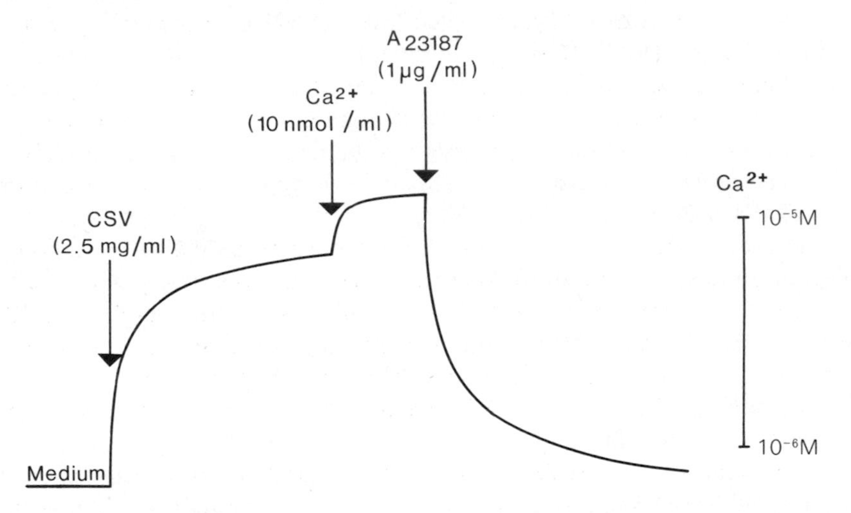

Fig. 3.1. Ca^{2+} by chromaffin secretory vesicles in the presence of ionophore A23187. Chromaffin secretory vesicles (CSV) were added as indicated to a medium containing 20 mM Mops, 40 mM KCl and sucrose to give a final osmolarity of 420 mOsm. The CSVs were prepared as described by Bulenda and Gratzl (1985). The free Ca^{2+} concentration was monitored by a microelectrode before and after addition of CSVs. Ca^{2+} was added as $CaCl_2$ causing the free Ca^{2+} concentration to increase. The subsequent addition of A23187 caused a rapid and substantial decrease in the free Ca^{2+} concentration. The total amount of Ca^{2+} taken up was 310 nmoles per mg of protein.

secretory vesicles remained stable and did not release their catecholamine content. As seen from the figure, the addition of the ionophore results in a dramatic decrease of the external Ca^{2+} concentration. The total amount of Ca^{2+} taken up in this experiment was 310 nmol per mg of protein. Presumably both the internal face of the membrane as well as the protein content of the vesicles can bind Ca^{2+}. The Ca^{2+} binding capacity of the osmotically released protein is about 150 nmol per mg of protein with a K_D of 50×10^{-6} mM (Reiffen and Gratzl 1986a,b). Binding to the large amounts of ATP found in these vesicles is also important (Reiffen and Gratzl 1986b). There seems to be a difference in binding capacity between chromaffin secretory vesicles and secretory vesicles from the posterior pituitary. Secretory vesicles from the posterior pituitary bind 3–4 times less Ca^{2+} than chromaffin secretory vesicles (D. Bulenda and T. Saermark, unpublished results), but they do not contain ATP (Gratzl *et al.* 1980; Poisner and Douglas 1968; Thorn, Russell, and Treiman 1982). However, it should be pointed out that Ca^{2+} binding by posterior pituitary secretory vesicles is still quite large and that the actual content of Ca^{2+} in the isolated vesicles from both tissues is very similar. Isolated posterior pituitary secretory vesicles were found to contain 25 nmol per mg of protein (Saermark *et al.* 1983) and chromaffin secretory vesicles contained 60 nmol per mg of protein (Bulenda and Gratzl 1985).

The high internal binding capacity indicates that the Ca^{2+} concentration gradient across the membrane may not be large. Since the Ca^{2+} gradient is coupled to the Na^+ gradient it follows (Aronson 1981) that a 10-fold gradient for Ca^{2+} is not likely because a similar gradient of Na^+ at external 130 mM Na^+ would result in internal Na^+ concentrations approaching 1 M. This would be incompatible with the stability of the vesicles. It should be pointed out that there is no difference in stability of secretory vesicles in KCl or NaCl medium (Saermark *et al.* 1983; Saermark, Thorn, and Gratzl 1983). On the other hand, it is quite clear that gradients do exist since osmotic swelling due to transport can be observed (Saermark, Thorn, and Gratzl 1983). What is then the driving force for this Ca^{2+} accumulation against a concentration gradient?

In the case of the chromaffin vesicles a mechanism involving a coupling of the exchanger to potentials generated by the H^+-transporting ATPase in these vesicles, or to the H^+ gradient created, may be imagined. However, the Ca^{2+} accumulation does not seem to be ATP dependent and is not very sensitive to pH (Krieger-Brauer and Gratzl 1982). In the case of the posterior pituitary no obvious energy source exists. Although the presence of a proton pump in posterior pituitary secretory vesicles has been suggested (Scherman, Nordmann, and Henry 1982; Russell 1984; Russell and Holz 1981) this does not seem to be the case for several reasons. ATPase activities are not enriched in secretory vesicle preparations (Gratzl *et al.* 1980; Saermark *et al.* 1986;

Saermark, Jones, and Robinson 1984; Saermark, Thorn, and Gratzl 1983; Vilhardt and Hope 1974) and analysis on discontinuous Percoll gradients (Saermark *et al.* 1986) and sucrose gradients (Saermark *et al.* 1986; Vilhardt and Hope 1974) with increasing densities show marked differences in density of membrane-bound ATPase activity and hormone markers. Furthermore, the processing enzymes in these vesicles seem to have neutral or basic pH optima (Eipper, Glembotski, and Mains 1983, Eipper, Mains, and Glembotski 1983) and hormone processing in intact secretory vesicles is inhibited by an internal pH below 7 (Saermark *et al.* 1986) making a pronounced pH gradient unlikely. Also secretory vesicles from the parotid gland have been shown not to contain a proton pump (Arvan, Rudnick, and Castle 1984). The presence or absence of proton pumps in posterior pituitary secretory vesicles is a controversial point and alternative views exist (Njus, Kelley, and Hardanek 1986).

It should be realized that at a pH of about 6 very few protons are actually present in the water space in each granule. In pituitary secretory vesicles between 5 and 10 free protons are present depending on the assumptions, hardly worth the presence of a proton pump unless a proton-consuming process takes place such as trapping of amines as in the chromaffin vesicles. No such process has been demonstrated in posterior pituitary secretory vesicles (Njus, Kelley, and Hardanek 1986; Thorn, Russell, and Treiman 1982). The regeneration of ascorbate, by electron transfer from external ascorbate, is necessary for the amidation of hormones (Eipper, Glembotski, and Mains 1983, Eipper, Mains, and Glembotski 1983), presumably also in posterior pituitary secretory vesicles. In the chromaffin secretory vesicles this transfer is driven by gradients created by the proton pump (Njus *et al.* 1981, 1983). In the posterior pituitary this may be driven by a proton gradient resulting from the known release of H^+ during binding of the processed hormones to their binding proteins (Breslow 1975). The absence of a proton pump in parotid secretory vesicles (Arvan, Rudnick, and Castle 1984) and posterior pituitary secretory vesicles (Saermark *et al.* 1986; Saermark, Thorn, and Gratzl 1983; Vilhardt and Hope 1974), although they both contain Ca^{2+} (Wick and Hepler 1982), and the relatively small effect of external pH variations on Ca^{2+} accumulation by chromaffin secretory vesicles (Krieger-Brauer and Gratzl 1982) having a proton pump (Apps and Reid 1977) indicates that a pH gradient is not the primary driving force. On the other hand, it seems clear that protons can exchange for external cations since NH_4^+ reduces Ca^{2+} uptake by secretory vesicles, although it does not abolish it completely (T. Saermark, unpublished data).

A diffusion potential across the secretory membrane may partly explain Na^+ accumulation by secretory vesicles. K^+ cannot act as a substitute for Na^+ in generating a Ca^{2+} efflux from secretory vesicles (Krieger-Brauer and Gratzl 1982; Saermark *et al.* 1983; Saermark, Thorn, and Gratzl 1983).

Although K^+ presumably can permeate the membrane of secretory vesicles, KCl may set up a diffusion potential across the secretory vesicle membrane. Secretory vesicles are known to have a potential across their membrane, the inside being negative in the absence of ATP (Russell 1984; Russell and Holz 1981; Scherman, Nordmann, and Henry 1982).

The contribution by a diffusion potential to the driving force for Na^+ accumulation can be demonstrated using the osmotic swelling of the secretory vesicles as an indication of Na^+ uptake. During Na^+ uptake posterior pituitary secretory vesicles change their osmotic content. This is revealed by a decrease in the optical density of a suspension of the vesicles (Saermark, Thorn, and Gratzl 1983). The optical density of a suspension of secretory vesicles is linearly dependent on the logarithm of the external KCl concentration. Figure 3.2 shows the effect of increasing KCl concentration on the optical density of a posterior pituitary secretory vesicle suspension in the presence of 10 mM Na^+. The KCl-induced decrease in optical density was completely abolished if Na^+ was omitted from the medium (Saermark, Thorn, and Gratzl 1983). It thus seems as if Na^+ uptake is increased by increasing KCl concentrations, an effect that would be expected if K^+ was less permeable than Cl^- and Na^+. The presence of a diffusion potential may at least partly explain the cation accumulation.

The nature of the driving forces that leads to Ca^{2+} and Na^+ accumulation

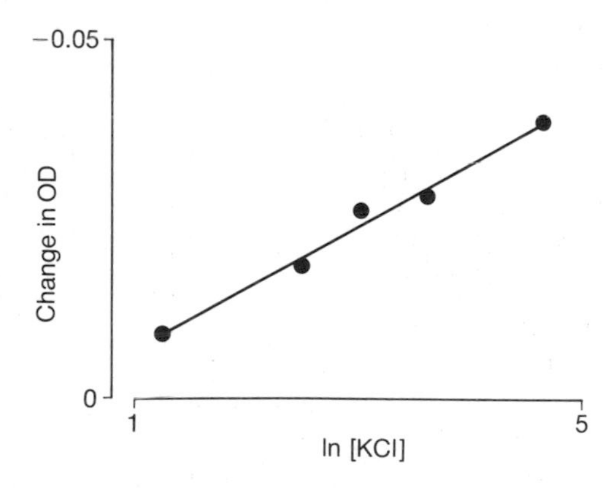

Fig. 3.2. Na^+-dependent osmotic swelling of posterior pituitary secretory vesicles is dependent on the external KCl concentration. Na^+-induced swelling of posterior pituitary secretory vesicles was measured at 10 mM NaCl as a decrease in optical density (OD) at 550 nm (Saermark *et al.* 1983). The change in OD after 15 minutes was plotted as a function of the natural logarithm of the (external) KCl concentration for separate samples incubated in parallel. The medium contained 20 mM TES pH 7.0 and KCl. Sucrose was added to obtain a final osmolarity of 320 mOsm. The vesicles remained stable during this treatment (Saermark, Thorn, and Gratzl 1983). No swelling was observed if NaCl was omitted from the medium.

by secretory vesicles remains to be discovered. It should be pointed out that the ability of the secretory vesicles to maintain ion gradients may not be important for their function. The presence of a large internal buffer (Reiffen and Gratzl 1986a,b) and an affinity for Ca^{2+} just above the resting Ca^{2+} concentration (Krieger-Brauer and Gratzl 1983; Saermark, Thorn, and Gratzl 1983) will make them function as a Ca^{2+} sink.

3.5 The cell biology of secretory vesicle Na–Ca exchange

Secretory vesicle proteins are temporarily inserted into the plasma membrane during exocytosis and are subsequently retrieved. The protein responsible for the observed Na–Ca transport across the secretory vesicle membrane must also be retrieved from the cell surface after exocytosis. Coated microvesicles may be involved in this process (Saermark and Gratzl 1986).

Rather confusing evidence exists concerning Ca^{2+} accumulation by coated microvesicles. Initial experiments suggested the presence of an ATP-dependent Ca^{2+} accumulation by these vesicles (Blitz, Fine, and Toselli 1977; Nordmann and Chevallier 1981; Torp-Petersen *et al.* 1980). Although part of this may be due to plasma membrane contamination (Saermark and Thorn 1982) it seems clear from data obtained on coated microvesicles from the posterior pituitary (Nordmann and Chevallier 1981; Saermark and Gratzl 1986) that the specific activity of Ca^{2+} accumulation by preparations of these vesicles is very high and even higher than Ca^{2+} accumulation by plasma membrane preparations. The V_{max} appears to be about 60 nmol per mg protein per hour (Saermark and Gratzl 1986). Posterior pituitary plasma membrane preparations accumulated 12 nmol per mg protein per 15 minutes and microsomes 7 nmol per mg per 15 minutes (Conigrave *et al.* 1981), a rate which is similar to that of the coated microvesicles. It thus seems as if coated microvesicle Ca^{2+} accumulation is too high to be explained by contamination only. It seems much more likely that coated microvesicles can accumulate Ca^{2+} by a mechanism involving the H^+-transporting ATPase known to be present in these vesicles (Forgac *et al.* 1983; Stone, Xie, and Racker 1983). We have shown that both ATP and loading with Na^+ stimulates Ca^{2+} accumulation by the microvesicles suggesting a functional coupling between the H^+ ATPase in these vesicles and an Na–Ca exchange mechanism (Saermark and Gratzl 1986). Na^+, but not K^+, can release the Ca^{2+} accumulated by coated microvesicles. This Ca^{2+}-induced Ca^{2+} release suggests the presence of an Na–Ca exchanger in coated microvesicles very similar to the one present in secretory vesicles (Saermark and Gratzl 1986). The Na–Ca exchanger from secretory vesicles may thus be retrieved by coated microvesicles after exocytosis.

3.6 The role of secretory vesicle Na–Ca exchange in intracellular calcium regulation and hormone secretion

The characteristics of secretory vesicle Na–Ca exchange allow it to operate at ion concentrations found in nerve endings (Table 3.1). It has been shown that ouabain releases vasopressin from the posterior pituitary and that this process is independent of external Ca^{2+} in the posterior pituitary (Dicker 1966) and the adrenal medulla (Banks 1967). Release of Ca^{2+} from secretory vesicles due to a ouabain-induced increase in intracellular Na^{+} may explain these results, although other explanations are possible. It has also been shown that the Ca^{2+} content in secretory vesicles is changed during secretion (Serck-Hanssen and Christiansen 1973). These experiments suggest that secretory vesicles may play a role in Ca^{2+} buffering in secretory cells.

The ion gradients across the secretory vesicle membrane may also be involved in the mechanism of exocytosis. During exocytotic fusion a swelling of secretory vesicles can be observed (Douglas 1974; Grinstein, VanderMeulen, and Furuga 1982). Extracellular Na^{+} is important for this process but exocytosis is not dependent on extracellular Na^{+} (Douglas 1974). Grinstein, VanderMeulen, and Furuga (1982) suggested that this swelling was necessary to open the secretory vesicles after fusion with the cell membrane and might be due to a process involving an H^{+} exchange mechanism. Alternatively Na–Ca exchange could be involved since Na^{+} uptake by these vesicles is accompanied by a swelling of isolated vesicles (Saermark, Thorn, and Gratzl 1983).

3.7 Conclusion

Secretory vesicles are able to accumulate Na^{+} and Ca^{2+} by an Na–Ca exchange mechanism that is not directly dependent on ATP. This uptake is favoured by a large intravesicular buffer capacity for Ca^{2+}, but concentration gradients are formed as revealed by osmotic swelling of the vesicles. The driving forces are not known. Diffusion potentials and H^{+} gradients may be involved.

The presence of the Na–Ca exchange mechanism in secretory vesicles suggests that there may be a stimulus-dependent regulation of the Ca^{2+}-transporting capacity across the cell membrane. It is difficult to assess the relative importance of this compared to other types of stimulus-dependent Ca^{2+} buffering systems. Several questions are unanswered at present, the most important being whether the exchanger in secretory vesicles is the same as the one in the plasma membrane or if the secretory vesicle exchanger is retrieved immediately after exocytosis.

Acknowledgements

The work was supported by the Danish Medical Research Council, the Nordisk Insulin Fond, Novo's Fond, Fonden til Lægevidenskabens Fremme and Diabetesforeningen. The author wishes to express his gratitude to M. Gratzl and N. A. Thorn for many fruitful discussions.

References

Apps, D. K. and Reid, A. G. (1977). *Biochem. J.* **167**, 297–300.
Aronson, P. S. (1981). *Am. J. Physiol.* **240**, F1–F11.
Arvan, P., Rudnick, G., and Castle, J. D. (1984). *J. Biol. Chem.* **259**, 13567–75.
Baker, P. F., Hodgkin, A. L., and Ridgway, E. B. (1971). *J. Physiol.* **218**, 709–55.
Banks, P. (1967). *J. Physiol.* **193**, 631–7.
Blaustein, M. P., Ratzlaff, R. N., Kendrick, N. C., and Schweitzer, E. S. (1978). *J. Gen. Physiol.* **72**, 15–42.
Blitz, A. L., Fine, R. E., and Toselli, P. A. (1977). *J. Cell Biol.* **75**, 135–47.
Borowitz, J. L., Fuwa, K., and Weiner, N. (1965). *Nature* **205**, 42–3.
Breslow, E. (1975). *Ann. NY Acad. Sci.* **248**, 423–41.
Bulenda, D. and Gratzl, M. (1985). *Biochemistry* **24**, 7760–5.
Conigrave, A. D., Treiman, M., Saermark, T., and Thorn, N. A. (1981). *Cell Calcium* **2**, 125–36.
Dicker, S. E. (1966). *J. Physiol.* **185**, 429–44.
Douglas, W. W. (1974). In: *Handbook of physiology* (eds R. P. Greep and E. B. Astwood), sect. VII Endocrinology, Vol. IV, pp. 191–224. American Physiological Society.
Eipper, B. A., Glembotski, C. C., and Mains, R. E. (1983). *J. Biol. Chem.* **258**, 7292–8.
Eipper, B. A., Mains, R. E., and Glembotski, C. C. (1983). *Proc. Natl. Acad. Sci. USA*, **80**, 5144–8.
Forgac, M., Cantley, L., Weidenmann, B., Altsteil, L., and Branton, D. (1983). *Proc. Natl. Acad. Sci. USA* **80**, 1300–3.
Grafenstein, H. R. K. V. and Neuman, E. (1983). *Biochem. Biophys. Res. Comm.* **117**, 245–51.
Gratzl, M., Torp-Petersen, C., Dartt, D., Treiman, M., and Thorn, N. A. (1980). *Hoppe-Seyler's Z. Physiol. Chem.* **361**, 1615–28.
Grinstein, S., VanderMeulen, J., and Furuga, W. (1982). *FEBS Lett.* **148**, 1–4.
Hausler, R., Burger, A., and Niedermaier, W. (1981). *Naunyn-Schmiedeberg's Arch. Pharmakol.* **315**, 225–67.
Howell, S. L. (1984a). *Diabetologia* **26**, 319–27.
Howell, S. L. (1984b). In: *Methods in diabetes research 1, part B* (eds J. Larner and S. Pohl) pp. 205–13. John Wiley and Sons, New York.
Howell, S. L., Montague, W., and Tyhurst, M. (1975). *J. Cell Sci.* **19**, 395–409.

Hutton, J. C. (1984). *Experientia* **40**, 1091–8.
Jones, P. M., Saermark, T., and Howell, S. L. (1987). *Anal. Biochem.* **166**, 142–9.
Kostron, H., Winkler, H., Geissler, D., and Koenig, P. (1977). *J. Neurochem.* **28**, 487–93.
Krieger-Brauer, H. I. and Gratzl, M. (1981). *FEBS Lett.* **133**, 244–6.
Krieger-Brauer, H. I. and Gratzl, M. (1982). *Biochem. Biophys. Acta* **691**, 61–70.
Krieger-Brauer, H. I. and Gratzl, M. (1983). *J. Neurochem.* **41**, 1269–76.
Murphy, E., Coll, K., Rich, T. L., and Williamson, J. R. (1980). *J. Biol. Chem.* **255**, 6600–8.
Niedermaier, W. and Burger, A. (1981). *Naunyn-Schmiedeberg's Arch. Pharmakol.* **316**, 69–80.
Njus, D., Kelley, P. M., and Hardanek, G. J. (1986). *Biochem. Biophys. Acta* **853**, 237–65.
Njus, D., Zallakian, M., and Knoth, J. (1981). In: *Chemiosmotic proton circuits in biological membranes* (eds V. P. Skulachev and P. C. Hinkle) pp. 365–74. Addison-Wesley, Reading, MA.
Njus, D., Knoth, J., Cook, C., and Kelley, P. M. (1983). *J. Biol. Chem.* **258**, 27–30.
Nordmann, J. J. and Chevallier, J. (1981). *Nature* **287**, 54–6.
Phillips, J. H. (1981). *Biochem. J.* **200**, 99–107.
Poisner, A. M. and Douglas, W. W. (1968). *Mol. Pharmacol.* **4**, 531–40.
Ravazolla, M. (1976). *Endocrinology* **98**, 950–3.
Reiffen, F. U. and Gratzl, M. (1986a). *FEBS Lett.* **195**, 327–30.
Reiffen, F. U. and Gratzl, M. (1986b). *Biochemistry* **25**, 4402–6.
Russell, J. T. (1984). *J. Biol. Chem.* **259**, 9496–507.
Russell, J. T. and Holz, R. W. (1981). *J. Biol. Chem.* **256**, 5950–3.
Russell, J. T. and Thorn, N. A. (1975). *Acta Physiol. Scand.* **93**, 364–77.
Saermark, T. and Gratzl, M. (1986). *Biochem. J.* **233**, 643–8.
Saermark, T. and Thorn, N. A. (1982). *Cell Calcium* **3**, 561–82.
Saermark, T., Jones, P. M., and Robinson, I. C. A. F. (1984). *Biochem. J.* **218**, 591–9.
Saermark, T., Thorn, N. A., and Gratzl, M. (1983). *Cell Calcium* **4**, 151–70.
Saermark, T., Krieger-Brauer, H. I., Thorn, N. A., and Gratzl, M. (1983). *Biochem. Biophys. Acta* **727**, 239–45.
Saermark, T., Andersen, N. M., Atke, A., Jones, P. M., and Vilhardt, H. (1986). *Biochem. J.* **236**, 77–84.
Scherman, D., Nordmann, J. J., and Henry, J. P. (1982). *Biochemistry* **21**, 687–94.
Serck-Hanssen, G. and Christiansen, E. G. (1973). *Biochem. Biophys. Acta* **307**, 404–14.
Simons, T. J. B. (1982). *J. Membr. Biol.* **66**, 235–47.
Stone, D. K., Xie, X.-S., and Racker, E. (1983). *J. Biol. Chem.* **258**, 4059–63.
Thorn, N. A., Russell, J. T., and Treiman, M. (1982). The neurosecretory granule. In: *The secretory granule* (eds A. Poisner and J. M. Trifaro) pp. 119–51. Elsevier, Amsterdam.
Torp-Petersen, C., Saermark, T., Bundgaard, M., and Thorn, N. A. (1980). *J. Neurochem.* **35**, 552–7.
Vilhardt, H. and Hope, D. B. (1974). *Biochem. J.* **143**, 181–9.
Whittaker, V. P. (1982). *Fed. Proc.* **41**, 2759–64.
Wick, S. M. and Hepler, P. K. (1982). *J. Histochem. Cytochem.* **30**, 1190–204.

4 Development and use of inhibitors to study sodium–calcium exchange

Gregory J. Kaczorowski, Maria L. Garcia, V. Frank King, and Robert S. Slaughter

4.1 Introduction

Although the existence of Na–Ca exchange in the plasma membrane of different types of electrically excitable and non-excitable cells is well-established, the precise physiological role of this transport system in the various tissues in which it is found is still controversial. A major reason for this dilemma is the lack of potent selective inhibitors with which to study this system in intact cells. Since Na–Ca exchange is a completely reversible mechanism, it can mediate bidirectional transmembrane Ca^{2+} flux, the vectorial nature of which depends on the magnitude and orientation of electrical and chemical ion gradients that directly control the operation of the transporter. Under conditions in which cells are polarized with a Na^{+} gradient directed inward, Na–Ca exchange probably functions in net Ca^{2+} efflux. However, on membrane depolarization, or with elevated intracellular Na^{+} concentrations, Na–Ca exchange can drive Ca^{2+} accumulation. Therefore, in electrically-active tissues which are undergoing cycles of membrane depolarization and repolarization (e.g. heart), it is difficult to be certain whether Na–Ca exchange mediates overall Ca^{2+} influx or efflux. Moreover, without the use of specific inhibitors, it is very difficult to study the action of an individual transport process in whole cells where a number of other competing reactions (e.g. other Ca^{2+} transport systems, sequestration organelles, and binding proteins) function simultaneously to regulate Ca^{2+} homeostasis. For example, even if one were certain that the main function of Na–Ca exchange were to clear Ca^{2+} from the cell, it would be difficult to ascribe the relative importance of Na–Ca exchange and the Ca^{2+}-ATPase, the other major plasmalemmal Ca^{2+} transport system, as efflux pathways because specific inhibitors do not exist for either process. Nonetheless, some progress has been made towards identifying different structural classes of

Na–Ca exchange inhibitors and determining their mechanisms of action. This chapter will review this progress with particular focus on the action of mechanism-based inhibitors which interact directly with the transport protein.

4.2 Use of membrane vesicles to detect and characterize Na–Ca exchange inhibitors

The Na–Ca exchange reaction has been investigated in many different preparations. These include intact tissues, single cells, organelles (e.g. mitochondria, where a distinct non-electrogenic Na–Ca exchange reaction is found), and isolated plasma membrane vesicles (see Chapters 2–4 for reviews). It is recognized that plasmalemmal Na–Ca exchange can be monitored in tissues and cells by $^{45}Ca^{2+}$ flux measurements, electrophysiological techniques (due to the electrogenic nature of the transport process), and functional cellular responses (e.g. contractility of cardiac preparations). However, the simplest model system for studying this reaction uses plasma membrane vesicles derived from the tissue of interest. In terms of inhibitor development, there are several advantages in employing this latter approach. The first is that the Na–Ca exchange reaction can be studied in isolation, removed from other mechanisms which modulate movement, sequestration, and binding of Ca^{2+}. Vesicular Na^{+}-dependent Ca^{2+} flux measurements also provide a rapid, efficient, and high-capacity assay by which to screen for agents that directly affect the operation of the transport protein. Furthermore, because the kinetic properties of the Na–Ca exchange reaction have been extensively delineated in isolated membrane vesicles, this preparation is an ideal system for defining the mechanisms of action of transport inhibitors once they are discovered.

Na–Ca exchange activity was first demonstrated in a vesicle preparation using cardiac sarcolemmal membranes (Reeves and Sutko 1979), and it is this source which possesses the greatest transport capacity (i.e. highest V_{max} values for Ca^{2+} movement) compared to preparations from different tissues. Although Na–Ca exchange has been extensively characterized in cardiac membranes, the activity is clearly present in plasma membranes derived from other excitable tissues including brain (Gill, Grollman, and Kohn 1981), pituitary (Kaczorowski *et al.* 1984) and smooth muscle (Slaughter, Welton, and Morgan 1987; Slaughter *et al.* 1989). In general, the properties of Na–Ca exchange in vesicles from various sources are quite similar, suggesting that transport inhibitors derived from studies using one source of membranes would be useful as probes for studying the physiological role of this reaction in a number of different systems. Because the characteristics of cardiac sarcolemmal Na–Ca exchange are the most completely described, and the role of Na–Ca exchange in this target tissue is of particular interest, most inhibitor studies have focused on the reaction in this preparation. Another reason that this is a convenient system to study is because a model has been proposed, based on flux studies, to describe the operation of the cardiac transporter (Reeves 1985; Reeves *et al.* 1984). This formulation, then, has provided a conceptual basis by which the mechanism of action of different inhibitors can be evaluated.

In a hypothetical scheme, two different classes of substrate binding sites are postulated to exist on the transport protein; a common site at which either one or two Na^+ or a single Ca^{2+} can bind (A-site) and a separate site at which the third Na^+ that is transported interacts (B-site). Occupation of both A- and B-sites by Na^+ on one surface of the carrier is required to bring about translocation of Ca^{2+} bound to an A-site on the opposite face of the transporter. It has also been suggested that when Ca^{2+} occupies the A-site on both surfaces of the carrier, broader substrate specificity exists at the B-site, and it is here that alkali metal ions interact to stimulate rates of Ca–Ca exchange, a non-productive transport mode of this system. A similar mechanism to the one proposed in heart has been described for the Na-Ca exchange reaction in squid axon (Blaustein 1977). Using such a scheme, it is possible to predict the kinetic behaviour of various inhibitors that interact at one or more of the ion binding sites on the carrier. Moreover, mechanistic studies of inhibitor action in cardiac vesicles have been useful in confirming different aspects of the proposed model (Garcia *et al.* 1988; Slaughter *et al.* 1988).

4.3 Identification of Na–Ca exchange modulators

It has been difficult to find agents which will inhibit Na–Ca exchange. Since no specific, natural product inhibitors have been described that will potently block this transport reaction, no compelling chemical leads exist for rationally designing selective agents to modulate Na–Ca exchange activity. Therefore, the search for inhibitors has concentrated mainly on studies of inorganic cations, drugs with known cardiovascular activities, and previously identified blockers of transport reactions. These methods have met with limited success.

There are a number of metal ions which have been shown to block Na-dependent Ca^{2+} fluxes in both vesicles and whole cells, presumably by interacting at various cation binding sites on the carrier protein. Vesicle studies suggest the following rank order of potency for various tri- and divalent cations as transport inhibitors in cardiac membranes; La^{3+}, Nd^{3+}, Tm^{3+}, Y^{3+}, Cd^{2+}, Sr^{2+}, Ba^{2+}, Mn^{2+}, and Mg^{2+} (Trosper and Philipson 1983). Several of these divalent cations (e.g. Ba^{2+} and Sr^{2+}) may serve as substrates for the cardiac transporter (Philipson 1985a). Of these cations, the most potent inhibitor is La^{3+} which effectively blocks Na–Ca exchange activity in a variety of vesicle systems (Gill, Grollman, and Kohn 1981; Kaczorowski *et al.* 1984; Reeves and Sutko 1979) as well as in whole cells such as the squid giant axon (Baker *et al.* 1969) and embryonic cardiac myocytes (Barry and Smith 1982). The usefulness of an inhibitor such as La^{3+} is limited because this agent blocks most Ca^{2+} transport reactions. However, in experimental settings where conditions are established to monitor exclusively Na–Ca exchange-mediated fluxes (Barry and Smith 1982), La^{3+} can be an effective probe of transporter activity. It is generally assumed that La^{3+} functions as a competitive inhibitor of Ca^{2+} binding at the A-site of the carrier, in analogy to the mechanism by which this ion blocks other Ca^{2+} transport systems. However, recent evidence suggests that the action of La^{3+} may be more complex. For example, the potency of La^{3+} as an inhibitor of cardiac Na–Ca exchange is markedly reduced when the carrier is solubilized and reconstituted into proteoliposomes (unpublished observations). This finding implies that La^{3+} may exert part of its inhibitory effect by interacting at domains other than the A-site which might be located on the transporter or on phospholipid constituents surrounding the carrier. Several divalent cations have also been used to probe Na–Ca exchange activity. In cultured, arterial, smooth muscle cells, these agents function either as carrier substrates (e.g. Sr^{2+}), or as non-transportable competitive blockers of Ca^{2+} binding to the carrier (e.g. Mg^{2+}; Smith, Cragoe, and Smith 1987). In squid axon, Mn^{2+}, Co^{2+}, and Ni^{2+} are effective inhibitors of both Na-dependent Ca^{2+} uptake and efflux reactions (Allen and Baker 1988). In addition to tri- and divalent cations, various monvalent cations including Na^{+}, Li^{+}, K^{+}, Rb^{+}, and choline are also known to inhibit Na–Ca exchange activity with different potencies.

Different organic agents representing a wide variety of structural types have been shown to block the Na–Ca exchange reaction, but most of these are membrane-active compounds which require rather high concentrations to be effective. These include chlorpromazine (Caroni, Reinlib, and Carafoli 1980), dibucaine, tetracaine, and ethanol (Allen and Baker 1988; Michaelis and Michaelis 1983; Michaelis *et al.* 1987), verapamil (Erdreich, Spanier and Rahamimoff 1983), quinidine (Ledvora and Hegyvary 1983; Mentrard, Vassort, and Fischmeister 1984), and polymyxin B (Philipson and Nishimoto 1982). Since limited mechanistic investigations have been carried out with these inhibitors, it is not clear whether they interact directly with the carrier protein or affect transport activity in a secondary fashion by perturbing lipid–protein interactions. Quin-2, a membrane permeant fluorescent probe widely used to monitor intracellular Ca^{2+} concentrations, has been identified as an inhibitor of Na–Ca exchange in squid axon where it blocks Na_i-dependent Ca^{2+} influx without affecting the back reaction (Allen and Baker 1985). This phenomenon appears to be related to the ability of quin-2 to chelate intracellular Ca^{2+}, rather than to be an effect at the level of the transporter or membrane, because Na_i-dependent Ca^{2+} influx in this preparation requires intracellular Ca^{2+} for activation of the carrier (Baker and Dipolo 1984). However, these findings raise concerns regarding the use of quin-2 or other Ca^{2+}-chelating indicators to study Ca^{2+} homeostasis in tissues where Na–Ca exchange functions and where carrier activity may be regulated by intracellular Ca^{2+} concentrations.

Other classes of membrane active agents are more potent modulators of Na–Ca exchange activity. For example, various cationic amphiphilic molecules, such as dodecylamine, are effective inhibitors of Na–Ca exchange in cardiac vesicles (Philipson 1984). These compounds are active at concentrations below that at which they disrupt the osmotic integrity of the membrane (e.g. the IC_{50} for Na–Ca exchange inhibition by dodecylamine is 20 μM). Interference of Na–Ca exchange activity by such agents may be related to their marked effect on contractility of isolated cardiac tissues (Philipson, Langer and Rich 1985). Dodecylamine has also been shown to block currents associated with Na–Ca exchange action in isolated myocytes (Bielefeld *et al.* 1986). Nonetheless, it is unlikely that amphiphilic cations could be used as specific probes of carrier action because they appear not to interact directly with the transporter. Inhibitors of this type will probably affect a number of other membrane-associated activities (e.g. dodecylamine is also known to block cardiac Ca^{2+} and K^+ channels; Bielefeld *et al.* 1986). Although it has been shown that dodecylamine is a non-competitive inhibitor with respect to Ca^{2+} (Philipson 1984), a pattern which is consistent with an interaction of inhibitor at a Na^+ binding site on the carrier (e.g. the B-site), more recent data indicate that kinetics of inhibition *vs.* Na^+ are not strictly competitive either (Slaughter *et al.* 1988). Thus, agents like dodecylamine, as well as many other structurally similar compounds that have also been found to block Na–Ca exchange (unpublished observations), probably act in a secondary fashion to alter carrier activity by disrupting the membrane environment surrounding the transport protein. This interpretation is consistent with other findings which demonstrate that Na–Ca exchange activity in vesicles is exquisitely sensitive to the types of phospholipids, fatty acids, and other amphiphilic molecules which comprise membrane bilayers (Philipson, Frank, and Nishimoto 1983; Philipson and Nishimoto 1984; Philipson and Ward 1985, 1987).

Another membrane-active agent that effectively blocks Na–Ca exchange in both vesicles (de la Pẽna and Reeves 1987) and cells (Bielefeld *et al.* 1986) is quinacrine. This compound inhibits all three modes of carrier action (i.e. Na–Ca, Ca–Ca, and Na–Na exchange) in cardiac sarcolemmal vesicles with a K_i of 50 μM. Inhibition is non-competitive *vs.* Ca^{2+}, but is reversed by high concentrations of Na^+. This pattern is qualitatively similar to the inhibitory profile produced by other classes of mechanism-based Na–Ca exchange inhibitors (e.g. amiloride derivatives and bepridil; see below). However, it is unlikely that quinacrine blocks carrier activity by interacting at either the A- or B-sites of the transporter because of the following lines of evidence: Dixon analyses of the inhibition of Na–Ca exchange in the presence of quinacrine and amiloride analogues suggest that both inhibitors can interact simultaneously with the transport protein (Pilar de la Pẽna and John P. Reeves, personal communication); and the Na^+ dependence for reversal of transport inhibition by quinacrine is very broad and incomplete (see below). These findings suggest that quinacrine binds to a site on the transporter which is distinct from the carrier's two classes of substrate binding sites. That quinacrine affects carrier activity through a direct interaction with the transport protein, rather than by a less specific mechanism involving perturbation of the membrane environment, has been suggested by the following observation: treatment of vesicles with quinacrine mustard, an alkylating agent, results in irreversible inactivation of carrier activity which persists after reconstitution of the soluble transporter into proteoliposomes (de la Pẽna and Reeves 1984). Moreover, pre-incubation of vesicles with quinacrine followed by assay of transport activity in quinacrine-free medium results in stimulation of Na–Ca exchange activity (de la Pẽna and Reeves 1987). Stimulation is the result of a decrease in the K_m for Ca^{2+}, and similar effects can be produced by other lipophilic cations (e.g. tetraphenylphosphonium ion). This phenomenon appears to involve an electrostatic interaction at the inner surface of the membrane whereby the hydrophobic cation alters the local charge distribution near the carrier to accelerate the rate-determining step of the transport reaction. Such a pattern has not been observed with other classes of mechanism-based Na–Ca exchange inhibitors. Although the interaction of quinacrine is mechanistically interesting, the use of this agent as a probe of Na–Ca exchange in physiological investigations is restricted because of its many other inhibitory actions.

A number of cardiovascular drugs have been tested for their ability to inhibit Na–Ca exchange. Amrinone, a positive inotropic agent, can block Na–Ca exchange in cardiac membranes, but it causes only partial inhibition at very high concentrations (Mallov 1983). Doxorubicin (Adriamycin), an anthracycline antibiotic which is known to be cardiotoxic, has been reported to be a potent inhibitor of Na–Ca exchange in cardiac sarcolemmal vesicles (Caroni, Villani, and Carafoli 1981). Unfortunately, this observation has not been confirmed by others who have studied the effects of doxorubicin on Na–Ca exchange in either cardiac (Reeves 1985) or pituitary (Kaczorowski *et al.* 1984) plasma membrane vesicles. It is also noteworthy that this agent was only moderately effective in blocking a current related to Na–Ca exchange in cardiac myocytes (Mentrard, Vassort and Fischmeister 1984) and had no effect on squid axon Na–Ca exchange (Allen and Baker 1988). The reason for this discrepancy is not clear at present.

Members of three different structural classes of organic Ca^{2+} entry blockers (i.e. benzothiazepines, aralkylamines, and dihydropyridines) that function therapeutically as L-type Ca^{2+} channel inhibitors are not, in general, very effective inhibitors of Na–Ca exchange (Johansson and Hellstrand 1987). The benzothiazepine diltiazem does not inhibit plasmalemmal (electrogenic) Na–Ca exchange, but is an effective inhibitor of the mitochondrial (electroneutral) Na–Ca exchange pathway present in organelles isolated from electrically excitable tissues (e.g. heart and brain; Vaghy *et al.* 1982). Indeed, various benzothiazepines and benzodiazepines have been found to inhibit Na–Ca exchange in cardiac mitochondria at μM concentrations (Matlib *et al.* 1983). Since these agents have no effect whatsoever on the sarcolemmal transport system, a determination of whether diltiazem blocks Na–Ca exchange in vesicles is a convenient means of assessing the purity of plasmalemmal membrane preparations. It is difficult to use diltiazem in physiological studies of cardiac mitochondrial Na–Ca exchange because this compound is a potent blocker of L-type Ca^{2+} channels. However, the benzodiazepine clonezapam inhibits the mitochondrial exchange process without possessing significant Ca^{2+} entry blocker activity (Matlib *et al.* 1985). Such an agent could be a relatively specific probe of mitochondrial Na–Ca exchange action in heart. Moreover, certain stereoisomers of diltiazem may be more useful in this type of study. A recent report presents data indicating that while D-*cis*-diltiazem is a good inhibitor of both Ca^{2+} channels and mitochondrial Na–Ca exchange, D-*trans*-diltiazem will block the latter transport process without any significant effect on the former system (Chiesi *et al.* 1987). In addition, these data suggest that diltiazem's ability to inhibit mitochondrial exchange occurs through a stereospecific interaction with the transport protein because L-*cis*- and L-*trans*-diltiazem are more than 10-fold weaker as inhibitors.

Although aralkylamines can block a number of different membranous ion flux pathways besides voltage-dependent Ca^{2+} channels (e.g. Na^+ channels, K^+ channels, sarcolemmal Ca^{2+}-ATPase), a phenomenon which could be related to their local anaesthetic-like properties, they are not potent inhibitors of Na–Ca exchange. In one study with cardiac sarcolemmal vesicles (Van Amsterdam and Zaagsma 1986), verapamil was shown to be a weak blocker of transport activity (IC_{50} of *c.* 200 μM), suggesting that at therapeutic concentrations this mode of action is unimportant. Other aralkylamines such as WB 4101, TMB-8, (+) and (−) D-600 (methoxyverapamil), AQA 39, lidoflazine, cinnarizine, flunarizine, cyproheptadine, and prenylamine are without inhibitory activity (unpublished observations). In the former study, the dihydropyridines nifedipine and nimodipine were also found to be without effect (Van Amsterdam and Zaagsma 1986). Others have reported that various dihydropyridines will inhibit Na–Ca exchange activity (Allen and Baker 1988; Takeo, Adachi, and Sakanashi 1985; Takeo, Elimban, and Dhalla 1985), but these agents have been found to function either at very high concentrations, or have peculiar inhibitory properties. For example, nicardipine is reported to inhibit Na–Ca exchange in sarcolemmal vesicles by interfering with Na_i-dependent Ca^{2+} uptake, but not Na_o-dependent Ca^{2+} efflux (actually a slight stimulation of efflux activity was reported; Takeo, Adachi, and Sakanashi 1985). Since the Na–Ca exchange reaction appears symmetric in cardiac membranes (Philipson 1985b), the specificity of an interaction of this sort is uncertain. Agents which interfere with Ca^{2+} binding to intravesicular sites will decrease Ca^{2+} uptake by decreasing the Ca^{2+} binding capacity of vesicles and this manifests itself as apparent inhibition of transport activity (unpublished observations). However, such compounds will not block Ca^{2+} movements via the efflux pathway, making them uninteresting as probes since they do not interact directly with the carrier protein. Of all the different Ca^{2+} entry blockers that have been investigated in our laboratory, only the anti-arrhythmic agent bepridil has been found to be both a potent blocker and a mechanism-based inhibitor of Na–Ca exchange (see below; Garcia *et al.* 1988).

Some agents which are known blockers of Na-dependent transport systems have also been identified as inhibitors of Na–Ca exchange. For example, the alkaloid harmaline is a well-known inhibitor of Na-dependent amino acid transport systems in eukaryotic cells (Young, Mason, and Fincham 1988). It has been postulated, from studies with ileum smooth muscle, that this agent can block Na–Ca exchange activity (Suleiman and Hider 1985). This was verified using cardiac sarcolemmal vesicles where harmaline inhibits both Ca^{2+} influx and efflux reactions with a K_i of 250 μM (Suleiman and Reeves 1987). The application of harmaline in physiological experiments is problematic, however, because, in addition to blocking various Na-dependent transport reactions, it has also been shown to possess significant Ca^{2+} entry blocker activity (e.g. the IC_{50} for the block of high K^+-induced contraction of smooth muscle is 460 μM; Karaki *et al.* 1986).

The pyrazinoyl–acylguanidine diuretic amiloride is a known inhibitor of a number of different Na^+ transport reactions including epithelial Na^+ channels and Na–H exchange (for reviews see Benos 1982; Kleyman and Cragoe 1988). It is also a weak inhibitor of Na–Ca exchange in murine erythroleukaemia cells (Smith *et al.* 1982), and in heart (Siegl *et al.* 1984), brain (Gill, Chueh, and Whitlow 1984; Schellenberg, Anderson, and Swanson 1983), and pituitary (Kaczorowski *et al.* 1984) plasma membrane vesicles. In all cases, amiloride functions at mM concentrations. Given the poor potency and selectivity of amiloride itself, it is not an attractive agent to use for investigating the Na–Ca exchange reaction. However, various analogues of amiloride exist which display better potency as inhibitors of transport (Kaczorowski *et al.* 1985; Schellenberg *et al.* 1985a). In addition to inhibiting plasmalemmal Na–Ca exchange, amiloride and amiloride derivatives block cardiac mitochondrial Na–Ca exchange (Jurkowitz *et al.* 1983; Sordahl *et al.* 1984). These agents will also inhibit Na–Ca exchange in mitochondria isolated from brain (Schellenberg *et al.* 1985b), but differences exist in the structure–activity relationship noted for this series of compounds. Although mechanisms of inhibition of plasmalemmal and mitochondrial Na–Ca exchange reactions are different, both transport systems can be effectively blocked by amiloride analogues. This fact must be considered if such agents are used in intact cell or tissue studies of cellular Ca^{2+} metabolism.

Various chemical reagents which can be used to modify specific amino acid residues in proteins have been shown to affect vesicular Na–Ca exchange activity. N-ethylmaleimide which alkylates free sulphydryl groups blocks Na–Ca exchange in cardiac membranes (Hazama, Hattori, and Kanno 1987). When this phenomenon was studied in detail, it was found that a number of different protein sulphydryl-modifying reagents (either alkylating or reducing agents) can affect carrier activity and that their mode of action depends on the presence of carrier substrates (e.g. Na^+) in the incubation medium (Pierce, Ward, and Philipson 1986). For example, reducing agents cause inhibition of Na–Ca exchange when the reaction is carried out in the presence of K^+, but Na^+ protects against carrier inactivation. Inhibition by N-ethylmaleimide is more marked in Na^+ than K^+ media, whereas a different alkylating reagent, methanethiosulphonate, actually causes stimulation of transporter activity through a process that could be blocked by Na^+. These data suggest that a sulphydryl residue is critically involved in the Na–Ca exchange transport mechanism and that this residue may be localized at, or coupled to, a Na^+ binding site(s) on the carrier. In contrast, other amino acid modifying reagents which interact with carboxyl, lysyl, histidinyl, tyrosyl, tryptophanyl, arginyl, or hydroxyl residues also affect carrier activity, but do so regardless of the presence of Na^+ in the reaction buffer (Pierce, Ward, and Philipson 1986). Chemical modification of phospholipid components can also cause inhibition of Na–Ca exchange in cardiac membranes. N-methylation of phosphatidylethanolamine causes 50 per cent inhibition of transport activity in sarcolemmal vesicles (Panagia *et al.* 1987), consistent with the hypothesis that the Na–Ca exchange mechanism is very sensitive to its membrane environment (Philipson and Ward 1987).

Besides small organic molecules, investigators have reported that certain antibody reagents will inhibit Na–Ca exchange in vesicles. A family of monoclonal antibodies (Quackenbush *et al.* 1986), one of which is known to cause an increase in the intracellular Ca^{2+} concentration of parathyroid cells (Posillico *et al.* 1985), have been tested to determine if they would interact with the Na–Ca exchange system. Data indicating that one of these antibodies, 44D7, specifically inhibits Na–Ca exchange activity in cardiac sarcolemmal vesicles by interacting directly with the transport protein were reported (Michalak 1986). Unfortunately, our laboratory, and two other laboratories (J. P. Reeves, personal communication; K. D. Philipson, personal communication), have tested this antibody and have not been able to confirm these findings. Recently, it has been observed that polyclonal antibodies raised against various fractions of membrane proteins isolated from cardiac sarcolemma by $NaDodSO_4$ polyacrylamide gel electrophoresis will inhibit Na–Ca exchange in vesicles (Longoni and Carafoli 1987). Maximal inhibition achieved was 50 per cent, perhaps indicating that the epitopes which are recognized by those antibodies that block carrier function are exposed only on one membrane surface, consistent with the fact that cardiac sarcolemmal vesicles are a mixture of about 50 per cent right-side-out and 50 per cent inside-out membranes. An excess of IgG over sarcolemmal membrane protein was required to attain maximum inhibitory activity, suggesting that the titer of active antibody is not very high. The production of antibodies which block function raises the possibility that such reagents can be used in intact tissue studies to assess the physiological role of Na–Ca exchange, as well as for the facile purification of the transport protein to homogeneity. In following this latter approach, other investigators have raised antibodies directed against the Na–Ca exchange transporter from brain (Barzilai, Spanier, and Rahamimoff 1987). Unfortunately the polyclonal antibody preparation did not block the transport activity in vesicles, but it effectively immunoprecipitated carrier activity from detergent-solubilized membranes.

In addition to compounds that inhibit Na–Ca exchange, several agents have been identified which stimulate transporter activity. One class of stimulatory agents include hydrophobic anions which probably perturb the membrane environment in which the carrier is located and thereby cause enhanced transport activity. Examples of effectors that are active in cardiac sarcolemmal membrane vesicles are anionic phospholipids (Philipson, Frank, and Nishimoto 1983; Philipson and Nishimoto 1984), negatively-charged amphiphilic molecules (Philipson 1984), fatty acids (Philipson and Ward 1985), and fatty acid derivatives (Philipson and Ward 1987). Stimulation of Na–Ca exchange by fatty acids may have relevance in cardiac tissue since such agents are produced in ischaemic myocardium. However, rather high concentrations of fatty acids relative to membrane phospholipid are sometimes required to elicit effects in vesicles making it difficult to know the exact physiological correlate of these observations *in vitro* (Ashavaid *et al.* 1985). It is also possible to stimulate Na–Ca exchange activity through a direct redox modification of the carrier protein (Reeves, Bailey, and Hale 1986). For this process both a reducing agent (e.g. dithiothreitol, reduced glutathione, Fe^{2+}, $O_2^{\bar{\cdot}}$) and an oxidizing agent (e.g. oxidized glutathione, H_2O_2, O_2) are required, and stimulation is substrate dependent (i.e. Ca^{2+} blocks the effect, but Na^+ is necessary to achieve enhancement of transport activity). Regulation of Na–Ca exchange by this redox pathway may be important in ischaemic situations in heart, but could also be a control mechanism for modulating carrier activity under normal conditions. Further investigation of Na–Ca exchange stimulators in whole cell and tissue preparations is needed to ascertain the physiological consequences of regulating transport activity in this fashion.

4.4 Mechanism-based inhibitors of Na–Ca exchange

Two different structural classes of Na–Ca exchange inhibitors (Fig. 4.1) have been extensively investigated in order to characterize their mechanisms of action (Garcia *et al.* 1988; Slaughter *et al.* 1988). The first class is comprised of analogues of amiloride. Agents such as these have been useful for studying

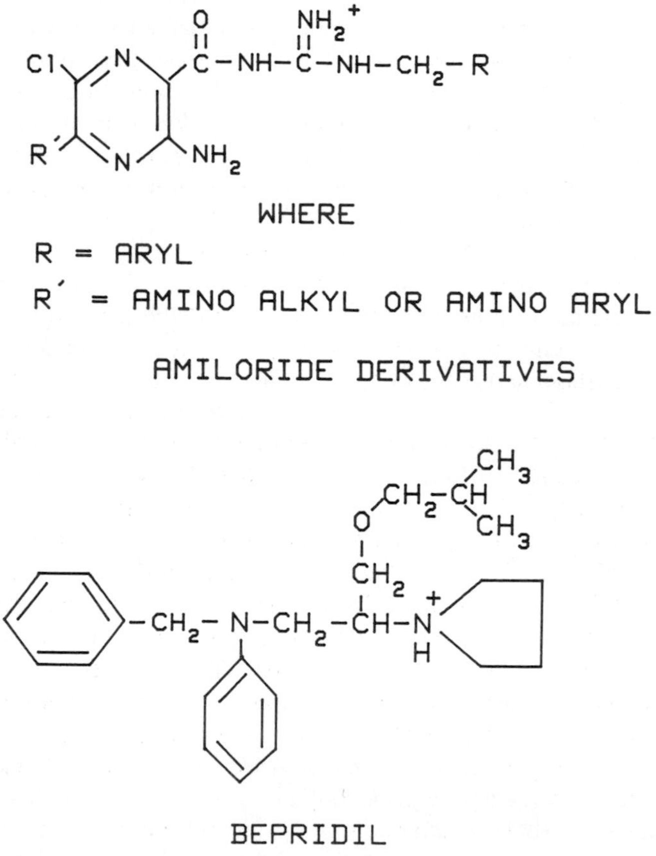

Fig. 4.1. Two different mechanism-based classes of Na–Ca exchange inhibitors. Structures of terminal guanidino-nitrogen and N^5-pyrazine ring-substituted derivatives of amiloride, as well as that of bepridil are shown.

a number of different Na-dependent transport systems. The second group of compounds is represented by the anti-arrhythmic agent bedpridil, which displays an inhibitory profile unique from that of amiloride derivatives. Together, these studies have demonstrated that amiloride analogues and bepridil have different modes of action. Nevertheless, both types of inhibitors are mechanism-based in that they block carrier activity by interacting at discrete sites on the transport protein. Studies of this type are useful because they provide a paradigm by which other Na–Ca exchange inhibitors can be evaluated. In addition, they form the basis for the development of more selective inhibitors.

Although amiloride is a weak inhibitor of Na–Ca exchange, a wide variety of amiloride derivatives have been prepared (for a review see Kleyman and Cragoe 1988), and some of these possess significantly greater potency than the parent compound as blockers of Na–Ca exchange activity (Kaczorowski *et al.* 1985; Siegl *et al.* 1984). Two major structural classes of analogues have been synthesized; molecules substituted either at the terminal nitrogen atom of the guanidino group, or at the five position amino group of the pyrazine ring. Examination of over 300 amiloride derivatives indicate that the substitution of hydrophobic groups at the guanidinium nitrogen increases inhibitory potency. For example, the addition of a benzyl group increases potency by an order of magnitude as compared with amiloride (e.g. the K_i of benzylamiloride (benzamil) is 100 μM). Further substitution of the benzyl ring with chlorine or methyl groups results in analogues that are 50- to 100-fold more potent than amiloride (e.g. the K_i of 3,4-dichlorobenzylamiloride (dichlorobenzamil) is 17 μM). The most potent member of this series is naphthylmethylamiloride ($K_i = 10$ μM). Although the nature of the hydrophobic substitution is crucial for enhancement of inhibitory activity (e.g. introduction of alkyl groups causes either very small potentiation in activity over that of amiloride, or can actually result in loss of potency), it does not appear that these substituents interact at a chiral pocket on the carrier since the two stereoisomers of 1-naphthylethylamiloride have equivalent potency. The guanidino moiety is essential for inhibitory activity in this series of compounds. Analogues in which the guanidino group is deleted or substituted in such a way that the pK_a of the protonated nitrogen is lowered, are inactive. These results agree with the observation that amiloride and its analogues function as Na^+ mimics in inhibiting many transport systems. Therefore, the cationic charge on the guanidino moiety would be expected to be critical for eliciting inhibitory activity.

Substitutions where hydrophobic groups are added at the 5-position nitrogen atom of the pyrazine ring also result in compounds with enhanced potency as blockers of Na–Ca exchange (Kaczorowski *et al.* 1985). In this case, however, the nature of the hydrophobic constituent is less crucial for gaining increases in inhibitory activity. Thus, both alkyl and aryl substitutions increase inhibitory potency, enhancement is a strict function of the hydrophobic nature of the substituent, and analogues that are 50- to 100-fold more potent than amiloride can be identified (e.g. the K_i of N^5-(propyl, butyl)amiloride is 20 μM; the K_i of N^5-(2,4-dichlorobenzyl)amiloride is 10 μM). The mechanism by which these 5-pyrazine substituted derivatives act appears identical to that by which terminal guanidino–nitrogen analogues function (Kaczorowski *et al.* 1985; unpublished observations). Several amiloride derivatives have been synthesized which possess both N^5- and terminal guanidino–nitrogen substituents. The most active of these compounds are approximately 500-fold more potent than amiloride (e.g. the K_i of N-(2,4-dichloro), N^5-(propyl, butyl)amiloride is 3 μM), but they have limited solubility in aqueous buffers making their study difficult. Of the other amiloride analogues that have been investigated, substitution at either the 5- or 6-positions of the pyrazine ring, or at the N^3 pyrazine nitrogen results in compounds which either inhibit no better than amiloride, or not at all. It has also been possible to replace the pyrazine ring of amiloride with other groups, while maintaining the acylguanidinium moiety, and still achieve inhibitory activity. For example, diphenylmethylacylguanidine is as active as benzamil in inhibiting Na–Ca exchange (K_i of 100 μM). This is noteworthy because the pyrazine function of amiloride is crucial for its interaction with epithelial Na^+ channels—the basis of this agent's diuretic activity. Most other types of guanidinium-containing compounds that have been tested do not effectively inhibit Na–Ca exchange.

Although amiloride analogues are mechanism-based inhibitors, it is important to emphasize that these agents, for the most part, lack the high degree of selectivity required for physiological studies of transport activity in intact cellular systems. While this does not invalidate their use for studying Na–Ca exchange in isolated vesicle preparations, it makes it very difficult to use these agents in studies *in vivo* where they may interfere with many different cellular reactions. For example, while both terminal guanidino–nitrogen and N^5-substituted analogues of amiloride do inhibit Na–Ca exchange, representatives of the former structural type are potent blockers of non-tetrodotoxin sensitive epithelial Na^+ channels, whereas members of the latter structural class are very effective inhibitors of Na–H exchange (Kleyman and Cragoe 1988; Lazdunski, Frelin, and Vigne 1985). In addition, certain of these compounds have been shown to inhibit Na^+, K^+-ATPase, other Na^+-coupled transport systems (e.g. amino acid and sugar transporters), voltage-gated Na^+, Ca^{2+}, and K^+ channels, protein kinases, adenylate cyclase, various receptor systems (e.g. nicotinic and muscarinic acetylcholine receptors, α-receptors, β-receptors, atrial naturetic factor receptors), and cellular RNA and DNA synthesis, and to do so with a variety of potencies. Although it was hoped that some terminal guanidino-nitrogen-substituted derivatives might be more selective as inhibitors of Na–Ca exchange than of other processes, all the data indicate that the windows of specificity are small. Perhaps the most troublesome property of these compounds, in terms of their use in investigations of Ca^{2+} homeostasis, is their ability to effectively block voltage-dependent Ca^{2+} channels. For example, a compound such as 3,4-dichlorobenzamil will inhibit L-type (Bielefeld *et al.* 1986; Garcia *et al.* 1987; Suarez-Kurtz and Kaczorowski 1988), T-type (Suarez-Kurtz and Kaczorowski 1988), and N-type (Feigenbaum, Garcia, and Kaczorowski 1988) Ca^{2+} channels more potently than it will block the Na–Ca exchange reaction. Therefore, while amiloride analogues can be successfully employed for studying Na–Ca exchange activity in intact cells and tissues using experimental protocols in which this system is isolated, their use in probing the physiological role of Na–Ca exchange under normal cellular conditions is completely unwarranted.

The other structural class of Na–Ca exchange inhibitor whose mechanism has been extensively characterized is the substituted pyrrolidineethanamine, bepridil, a cardiovascular agent with anti-anginal and anti-arrhythmic properties (Alpert *et al.* 1985; Flaim and Cummings 1986). Bepridil is known to inhibit voltage-dependent Ca^{2+} channels (Galizzi *et al.* 1986; Mras and Speralakis 1981), and sarcolemmal Ca^{2+}-ATPase (Agre, Virshup, and Bennett 1984; Lamers, Cysouw, and Verdouw 1985) through its interaction with calmodulin, and cardiac K^{+} channels (Reiser and Sullivan 1986). It also inhibits certain mitochondrial functions (Matlib 1985). However, because bepridil is a potent inhibitor of Na–Ca exchange ($K_i = 20\ \mu M$), and because significant differences exist between bepridil and the amiloride series of inhibitors (both in terms of their structures and their characteristic patterns of inhibition), the mechanism of action of bepridil has been investigated and compared with that of amiloride derivatives. Again, since this agent interferes with other cellular Ca^{2+} transport systems, its use in physiological studies of Na–Ca exchange is limited.

The manner in which amiloride analogues and bepridil interact with the Na–Ca exchange carrier has been characterized by assessing their ability to modulate Na–Ca, Ca–Ca, and Na–Na exchange activities in vesicles using kinetic measurements of ion fluxes. For such mechanistic investigations, purified bovine or porcine cardiac sarcolemmal membrane vesicles have been employed. However, in many cases, quantitatively similar data have been obtained with sarcolemmal membranes derived from smooth muscle (both respiratory and vascular), as well as plasma membranes prepared from brain and pituitary cells. This indicates that the mechanisms by which these agents block Na–Ca exchange activity in different tissues are fundamentally the same. In all vesicle preparations tested, amiloride derivatives produce concentration-dependent inhibition of Na_i-mediated $^{45}Ca^{2+}$ uptake, with complete block of transport activity occurring at elevated inhibitor concentration. An example of this behaviour is illustrated in Fig. 4.2 where the effects of 3,4-

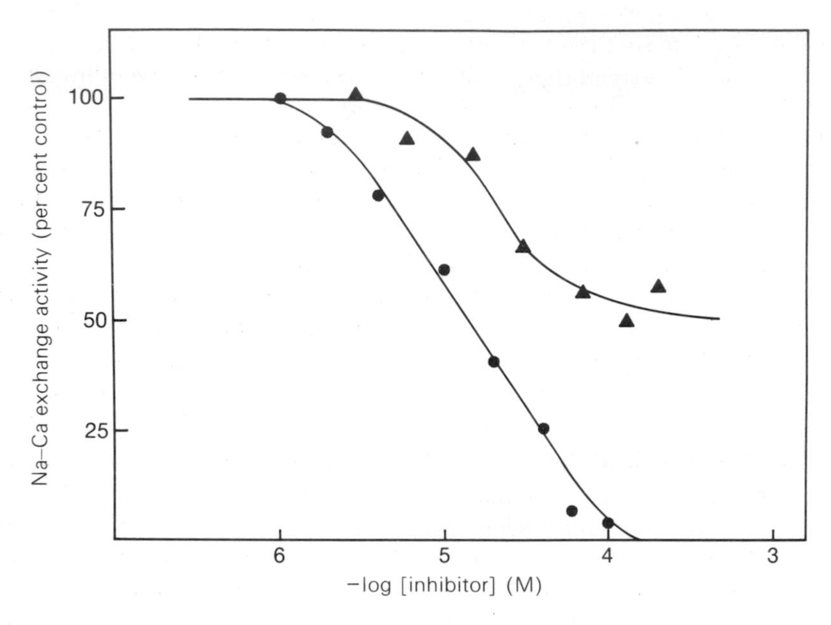

Fig. 4.2. Inhibition of Na–Ca exchange activity in sarcolemmal membrane vesicles prepared from bovine tracheal smooth muscle. Effects of 3, 4-dichlorobenzamil (●) and bepridil (▲) are represented.

dichlorobenzamil (DCB) are monitored on Na–Ca exchange activity in sarcolemmal membranes isolated from tracheal smooth muscle. The K_i of DCB in this experiment is 17 μM, a value identical to that observed for the inhibition of Na–Ca exchange in heart, brain, and pituitary plasma membranes (Siegl *et al.* 1984; Kaczorowski *et al.* 1985; Slaughter *et al.* 1988). Interestingly, when bepridil is tested in this preparation, a different inhibitory profile is observed (Fig. 4.2). Bepridil reproducibly causes only 50 per cent inhibition of Na_i-dependent Ca^{2+} uptake with an IC_{50} of 20 μM. An identical pattern has been observed with vesicles from heart (Garcia *et al.* 1988) and vascular smooth muscle (Slaughter *et al.* 1989). However, if vesicles are preincubated with bepridil before beginning the uptake reaction, inhibition of Na–Ca exchange can go to completeness. Therefore, partial inhibition of transport activity under certain experimental conditions is not the result of some spurious physical property of this inhibitor (e.g. limited solubility in the buffers employed in these experiments), but rather is mechanistically meaningful (see below). One way of demonstrating that blocking actions of inhibitors occur by direct intervention with the carrier is to show that these agents inhibit Na_o-dependent Ca^{2+} efflux. Both amiloride analogues and bepridil block Ca^{2+} efflux and the potency of block for each is the same as that determined for Na_i-dependent Ca^{2+} uptake (i.e. under similar ionic conditions). Bepridil causes complete inhibition of Na–Ca exchange activity in this protocol.

To ascertain at which, if any, substrate binding sites on the transporter amiloride analogues and bepridil interact, the kinetics of Na–Ca exchange inhibition were monitored with respect to either Ca^{2+} or Na^+ concentration. The kinetics of Na–Ca exchange *vs.* Ca^{2+} are displayed for the cardiac system in the Eadie-Hofstee representation of Fig. 4.3A. Addition of increasing concentrations of DCB to the K^+ containing medium used for these transport experiments results in a family of parallel lines indicating that inhibition by DCB is strictly non-competitive as a function of Ca^{2+} concentration. In contrast, if the experiment is repeated using an extravesicular buffer in which sucrose replaces K^+, the kinetic patterns are mixed. This indicates that in the absence of K^+, DCB possesses both competitive and non-competitive components of inhibition *vs.* Ca^{2+}. Bepridil, on the other hand, displays only non-competitive kinetics *vs.* Ca^{2+} regardless of whether K^+ or sucrose media are used for the measurements. Results obtained in K^+ medium are illustrated by the Dixon plots of Fig. 4.3B. Bepridil inhibition produces linear data which

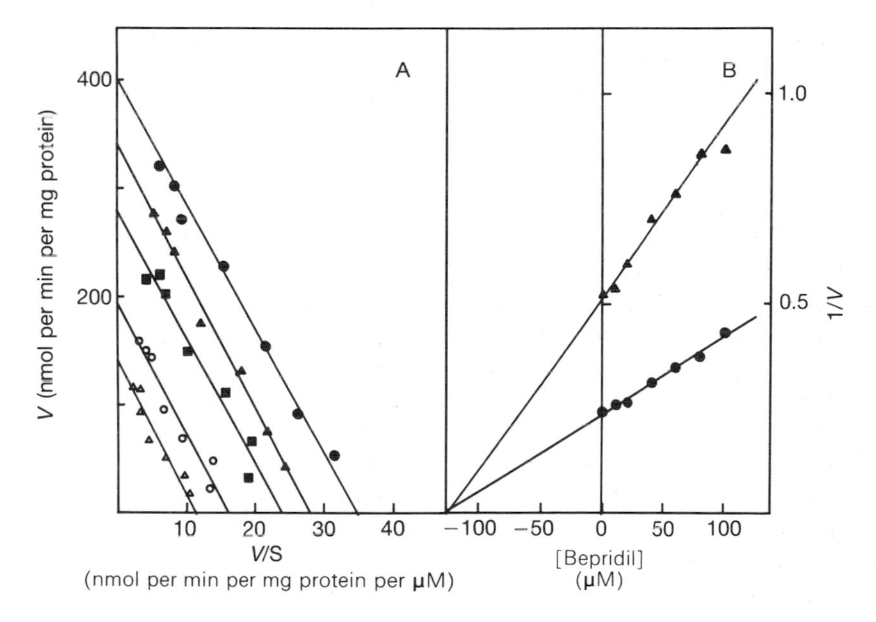

Fig. 4.3. Kinetics of cardiac sarcolemmal Na–Ca exchange inhibition by 3,4-dichlorobenzamil and bepridil. (A) Eadie-Hofstee representation of Na_i-dependent Ca^{2+} uptake kinetics in the presence of increasing concentrations of 3,4-dicholorobenzamil (control, ●; 10, ▲; 25, ■; 50, ○; or 70 μM, △; inhibitor). (B) Dixon plot of the effects of increasing concentrations of bepridil on Na_i-dependent Ca^{2+} uptake (7.5, ● or 75 μM, ▲, $^{45}Ca^{2+}$.

intersect on the x-axis at two different Ca^{2+} concentrations, indicating that inhibition with respect to Ca^{2+} is non-competitive. When the interaction between each of these inhibitors and Na^+ is investigated by monitoring the ability of extravesicular Na^+ to reverse inhibition of Na_o-dependent Ca^{2+} efflux in cardiac membranes, data shown in Fig. 4.4 are obtained. In both cases, Na^+ is able to alleviate the DCB or bepridil inhibition of efflux activity, and reversal of block occurs over a Na^+ concentration range which corresponds to the saturation of binding sites on the carrier by this substrate (Reeves and Sutko 1983). In contrast, data obtained with either dodecylamine or quinacrine (Fig. 4.4) do not support the idea that Na^+ can reverse the blockage of transport activity caused by these inhibitors over the same concentration range. Taken together, these data indicate that amiloride analogues and bepridil are mechanism-based inhibitors of Na–Ca exchange because they interact with discrete substrate binding sites on the transport protein: the interaction of both agents is competitive with Na^+; the interaction of bepridil is strictly non-competitive with Ca^{2+}; amiloride derivatives are either non-competitive or mixed inhibitors *vs.* Ca^{2+}, depending on whether K^+ is present in the solution bathing the vesicles.

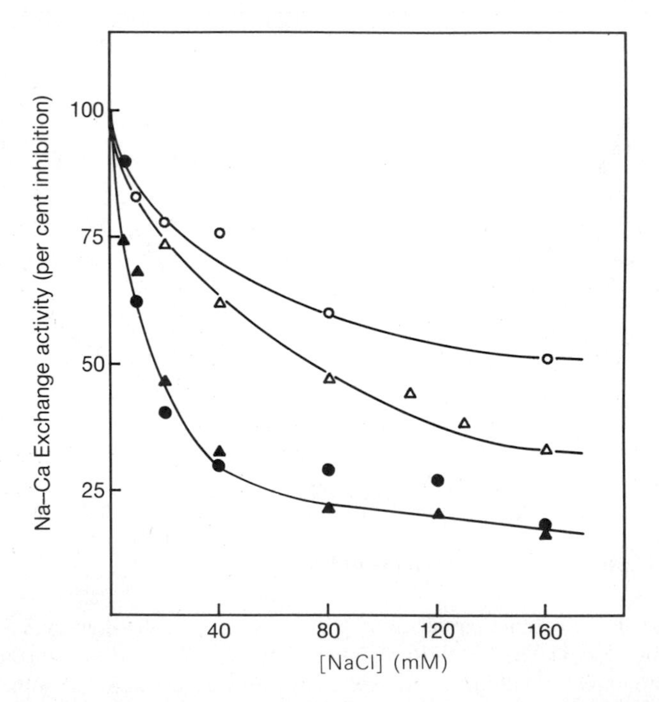

Fig. 4.4. Sodium reversal of Na_o-dependent Ca^{2+} efflux inhibition produced by various inhibitors. The patterns obtained with 3,4-dichlorobenzamil (20 μM, ▲), bepridil (20 μM●), quinacrine (50 μM, ○), and dodecylamine (20 μM, △) are shown.

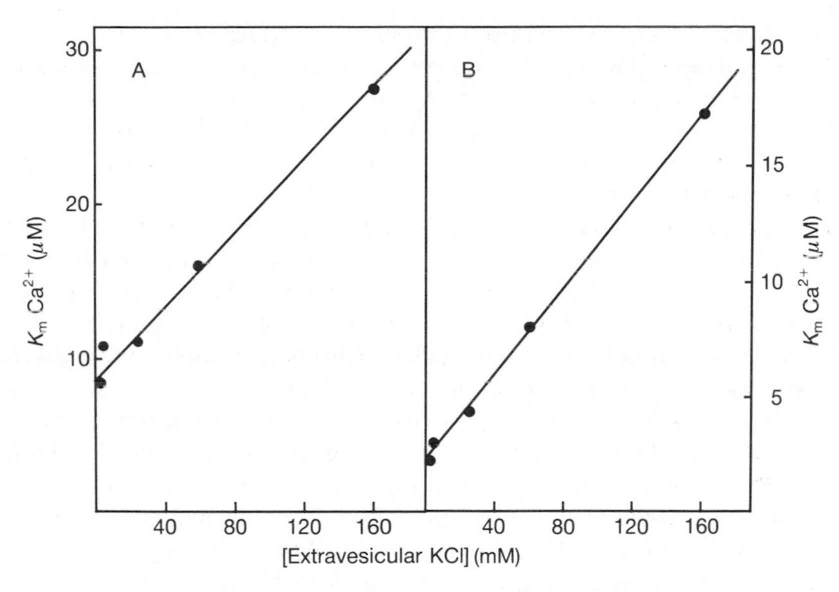

Fig. 4.5. Relationship between the Michaelis constant for Ca^{2+} and extravesicular K^+ concentration during Na–Ca and Ca–Ca exchange. Data obtained for carrier activity in the (A) Na–Ca exchange mode, and (B) Ca–Ca exchange mode are represented.

The effects of K^+ on operation of the Na–Ca exchange carrier in cardiac membranes may provide a clue as to the mechanism by which amiloride analogues block this reaction. As shown by the data in Fig. 4.5A, increasing extravesicular K^+ results in a linear increase in the K_m of Ca^{2+} in Na_i-dependent Ca^{2+} uptake. This occurs without any significant change in the V_{max} of the transport reaction over this range of K^+ concentrations. The action of K^+ is different when its effects are monitored on Ca–Ca exchange kinetics (Slaughter, Sutko, and Reeves 1983). In this case, K^+, and other alkali metal ions, can increase the V_{max} of Ca–Ca exchange, presumably by interacting at a monovalent regulatory site on the carrier. None the less, similar to its effects on Na–Ca exchange, K^+ also effects a linear increase in the K_m measured for Ca^{2+} in the Ca–Ca exchange mode (Fig. 4.5B). These data regarding the interaction between K^+ and Ca^{2+} suggest that both ions may bind in a mutually exclusive fashion at the A-site of the carrier. Thus, K^+ could lessen the ability of amiloride derivatives to bind at the A-site. Under conditions of high K^+ then, the interaction of these inhibitors would be primarily directed at the B-site of the transporter. This interpretation is consistent with non-competitive kinetics for amiloride inhibition *vs.* Ca^{2+}, but with competitive kinetics of inhibition *vs* Na^+. However, decreasing K^+ would allow inhibitor to gain access to the A-site resulting in mixed kinetic patterns of inhibition *vs.* Ca^{2+}, the profile that is actually observed. Therefore, it appears that amiloride derivatives can interact either at one or both classes of substrate binding sites on the transporter, depending on ambient ionic conditions, but that binding occurs preferentially at the B-site. Bepridil probably only interacts at the B-site of the carrier since it never displays competitive inhibition kinetics *vs.* Ca^{2+}.

It is expected that both amiloride analogues and bepridil would block Ca–Ca exchange if the B-site is accessible during operation of the carrier in this transport mode. This result is obtained when Ca–Ca exchange is monitored in cardiac membranes either in sucrose or in the presence of the monovalent stimulatory cation, K^+. The kinetics of benzamil (BNZ) inhibition of transport activity measured with respect to Ca^{2+} are markedly mixed in sucrose medium (Fig. 4.6A), but become more non-competitive in the presence of K^+ (Fig. 4.6B). Bepridil, in contrast, displays purely non-competitive kinetics for Ca–Ca exchange inhibition *vs.* substrate in either media (Fig. 4.6). These findings are consistent with the postulate that amiloride analogues interact at both carrier A- and B-sites in sucrose, but bind more strongly at the B-site in K^+, while bepridil associates exclusively with the B-site of the transporter regardless of ionic conditions. However, because of multiple effects of K^+ on the Ca–Ca exchange process, interactions between amiloride derivatives or bepridil and K^+ have been further characterized. When a BNZ block of Ca–Ca

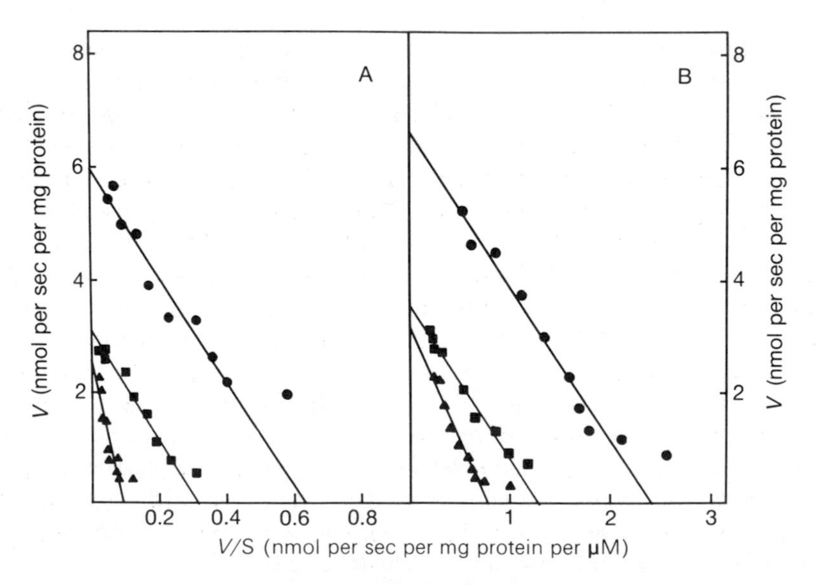

Fig. 4.6. Inhibition of Ca–Ca exchange by benzamil or bepridil. Eadie-Hofstee representations of Ca–Ca exchange kinetics in sucrose (A) or K^+ (B) media. (Control, ●; benzamil (200 μM, ▲; bepridil (20 μM and 100 μM respectively), ■.)

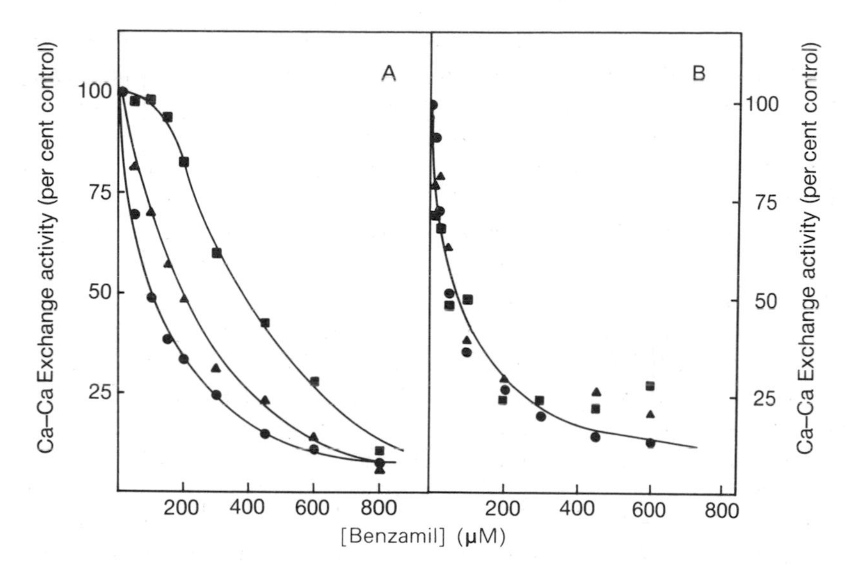

Fig. 4.7. Effect of monovalent cations on benzamil inhibition of Ca–Ca exchange. (A) Data obtained in the presence of sucrose, ●; 40 mM K^+, ▲; or 80 mM K^+, ■, are presented. (B) Data obtained in the presence of sucrose, ●; 40 mM choline$^+$, ▲; or 80 mM choline$^+$, ■, are represented.

exchange is recorded as a function of fixed extravesicular K^+ concentration, it appears that increasing K^+ will decrease inhibitor potency (Fig. 4.7A). This phenomenon is not simply an ionic strength effect because similar increases in choline concentrations fail to alter the efficacy of BNZ as a Ca–Ca exchange inhibitor (Fig. 4.7B). It is noteworthy that K^+, but not choline, stimulates Ca–Ca exchange by binding at a monovalent cation regulatory site (Slaughter, Sutko, and Reeves 1983) which has been hypothesized to be identical to the carrier's B-site (Reeves 1985; Reeves *et al.* 1984). The data of Fig. 4.7 are consistent with, but do not prove, this hypothesis. Investigation of the action of K^+ on bepridil inhibition of Ca–Ca exchange has revealed two effects: K^+ decreases the potency of bepridil as an inhibitor, as well as limits the maximal extent of block that can occur. For example, in sucrose, bepridil produces 100 per cent inhibition of Ca–Ca exchange activity, while with 160 mM intra- and extravesicular K^+ present, bepridil maximally inhibits transport activity 65 per cent (Garcia *et al.* 1988). These latter findings would suggest that monovalent stimulatory cations modulate Ca–Ca exchange by binding at other sites on the carrier in addition to the B-site.

It is predicted that inhibitors which interact at carrier Na^+ binding sites will also block Na–Na exchange. This prediction is verified by the data of Fig. 4.8 which have been obtained with BNZ. For these experiments, cardiac

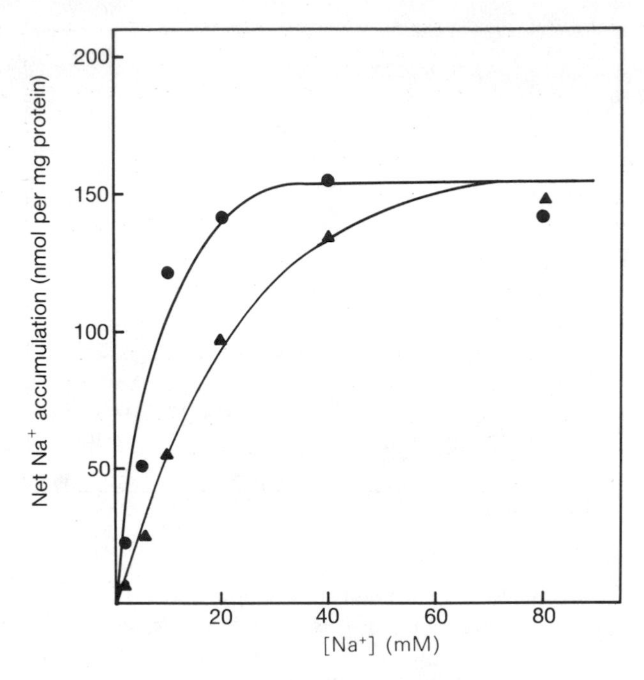

Fig. 4.8. Sodium reversal of Na–Ca exchange inhibition produced by benzamil. Data obtained in a control, ●, and with 200 μM benzamil present, ▲, are shown.

membranes were loaded with either KCl or NaCl and then diluted into media containing varying concentrations of $^{22}Na^+$. Uptake of $^{22}Na^+$ is greater in Na^+- than in K^+-loaded vesicles, and this stimulated accumulation of ion is ascribed to the operation of the transporter in the Na–Na exchange mode. In the absence of inhibitor, extravesicular Na^+ displays saturation behaviour in this Na–Na exchange mode with a concentration dependence that mirrors the ability of Na^+ to bind to the carrier. At a fixed concentration of BNZ, Na–Na exchange is blocked at low Na^+ concentration, but inhibition is completely alleviated as concentrations of Na^+ are increased. These findings provide a compelling argument that amiloride derivatives interact competitively with Na^+ at common binding sites on the transporter.

Several other analyses were employed to confirm the initial findings that amiloride derivatives bind at both A- and B-sites of the carrier, while the interaction of bepridil is restricted to the B-site. Given that strongly inhibitory terminal guanidino–nitrogen derivatives of amiloride are all structurally similar, it is likely that such agents will interact in an identical fashion with the transporter. Therefore, if these inhibitors compete for a single site on the carrier, Dixon analysis of inhibition by one amiloride analogue in the presence of fixed concentrations of a second compound should yield parallel line kinetics. This is behaviour expected for mutually exclusive inhibitors interacting at only one locus. As shown in Fig. 4.9A, a Dixon plot of Na–Ca exchange inhibition by naphthylmethylamiloride (NMA) alone does not yield linear data. Moreover, repetition of the experiment in the presence of increasing concentrations of DCB does not produce data displaying parallel line behaviour. Rather, a family of curves are generated which intersect on the x-axis and curve upward at elevated inhibitor concentrations. If these data are replotted as a function of the square of NMA concentration, the plots become linear suggesting that amiloride analogues may interact at two distinct sites on the carrier (Fig. 4.9B).

This idea was confirmed by Hill analyses of inhibition. The Hill plot for Na–Ca exchange inhibition by NMA is biphasic with a Hill coefficient of 1 at low inhibitor concentration and a Hill coefficient of 2 at higher concentrations of NMA. The Hill plot for inhibition of Na–Ca exchange by NMA in the presence of DCB is monophasic with a Hill coefficient of 2. Furthermore, in the former case, the break point of the Hill plot can be shifted by the concentration of Ca^{2+} used in the experiment; as Ca^{2+} is increased, the break occurs at a higher concentration of inhibitor. This type of kinetic behaviour is displayed

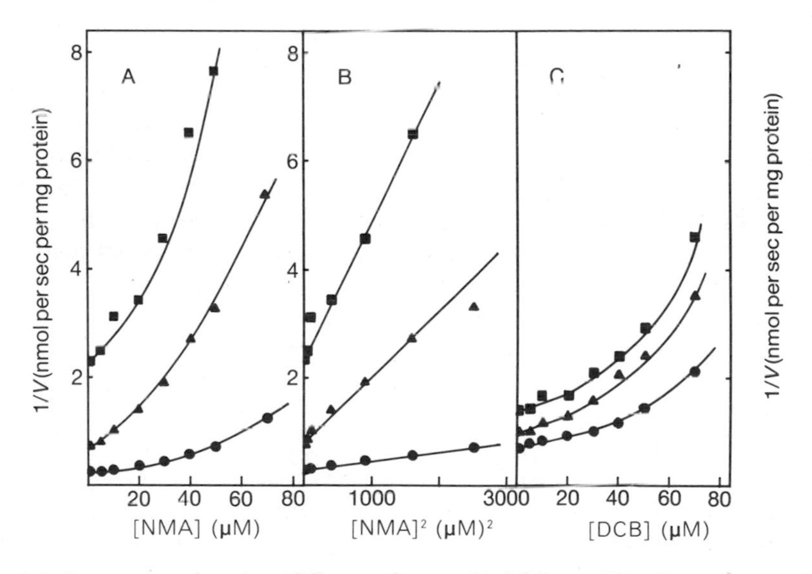

Fig. 4.9. Interactions between different classes of inhibitors of Na–Ca exchange. (A) Dixon plot of Na–Ca exchange inhibition produced by naphthylmethylamiloride (NMA) alone, ●, or in the presence of 30, ▲, or 60 μM, ■, 3,4-dichlorobenzamil (DCB). (B) Data from (A) are replotted as a function of $[NMA]^2$. (C) Dixon plot of inhibition of Na–Ca exchange produced by DCB alone, ●, or in the presence of 50, ▲, or 250 μM, ■, bepridil.

by all amiloride inhibitors tested (i.e. DCB, NMA, and BNZ), as well as by mixtures of these agents. Taken together, these data strongly suggest that amiloride derivatives can interact at two sites on the carrier, but that they have a preference for a single site at low inhibitor concentration, the B-site.

Investigation of the effects of bepridil on the pattern of DCB-mediated Na–Ca exchange inhibition using a Dixon analysis (Fig. 4.9C) reveals that in the presence of increasing bepridil, parallel line kinetics can be observed at low concentrations of DCB. Conversely, a Dixon analysis of bepridil inhibition with increasing DCB present also results in parallel line data at low amiloride analogue concentrations (Garcia *et al.* 1988). These results indicate that bepridil and amiloride derivatives function as mutually exclusive inhibitors when the concentration of the latter agents is low. This would imply that bepridil interacts at a single site on the carrier and that this binding site is the same as the one for which amiloride derivatives have a preference. That bepridil interacts at a single binding site was confirmed by Hill analyses of inhibition. These results illustrate that bepridil blocks both Na–Ca and Ca–Ca exchange with monophasic behaviour in Hill representations, the slopes of which display Hill coefficients of 1.

The findings of these mechanistic studies are summarized in Fig. 4.10. The proposed model suggests that the protonated acyl-guanidinium moiety of amiloride functions as a Na^+ mimic to block carrier activity by interacting at two classes of Na^+ binding sites (i.e. both A- and B-sites, but with a preference for the B-site). A similar mechanism is envisioned for protonated bepridil, except that its interaction is restricted to only one type of Na^+ binding site (the B-site). This mode of action is in marked contrast to that of other structural classes of Na–Ca exchange inhibitors which appear to block carrier activity by modifying the membrane environment surrounding the protein. In addition, there is a striking difference between the inhibitory activity of

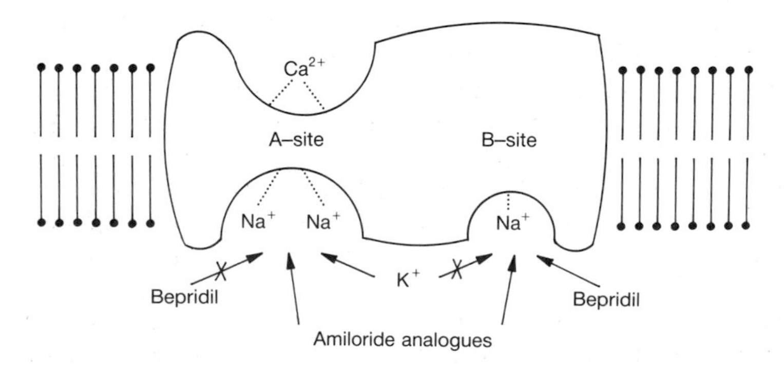

Fig. 4.10. Schematic representation of the mechanism of Na–Ca exchange inhibition by amiloride analogues and bepridil.

amiloride analogues and bepridil; at higher concentrations, the latter agent causes either partial or complete inhibition of carrier function depending on experimental conditions, while the former group of compounds always elicit complete block of transport regardless of the experimental protocols employed.

Bepridil is able to block completely Na–Ca exchange activity when it is added simultaneously with Na^+ (e.g. as in Na_o-dependent Ca^{2+} efflux), or if membranes are first pre-incubated with inhibitor. However, transport activity is maximally inhibited 50 per cent if bepridil is added at zero-time during Na_i-dependent Ca^{2+} uptake. This profile is mechanistically meaningful and may be related to the fact that bepridil binds only at the B-site of the carrier. A possible explanation of this kinetic behaviour is that during carrier turnover in the Na_i-dependent Ca^{2+} uptake mode, release of the three sodium ions from A- and B-sites to the extravesicular medium is random. Assuming that bepridil functions under these conditions by interacting with the carrier exclusively at external substrate binding sites, and knowing that inhibitor binds solely to the B-site when this site is unoccupied, the transporter may be accessible to inhibitor only half of the time during turnover. Given that bepridil inhibition is completely reversible, a scenario could be envisioned, using appropriate rate constants for binding of inhibitor at the B-site and recycling of the carrier, where partitioning occurs between blocked and free carrier to yield, at most, 50 per cent inhibition of transport activity. However, if bepridil and Na^+ are presented simultaneously at the same face of the transporter, the equilibrium between free and blocked carrier can be driven to the completely blocked state by increasing inhibitor concentration. This would result in complete abolition of Na–Ca exchange activity. Other kinetic schemes that explain the action of bepridil may also be possible. Yet this inhibitory pattern must somehow be related to bepridil's unique ability to interact at only one substrate binding site on the transport protein.

The results obtained from studies of amiloride analogues and bepridil support many aspects of the model proposed for the cardiac Na–Ca exchange reaction (Reeves 1985; Reeves *et al.* 1984). Clearly, data indicating a preferential interaction of inhibitors with various Na^+ sites on the carrier are consistent with the hypothetical scheme where separate A- and B-sites are thought to be involved in the transport mechanism. Because the B-site is more accessible to amiloride derivatives than the A-site, it may be located in close proximity to the membrane surface or in a hydrophobic pocket on the protein, whereas the A-site appears to be exposed to the aqueous environment.

Several lines of evidence are consistent with this idea. The structure–activity relationship observed with both terminal guanidino-nitrogen and N^5-pyrazine derivatives of amiloride indicate that potency of these agents as Na–Ca exchange inhibitors is strongly coupled to the hydrophobic nature of the analogue. Furthermore, block of carrier activity by less lipophilic amiloride derivatives (e.g. amiloride or BNZ) displays significantly greater mixed kinetic patterns of inhibition with respect to Ca^{2+} than is observed with other more hydrophobic molecules (e.g. DCB or NMA), indicating that reducing the ability of inhibitor to partition into the membrane may lessen an interaction at the B-site. Moreover, when binding of [^{3}H]N^5-(methyl,isobutyl)amiloride to cardiac sarcolemmal membrane vesicles was studied, it was observed that this agent binds to membranes with an apparent K_D which is nearly equivalent to its K_i for blocking Na–Ca exchange (i.e. 35 μM; unpublished observations). However, the number of binding sites determined for this probe is far in excess of the possible number of transporters present. Therefore, binding of [^{3}H]N^5-(methyl,isobutyl)amiloride to vesicles occurs through an interaction with lipid rather than membraneous protein components. Many different Na–Ca exchange inhibitors (e.g. amiloride derivatives, as well as structurally distinct inhibitors) prevent association of this compound with vesicles and in doing so display the same rank order of potency as noted for their ability to inhibit Na–Ca exchange activity. Similar findings have been obtained by studying binding of this radiolabelled amiloride analogue to protein-free liposomes. These results suggest that potent amiloride analogues are effective because they easily partition into the membrane, are anchored there by the hydrophobic portions of their structure, diffuse along the surface of the membrane into the B-site, and present the carrier with a high local concentration of cationic inhibitor which effectively competes with Na^+ for binding to the protein.

The data obtained with bepridil, another very lipophilic agent which inhibits a number of different membrane-associated systems and which interacts only with the B-site of the transporter, are also consistent with the proposed hypothesis. In addition, it might be predicted that K_i values observed for inhibition of Na–Ca exchange by amiloride derivatives would be dependent on the lipid composition of the membrane. Consistent with this prediction, is the observation that the effectiveness of DCB as a Na–Ca exchange inhibitor in proteoliposomes, (i.e. in a system where the carrier protein is in solution and subsequently reconstituted into artificially formed vesicles), is markedly dependent on lipid content, although the mechanism of inhibition (i.e. an interaction at two ion binding sites) is not altered (unpublished observations). Thus, little is revealed about the molecular structure of substrate binding sites on the transporter from structure–activity patterns of currently available Na–Ca exchange inhibitors. Rather, these data simply reflect the ability of agents to produce high local concentrations of cationic inhibitors at the surface of the carrier. It is not surprising, therefore, that the inhibitors described to date display little selectivity as blockers of Na–Ca exchange because they can interact with cation binding sites present on a number of different membrane-associated proteins.

4.5 Pharmacology of Na–Ca exchange inhibitors

Although currently available Na–Ca exchange inhibitors lack the specificity required for probing the physiological role of this transport reaction in cells and tissues, they can nevertheless be used in intact systems under defined experimental conditions where Na–Ca exchange activity is isolatable. For example, DCB will inhibit Na–Ca exchange driven $^{45}Ca^{2+}$ uptake in normal and Na-loaded chick embryonic myocytes. This occurs with apparently greater efficacy than inhibition of transport activity in vesicles (e.g. the K_i of DCB in the myocyte system is *c.* 1 μM; Kim and Smith 1986; William H. Barry, personal communication). However, it is difficult to compare potencies of inhibitor action in cells *vs.* vesicles because markedly different experimental protocols are used in the two types of studies (e.g. 1 second initial rate measurements are used for calculating K_i values in vesicles, while much longer time-point determinations are employed in intact cell experiments and these result in the lowering of apparent K_i values). Amiloride derivatives also block Na-dependent Ca^{2+} fluxes in cultural arterial smooth muscle cells (Smith, Cragoe, and Smith 1987). In this case, 2,4-dimethylbenzylamiloride (20 μM) completely inhibits $^{45}Ca^{2+}$ uptake and $^{22}Na^{+}$ efflux from Na-loaded preparations. It has been demonstrated that certain electrical currents can be associated with Na–Ca exchange activity in cardiac cells due to electrogenic operation of the transporter (Kimura, Noma, and Irisawa 1986; Mechmann and Pott 1986). Thus, it is predicted that amiloride derivatives would block such currents in whole cells. DCB has been shown to inhibit 'creep currents' (Hume and Uehara 1986a,b) resulting from action of Na–Ca exchange in Na-loaded frog atrial cells with a K_i of 4 μM (Bielefeld *et al.* 1986). Dodecylamine and quinacrine are also effective in this experimental protocol. DCB blocks a transient inward current and an outwardly-rectifying background current in single guinea-pig atrial cells believed to be mediated by Na–Ca exchange (Lipp and Pott 1987). In addition, DCB and NMA will block spontaneous inward currents in Na-loaded guinea-pig myocytes that have been linked to Na–Ca exchange activity (Cragoe, Raveno, and Wettwer 1987). Therefore, transport inhibitors developed from vesicle studies have proven useful in blocking Na–Ca exchange in whole cells. Given the caveat regarding lack of selectivity of the present inhibitors, it is still useful to review the pharmacology of these agents, especially in experimental settings where the Na–Ca exchange mechanism is thought to be very important.

High concentrations of amiloride are known to produce a positive inotropic and negative chronotropic effect in isolated guinea-pig atrial muscle (Pousti and Khoyi 1979). This pharmacological profile is non-catechol related and has been attributed directly to inhibition of Na–Ca exchange (Floreani and Luciani 1984). However, amiloride will inhibit other ion flux pathways in heart, such as voltage-dependent K^+ channels (Bielefeld *et al.* 1986; Yamashita, Motomura, and Taira 1981), which would produce the same pharmacological pattern. DCB causes a small positive inotropic effect in guinea-pig atria (Siegl *et al.* 1983), but a negative inotropic effect in guinea-pig papillary muscle (Siegl *et al.* 1984). Amphiphilic cation inhibitors of Na–Ca exchange (e.g. dodecylamine) also cause a negative inotropic effect in rabbit papillary, while amphiphilic anions that stimulate Na–Ca exchange activity produce a positive inotropic effect in this muscle (Philipson, Langer and Rich 1985). Since both DCB (Bielefeld *et al.* 1986; Garcia *et al.* 1987; Suarez-Kurtz and Kaczorowski 1988) and dodecylamine (Bielefeld *et al.* 1986) block voltage-dependent Ca^{2+} channels in heart, it is not clear whether the negative inotropic activity observed in papillary muscle is due to Na–Ca exchange inhibition, or to Ca^{2+} entry blocker pharmacology, or to a combination of both activities. On theoretical grounds, it is possible that Na–Ca exchange could mediate net Ca^{2+} influx in some cardiac preparations (e.g. papillary muscle), while providing mainly a Ca^{2+} efflux pathway in other tissues (e.g. atrial muscle). Such tissue-dependent vectorial operation of the transporter would explain how Na–Ca exchange inhibition could manifest itself in both positive and negative inotropic activity. Because blockers of Na–Ca exchange display positive inotropism under some conditions, it has been postulated that selective transport inhibitors would form a useful new class of inotropic agents (Luciani and Floreani 1985).

Perhaps the best correlation between pharmacology of Na–Ca exchange inhibitors and block of the transport reaction comes from investigations of Na-loaded preparations. Treatment of papillary muscle with agents that promote intracellular Na-loading (e.g. use of ouabain to block Na^+, K^+-ATPase activity or application of veratridine to increase Na^+ channel open times), results in a positive inotropic effect, presumably due to Ca^{2+} influx via the Na–Ca exchange mechanism. DCB effectively blocks positive inotropism elicited by such treatments (Siegl *et al.* 1984). Moreover, the positive inotropic response due to increases in frequency of electrical stimulation of a muscle preparation (i.e. a Treppe response) is also blocked by DCB (P. K. S. Siegl, personal communication). Arrhythmias that ensue during exposure of heart tissue to toxic levels of ouabain are believed to result in part from Ca^{2+} overloading of cardiac cells through Na–Ca exchange action (Baker *et al.* 1969). Subsequent alterations in the electrical properties of the sarcolemmal membrane are thought to occur because of spontaneous asynchronous release of Ca^{2+} from overloaded sarcoplasmic reticulum which produces local elevations in intracellular Ca^{2+} concentrations that interfere with normal activities of ion channels (Orchard, Eisner and Allen 1983; ter Keurs *et al.* 1987; Wier *et al.* 1983). DCB prevents ouabain-induced arrhythmogenesis in both *in vitro* and *in vivo* models as illustrated with data obtained from isolated guinea-pig papillary muscles and anaesthetized dogs, respectively (Bush,

Kaczorowski, and Siegl 1985). Taken together, these results are consistent with the hypothesis that elevation of intracellular Na^+ leads to Na–Ca exchange-mediated Ca^{2+} influx which is blockable by amiloride derivatives. It is unlikely that the pharmacological activity of DCB in these protocols is due to inhibition of other transport systems besides Na–Ca exchange because neither Ca^{2+} entry blockers nor Na^+ channel inhibitors display activity in all these different experimental paradigms.

There are other experimental protocols where Na–Ca exchange is not easily studied in intact tissues with amiloride analogues. One of these involves investigation of the Ca^{2+} paradox in cardiac muscle. It has been observed that exposure of cardiac cells to Ca^{2+}-free solution followed by re-perfusion with Ca^{2+}-containing medium causes large quantities of Ca^{2+} to be accumulated, cell contracture, loss of excitability, and significant intracellular damage (Bonvallet, Rougier, and Tourneur 1984; Nayler *et al.* 1984). Ca^{2+} influx mediated by Na–Ca exchange has been implicated in these events (Chapman 1983; Nayler *et al.* 1984). Influx of Ca^{2+} is driven by elevated intracellular Na^+ resulting from entry of Na^+ through Ca^{2+} channels during the Ca^{2+}-free exposure period (Chapman *et al.* 1984). DCB effectively inhibits the Ca^{2+} paradox in frog atrial muscle, but the action of this agent correlates with the block of Ca^{2+} channels, thereby preventing Na^+ gain by cells (Sollero, Suarez-Kurtz, and Kaczorowski 1987). DCB does not block Na^+ withdrawal contractures of either frog atria (G. Suarez-Kurtz, personal communication) or embryonic chick myocytes (Kim, Okada, and Smith 1987) which have traditionally been ascribed to action of Na–Ca exchange, although La^{3+} is effective in this regard. However, recent studies of processes that control cytosolic Ca^{2+} concentrations during low Na^+ exposure in chick ventricular cells indicate that other mechanisms besides Na–Ca exchange are involved in the development of Na-free contractures (Kim, Okada, and Smith 1987). Thus, while some pharmacology exists with currently available Na–Ca exchange inhibitors, their use is restricted to experimental settings where Na–Ca exchange activity is clearly expressed. Even in those situations, the interpretation of data should be carried out with caution.

4.6 Conclusions

In order to develop a better understanding of the role that Na–Ca exchange plays in regulating Ca^{2+} homeostasis in various tissues, more specific transport inhibitors must be designed. Although the selectivity of currently available inhibitors is not great enough for definitive physiological studies, it is important to understand how these agents function and to identify those classes of compounds which act as mechanism-based inhibitors so that more selective blockers may eventually be developed. The present studies indicate that specificity may be difficult to attain with hydrophobic cations which interact at substrate binding sites on the carrier. It is likely that such inhibitors will block a number of transport systems because they may be at high local concentrations in the membrane. Moreover, because similarities undoubtedly exist in the structure of ion binding sites localized on different transporters and channels, selectivity may be difficult to attain using this approach. It is not surprising that hydrophobic cations which block one transport system will also interfere with the operation of another because metal ion inhibitors which interact at substrate binding sites on transporters typically show little discrimination in their action. Therefore, molecules which act at sites on the carrier distal to substrate binding sites may be a better approach for developing highly specific inhibitors. This is in analogy to the interaction of ouabain with the Na^+, K^+-ATPase. Unfortunately, without knowledge of the molecular structure of the Na–Ca exchange carrier, the identification of such agents must be serendipidous. Nevertheless, the search should continue with small organic molecules, natural product isolates, and antibody reagents directed against the transport protein. Hopefully, this approach will eventually lead to the development of selective inhibitors, not only for probing the physiological role of Na–Ca exchange, but also for determining if modulation of this transport mechanism is therapeutically useful in various pathophysiological conditions (e.g. congestive heart failure, cardiac ischaemia and associated re-perfusion injury, or some forms of essential hypertension).

Acknowledgements

The authors would like to thank John P. Reeves for useful discussions and suggestions during the course of the mechanistic experiments that have been performed with amiloride analogues and bepridil, part of which were carried out in his laboratory. We would like to thank Mary Coward for her help in preparing this manuscript.

References

Agre, P., Virshup, D., and Bennett, V. (1984). Bepridil and cetiedil. Vasodilators which inhibit Ca^{2+}-dependent calmodulin interactions with erythrocyte membranes. *J. Clin. Invest.* **74**, 812–20.

Allen, T. J. A. and Baker, P. F. (1985). Intracellular calcium indicator quin2 inhibits calcium influx via sodium–calcium exchange in squid axon. *Nature* **315**, 755–6.

Allen, T. J. A. and Baker, P. F. (1988). The effects of some putative calcium antagonists on sodium–calcium exchange in the squid axon. *Mol. Pharmacol.* (in press).

Alpert, J. S., Coumel, P., Greeff, K., Kirkler, D. M., Remme, W. J., Schoenbaum, E., and Verduyn, C. W. (1985). Bepridil: a review of its pharmacology and clinical efficacy as an anti-anginal agent with anti-arrhythmic properties. *Pharmatherapeutica* **4**, 195–222.

Ashavaid, T. F., Colvin, R. A., Messineo, F. C., MacAlister, T., and Katz, A. M. (1985). Effects of fatty acids on sodium–calcium exchange in cardiac sarcolemmal membranes. *J. Mol. Cell. Cardiol.* **17**, 851–62.

Baker, P. F. and DiPolo, R. (1984). Axonal calcium and magnesium homeostasis. *Curr. Top. Membr. Trans.* **22**, 195–247.

Baker, P. F., Blaustein, M. P., Hodgkin, A. L., and Steinhardt, R. A. (1969). The influence of calcium on sodium efflux in squid axons. *J. Physiol.* **200**, 431–58.

Barry, W. H. and Smith, T. W. (1982). Mechanisms of transmembrane calcium movement in cultured chick embryo ventricular cells. *J. Physiol.* **325**, 243–60.

Barzilai, A., Spanier, R., and Rahamimoff, H. (1987). Immunological identification of the synaptic plasma membrane sodium–calcium exchanger. *J. Biol. Chem.* **262**, 10315–20.

Benos, D. J. (1982). Amiloride: A molecular probe of sodium transport in tissues and cells. *Am. J. Physiol.* **242**, C131–45.

Bielefeld, D. R., Hadley, R. W., Vassilev, P. M., and Hume, J. R. (1986). Membrane electrical properties of vesicular sodium–calcium exchange inhibitors in single atrial myocytes. *Circ. Res.* **59**, 381–389.

Blaustein, M. P. (1977). Effects of internal and external cations and of ATP on sodium–calcium and calcium–calcium exchange in squid axons. *Biophys. J.* **20**, 79–111.

Bonvallet, R., Rougier, O., and Tourneur, Y. (1984). Role of sodium–calcium exchange in the calcium paradox in frog auricular trabeculae. *J. Mol. Cell. Cardiol.* **16**, 623–32.

Bush, L. R., Kaczorowski, G. J., and Siegl, P. K. S. (1985). Anti-arrhythmic properties of dichlorobenzamil, a sodium–calcium exchange inhibitor. *Circulation* **72-III**, 313.

Caroni, P., Reinlib, L., and Carafoli, E. (1980). Charge movements during the sodium–calcium exchange in heart sarcolemmal vesicles. *Proc. Natl. Acad. Sci. USA* **77**, 6354–8.

Caroni, P., Villani, F., and Carafoli, E. (1981). The cardiotoxic antibiotic doxorubicin inhibits the sodium–calcium exchange of dog heart sarcolemmal vesicles. *FEBS Lett.* 130, 184–6.

Chapman, R. A. (1983). Control of cardiac contractility at the cellular level. *Am. J. Physiol.* **245**, H535–52.

Chapman, R. A., Rodrigo, G. C., Tunstall, J., Yates, R. J., and Busselen, P. (1984). Calcium paradox of the heart: a role for intracellular sodium ions. *Am. J. Physiol.* **247**, H874–9.

Chiesi, M., Rogg, H., Eichenberger, K., Gazzotti, P., and Carafoli, E. (1987). Stereospecific action of diltiazem on the mitochondrial sodium–calcium exchange system and on sarcolemmal calcium channels. *Biochem. Pharmacol.* **36**, 2735–40.

Cragoe, E., Ravens, U., and Wettwer, E. (1987). Amiloride derivatives that block sodium–calcium exchange inhibit spontaneous inward currents in sodium-loaded myocytes. *Eur. J. Pharmacol.* **140**, 113–16.

de la Pẽna, P. and Reeves, J. P. (1984). Inactivation of the cardiac sodium–calcium exchange system with quinacrine mustard. *Biophys. J.* **45**, 81a.

de la Pēna, P. and Reeves, J. P. (1987). Inhibition and activation of sodium–calcium exchange activity in cardiac sarcolemmal vesicles by quinacrine. *Am. J. Physiol.* **252**, C24–9.

Erdreich, A., Spanier, R., and Rahamimoff, H. (1983). The inhibition of sodium-dependent calcium uptake by verapamil in synaptic plasma membrane vesicles. *Eur. J. Pharmacol.* **90**, 193–202.

Feigenbaum, P., Garcia, M. L., and Kaczorowski, G. J. (1988). Evidence for distinct sites coupled to high affinity omega-conotoxin receptors in rat brain synaptic plasma membrane vesicles. *Biochem. Biophys. Res. Comm.* **154**, 298–305.

Flaim, S. F. and Cummings, D. M. (1986). Bepridil hydrochloride—A review of its pharmacologic properties. *Curr. Ther. Res.* **39**, 568–97.

Floreani, M. and Luciani, S. (1984). Amiloride: Relationship between cardiac effects and inhibition of sodium–calcium exchange. *Eur. J. Pharmacol.* **105**, 317–322.

Galizzi, J. P., Borsotto, M., Barhanin, J., Fosset, M., and Lazdunski, M. (1986). Characterization and photoaffinity labeling of receptor sites for the Ca^{2+} channel inhibitors D-*cis*-diltiazem, (+/−)-bepridil, desmethoxyverapamil, and (+)-PN 200-110 in skeletal muscle transverse tubule membranes. *J. Biol. Chem.* **261**, 1393–7.

Garcia, M. L., King, V. F., Cragoe, E. J., and Kaczorowski, G. J. (1987). Interaction of amiloride derivatives with the cardiac calcium entry blocker receptor complex. *Biophys. J.* **51**, 428a.

Garcia, M. L., Slaughter, R. S., King, V. F., and Kaczorowski, G. J. (1988). Inhibition of sodium–calcium exchange in cardiac sarcolemmal membrane vesicles: II. Mechanism of inhibition by bepridil. *Biochemistry* **27**, 2410–15.

Gill, D. L., Grollman, E. F., and Kohn, L. D. (1981). Calcium transport mechanisms in membrane vesicles from guinea pig brain synaptosomes. *J. Biol. Chem.* **256**, 184–92.

Gill, D. L., Chueh, S. H., and Whitlow, C. L. (1984). Functional importance of the synaptic plasma membrane calcium pump and sodium–calcium exchanger. *J. Biol. Chem.* **259**, 10807–13.

Hazama, S., Hattori, Y., and Kanno, M. (1987). Influence of N-ethylmaleimide on sodium–calcium exchange across cardiac sarcolemma. *Asia Pac. J. Pharmacol.* **2**, 1–6.

Hume, J. R. and Uehara, A. (1986a). Properties of 'creep currents' in single frog atrial cells. *J. Gen. Physiol.* **87**, 833–55.

Hume, J. R. and Uehara, A. (1986b). 'Creep currents' in single frog atrial cells may be generated by electrogenic sodium–calcium exchange. *J. Gen. Physiol.* **87**, 857–84.

Johansson, B. and Hellstrand, P. (1987). Contractures induced by reversed sodium–calcium exchange in rat portal vein: Effects of calcium antagonists. *J. Cardiovas. Pharmacol.* **10**, 575–81.

Jurkowitz, M. S., Altschuld, R. A., Brierley, G. P., and Cragoe, E. J. (1983). Inhibition of sodium-dependent calcium efflux from heart mitochondria by amiloride analogs. *FEBS Lett.* **162**, 262–5.

Kaczorowski, G. J., Costello, L., Dethmers, J., Trumble, M. J., and Vandlen, R. L. (1984). Mechanisms of calcium transport in plasma membrane vesicles prepared from cultured pituitary cells: I Characterization of sodium–calcium exchange activity. *J. Biol. Chem.* **259**, 9395–403.

Kaczorowski, G. J., Barros, F., Dethmers, J. K., Trumble, M. J., and Cragoe, E. J. (1985). Inhibition of sodium–calcium exchange in pituitary plasma membrane vesicles by analogs of amiloride. *Biochemistry* **24**, 1394–403.

Karaki, H., Kishimoto, T., Ozaki, H., Sakata, K., Umeno, H., and Urakawa, N. (1986). Inhibition of calcium channels by harmaline and other harmala alkaloids in vascular and intestinal smooth muscle. *Br. J. Pharmacol.* **89**, 367–75.

Kim, D. and Smith, T. W. (1986). Inhibition of multiple trans-sarcolemmal cation flux pathways by dichlorobenzamil in cultured chick heart cells. *Mol. Pharmacol.* **30**, 164–70.

Kim, D., Okada, A., and Smith, T. W. (1987). Control of cytosolic calcium activity during low sodium exposure in cultured chick heart cells. *Circ. Res.* **61**, 29–41.

Kimura, J., Noma, A., and Irisawa, H. (1986). Sodium–calcium exchange currents in mammalian heart cells. *Nature* **319**, 596–7.

Kleyman, T. R. and Cragoe, E. J. (1988). Amiloride and its analogs as tools in the study of ion transport. *J. Membr. Biol.* **105**, 1–21.

Lamers, J. M. J., Cysouw, K. J., and Verdouw, P. D. (1985). Slow calcium channel blockers and calmodulin. Effect of felodipine, nifedipine, prenylamine and bepridil on cardiac sarcolemmal calcium pumping ATPase. *Biochem. Pharmacol.* **34**, 3837–42.

Lazdunski, M., Frelin, C., and Vigne, P. (1985). The sodium–hydrogen exchange system in cardiac cells: Its biochemical and pharmacological properties and its role in regulating internal concentrations of sodium and internal pH. *J. Mol. Cell. Cardiol.* **17**, 1029–42.

Ledvora, R. F. and Hegyvary, C. (1983). Dependence of sodium–calcium exchange and calcium–calcium exchange on monovalent cations. *Biochem. Biophys. Acta* **729**, 123–36.

Lipp, P. and Pott, L. (1987). Transient inward current and outward-rectifying background current in single guinea pig cardiac cells are blocked by vesicular sodium–calcium inhibitor. *J. Physiol.* **394**, 28.

Longoni, S. and Carafoli, E. (1987). Identification of the sodium–calcium exchanger of calf heart sarcolemma with the help of specific antibodies. *Biochem. Biophys. Res. Comm.* **145**, 1059–63.

Luciani, S. and Floreani, M. (1985). Sodium–calcium exchange as a target for inotropic drugs. *Trends Pharmacol. Sci.* **6**, 316.

Mallov, S. (1983). Effect of amrinone on sodium–calcium exchange in cardiac sarcolemmal vesicles. *Res. Comm. Chem. Pathol. Pharmacol.* **41**, 197–210.

Matlib, M. A., Lee, S. W., Depover, A., and Schwartz, A. (1983). A specific inhibitory action of certain benzothiazepines and benzodiazepines on the sodium–calcium exchange process of heart and brain mitochondria. *Eur. J. Pharmacol.* **89**, 327–8.

Matlib, M. A. (1985). Action of bepridil, a new calcium channel blocker on oxidative phosphorylation, oligomycin-sensitive adenosine triphosphatase activity, swelling, Ca^{2+} uptake and Na^{+}-induced Ca^{2+} release processes of rabbit heart mitochondria *in vitro*. *J. Pharmacol. Exp. Ther.* **233**, 376–81.

Matlib, M. A., Doane, K. D., Sperelakis, N., and Riccippo-Neto, F. (1985). Clonazepam and diltiazem both inhibit sodium–calcium exchange of mitochondria but only diltiazem inhibits the slow action potentials of cardiac muscle. *Biochem. Biophys. Res. Comm.* **128**, 290–6.

Mechmann, S. and Pott, L. (1986). Identification of sodium–calcium exchange current in single cardiac myocytes. *Nature* **319**, 597–9.

Mentrard, D., Vassort, G., and Fischmeister, R. (1984). Changes in external sodium induce a membrane current related to the sodium–calcium exchange in cesium-loaded frog heart cells. *J. Gen. Physiol.* **84**, 201–20.

Michaelis, M. L. and Michaelis, E. K. (1983). Alcohol and local anesthetic effects on sodium-dependent calcium fluxes in brain synaptic membrane vesicles. *Biochem. Pharmacol.* **32**, 963–9.

Michaelis, M. L., Michaelis, E. K., Nunley, E. W., and Galton, N. (1987). Effects of chronic alcohol administration on synaptic membrane sodium–calcium exchange activity. *Brain Res.* **414**, 239–44.

Michalak, M., Quackenbush, E. J., and Letarte, M. (1986). Inhibition of sodium–calcium exchanger activity in cardiac and skeletal muscle sarcolemmal vesicles by monoclonal antibody 44D7. *J. Biol. Chem.* **261**, 92–5.

Mras, S. and Sperelakis, N. (1981). Bepridil (CERM-1978) blockade of action potentials in cultured rat aortic smooth muscle cells. *Eur. J. Pharmacol.* **71**, 13–19.

Nayler, W. G., Perry, S. E., Elz, J. S., and Daly, M. J. (1984). Calcium, sodium and the calcium paradox. *Circ. Res.* **55**, 227–237.

Orchard, C. H., Eisner, D. A., and Allen, D. G. (1983). Oscillations of intra-cellular calcium in mammalian cardiac muscle. *Nature* **304**, 735–8.

Panagia, V., Makino, N., Ganguly, P. K., and Dhalla, N. S. (1987). Inhibition of sodium–calcium exchange in heart sarcolemmal vesicles by phosphatidylethanolamine N-methylation. *Eur. J. Biochem.* **166**, 597–604.

Philipson, K. D. (1984). Interaction of charged amphiphiles with sodium–calcium exchange in cardiac sarcolemmal vesicles. *J. Biol. Chem.* **259**, 13999–4002.

Philipson, K. D. (1985a). Sodium–calcium exchange in plasma membrane vesicles. *Ann. Rev. Physiol.* **47**, 561–71.

Philipson, K. D. (1985b). Symmetry properties of the sodium–calcium exchange mechanism in cardiac sarcolemmal vesicles. *Biochem. Biophys. Acta* **821**, 367–76.

Philipson, K. D. and Nishimoto, A. Y. (1982). Stimulation of sodium–calcium exchange in cardiac sarcolemmal vesicles by proteinase pretreatment. *Am. J. Physiol.* **243**, C191–5.

Philipson, K. D. and Nishimoto, A. Y. (1984). Stimulation of sodium–calcium exchange in cardiac sarcolemmal vesicles by phospholipase D. *J. Biol. Chem.* **259**, 16–19.

Philipson, K. D. and Ward, R. (1985). Effects of fatty acids on sodium–calcium exchange and calcium permeability of cardiac sarcolemmal vesicles. *J. Biol. Chem.* **260**, 9666–71.

Philipson, K. D. and Ward, R. (1987). Modulation of sodium–calcium exchange and calcium permeability in cardiac sarcolemmal vesicles by doxylstearic acids. *Biochem. Biophys. Acta* **897**, 152–8.

Philipson, K. D., Frank, J. S., and Nishimoto, A. Y. (1983). Effects of phospholipase C on the sodium–calcium exchange and calcium permeability of cardiac sarcolemmal vesicles. *J. Biol. Chem.* **258**, 5905–10.

Philipson, K. D., Langer, G. A., and Rich, T. L. (1985). Regulation of myocardial contractility and sarcolemmal calcium binding and transport by charged amphiphiles. *Am. J. Physiol.* **248**, H147-150.

Pierce, G. N., Ward, R., and Philipson, K. D. (1986). Role for sulfur-containing groups in the sodium–calcium exchange of cardiac sarcolemmal vesicles. *J. Membr. Biol.* **94**, 217–26.

Posillico, J. T. Sprikant, S., Brown, E. M., and Eisanbarth, G. S. (1985). The 4F2 cell surface protein modulates intracellular calcium. *Clin. Res.* **33**, 385a.

Pousti, A. and Khoyi, M. A. (1979). Effect of amiloride on isolated guinea-pig atrium. *Arch. Intl. Pharmacodyn.* **242**, 222–9.

Quackenbush, E. J., Gougos, A., Baumal, R., and Letarte, M. (1986). Differential localization within human kidney of five membrane proteins expressed on acute lymphoblastic leukemia cells. *J. Immunol.* **136**, 118–24.

Reeves, J. P. (1985). The sarcolemmal sodium–calcium exchange system *Curr. Top Membr. Trans.* **25**, 77–127.

Reeves, J. P. and Sutko, J. L. (1979). Sodium–calcium ion exchange in cardiac membrane vesicles. *Proc. Natl. Acad. Sci. USA* **76**, 590–4.

Reeves, J. P. and Sutko, J. L. (1983). Competitive interactions of sodium and calcium with the sodium–calcium exchange system of cardiac sarcolemmal vesicles. *J. Biol. Chem.* **258**, 3178–82.

Reeves, J. P., Hale, C. C., de la Pẽna, P., Slaughter, R. S., Boulware, T., Froehlich, J., Kaczorowski, G. J., and Cragoe, E. J. (1984). A model for the cardiac sodium–calcium exchange carrier. In: *Epithelial calcium and phosphate transport; molecular and cellular aspects* (eds F. Bronner and M. Peterlik) pp. 77–82. Alan R. Liss, New York.

Reeves, J. P., Bailey, C. A., and Hale, C. C. (1986). Redox modification of sodium–calcium exchange activity in cardiac sarcolemmal vesicles. *J. Biol. Chem.* **261**, 4948–55.

Reiser, H. J. and Sullivan, M. E. (1986). Anti-arrhythmic drug therapy: new drugs and changing concepts. *Fed. Proc.* **45**, 2206–12.

Schellenberg, G. D., Anderson, L., and Swanson, P. D. (1983). Inhibition of sodium–calcium exchange in rat brain by amiloride. *Mol. Pharmacol.* **24**, 251–58.

Schellenberg, G. D., Anderson, L., Cragoe, E. J., and Swanson, P. D. (1985a). Inhibition of synaptosomal membrane sodium–calcium exchange transport by amiloride and amiloride analogs. *Mol. Pharmacol.* **27**, 537–43.

Schellenberg, G. D., Anderson, L., Cragoe, E. J., and Swanson, P. D. (1985b). Inhibition of brain mitochondrial calcium transport by amiloride analogs. *Cell Calcium* **6**, 431–47.

Siegl, P. K. S., Kaczorowski, G. J., Trumble, M. J., and Cragoe, E. J. (1983). Inhibition of sodium–calcium exchange in guinea-pig heart sarcolemmal vesicles and mechanical response by isolated atria and papillary muscle to 3,4-dichlorobenzamil (DCB) *J. Mol. Cell. Cardiol.* **14**[suppl.1], 363a.

Siegl, P. K. S., Cragoe, E. J., Trumble, M. J., and Kaczorowski, G. J. (1984). Inhibition of sodium–calcium exchange in membrane vesicle and papillary muscle preparations from guinea pig heart by analogs of amiloride. *Proc. Natl. Acad. Sci. USA* **81**, 3238–42.

Slaughter, R. S., Sutko, J. L., and Reeves, J. P. (1983). Equilibrium calcium–calcium exchange in cardiac sarcolemmal vesicles. *J. Biol. Chem.* **258**, 3183–90.

Slaughter, R. S., Welton, A. F., and Morgan, D. W. (1987). Sodium–calcium exchange in sarcolemmal vesicles from tracheal smooth muscle. *Biochem. Biophys. Acta* **904**, 92–104.

Slaughter, R. S., Garcia, M. L., Cragoe, E. J., Reeves, J. P., and Kaczorowski, G. J. (1988). Inhibition of sodium–calcium exchange in cardiac sarcolemmal membrane vesicles: I Mechanism of inhibition by amiloride analogs. *Biochemistry* **27**, 2403–9.

Slaughter, R. S., Shevel, J. L., Felix, J. P., Garcia, M. L., and Kaczorowski, G. J. (1989). High levels of sodium–calcium exchange in vascular smooth muscle sarcolemmal membrane vesicles. *Biochemistry* (in press).

Smith, J. B., Cragoe, E. J., and Smith, L. (1987). Sodium–calcium antiport in cultural arterial smooth muscle cells. *J. Biol. Chem.* **262**, 11988–94.

Smith, R. L., Macara, I. G., Levenson, R., Housman, D., and Cantley, L. (1982). Evidence that a sodium–calcium antiport system regulates murine erythroleukemia cell differentiation. *J. Biol. Chem.* **257**, 773–80.

Sollero, T., Suarez-Kurtz, G., and Kaczorowski, G. J. (1986). Effects of an amiloride analog on the contractile responses of isolated frog atria. *Braz. J. Med. Biol. Res.* **19**, 581a.

Sordahl, L. A., LaBelle, E. F., and Rex, K. A. (1984). Amiloride and diltiazem inhibition of microsomal and mitochondrial sodium and calcium transport. *Am. J. Physiol.* **246**, C172–6.

Suarez-Kurtz, G. and Kaczorowski, G. J. (1988). Effects of dichlorobenzamil on calcium currents in clonal GH_3 pituitary cells. *J. Pharmacol. Exp. Ther.* **247**, 248–53.

Suleiman, M. S. and Hider, R. C. (1985). The influence of harmaline on the movements of sodium ions in smooth muscle of the guinea pig ileum. *Mol. Cell Biochem.* **67**, 145–50.

Suleiman, M. S. and Reeves, J. P. (1987). Inhibition of sodium–calcium exchange mechanism by harmaline. *Comp. Biochem. Physiol. C. Comp. Pharmacol. Toxicol.* **88**, 197–200.

Takeo, S., Adachi, K. and Sakanashi, M. (1985). A possible action of nicardipine on the cardiac sarcolemmal sodium–calcium exchange. *Biochem. Pharmacol.* **34**, 2303–8.

Takeo, S., Elimban, V., and Dhalla, N. S. (1985). Modification of cardiac sarcolemmal Na–Ca exchange by diltiazem and verapamil. *Exp. Cardiol.* **1**, 131–8.

ter Keurs, H. E. D. J., Schouten, V. J. A, Bucx, J. J., Mulder, B. M., and de Tombe, P. P. (1987). Excitation–contraction coupling in myocardium: Implications of calcium release and sodium–calcium exchange. *Can. J. Physiol. Pharmacol.* **65**, 619–26.

Trosper, T. L. and Philipson, K. D. (1983). Effects of divalent and trivalent cations on sodium–calcium exchange in cardiac sarcolemmal vesicles. *Biochem. Biophys. Acta* **731**, 63–8.

Vaghy, P. L., Johnson, J. D., Matlib, M. A., Wang, T., and Schwartz, A. (1982). Selective inhibition of sodium-induced calcium release from heart mitochondria by diltiazem and certain other calcium antagonist drugs. *J. Biol. Chem.* **257**, 6000–2.

Van Amsterdam, F. T. H. M. and Zaagsma, J. (1986). Modulation of ATP-dependent calcium extrusion and sodium–calcium exchange across rat cardiac sarcolemma by calcium antagonists. *Eur. J. Pharmacol.* **123**, 441–9.

Wier, W. G., Kort, A. A., Stern, M. D., Lakatta, E. G., and Marban, E. (1983). Cellular calcium fluctuations in mammalian heart: Direct evidence from noise analysis of aequorin signals in Purkinje fibers. *Proc. Natl. Acad. Sci. USA* **80**, 7367–71.

Yamashita, S., Motomura, S., and Taira, N. (1981). Cardiac effects of amiloride in the dog. *J. Cardiovasc. Pharmacol.* **3**, 704–15.

Young, J. D., Mason, D. K., and Fincham, D. A. (1988). Topographical similarities between harmaline inhibition sites on sodium-dependent amino acid transport system ASC in human erythrocytes and sodium-independent system ASC in horse erythrocytes. *J. Biol. Chem.* **263**, 140–3.

5 Sodium–calcium exchange in the heart

R. A. Chapman and D. Noble

5.1 Introduction

Observation of the acute sensitivity of the strength of the heart beat to the bathing concentrations of *both* calcium and sodium, while providing the impetus which led to the discovery of the Na–Ca exchange in cardiac muscle, is still persuasive of a major function for this process in the control of contractility of cardiac muscle.

Early experiments showed that the steady level of the heart beat varied according to the quotient $[Ca]_o/[Na]_o^2$ (Lüttgau and Niedergerke 1958; Niedergerke 1963; von Willbrandt and Koller 1948). The notion of an exchange of $2Na^+$ for $1Ca^{2+}$ across the cell membrane was proposed when it was established that the Ca-efflux was sensitive to intracellular [Na], (Glitsch, Reuter, and Scholz 1970; Reuter and Seitz 1968). From the outset, it was emphasized that a simple 2Na–1Ca exchange would carry an influx of Ca^{2+} at the measured concentrations of intracellular Na^+ and Ca^{2+} in squid axons (Baker *et al.* 1969; Blaustein and Hodgkin 1969). Evidence that more than $2Na^+$ were exchanged per Ca^{2+}, in cardiac tissue, was initially indirect but tended to favour a 3Na–Ca stoichiometry (Bers and Ellis 1982; Bridge *et al.* 1981; Busselen 1982; Chapman and Tunstall 1980; Chapman, Coray, and McGuigan 1983; Chapman, Tunstall and Yates 1981; Lee, Uhm and Dresdner, 1980; Sheu and Fozzard 1982). However, this was subsequently and unequivocally established by measurements of the coupled movement of Ca and Na ions across the membranes of isolated cardiac sarcolemmal vesicles (Philipson and Nishimoto 1981; Pitts 1979; Reeves and Sutko 1980; as discussed in Chapter 2).

If the exchange exhibits a fixed coupling ratio of 3Na–1Ca, then at the measured levels of intracellular activities of Na and Ca ions, the ionic gradients of Na^+ and Ca^{2+} across the cell membrane would approach, but not reach equilibrium at the resting potential. This could mean that either the exchange is sufficiently powerful to equate the gradients of Na^+ and Ca^{2+}, or that it is a weak system driven near to equilibrium by the activity of other more active systems such as the metabolically driven Na- and Ca-pumps. Indeed the action of the Na- and Ca-ATP-dependent pumps could be considered to act antagonistically on a Na–Ca exchange when it is near to equilibrium. The sarcolemmal systems, the Na–Ca exchange and the uncoupled Ca-pump, because they expel Ca^{2+} from the cell, unlike the intracellular systems, will be able to provide long-term regulation of cellular Ca^{2+}. It is therefore, important to know the relative contributions of these two systems under physiological conditions. The data available for cardiac muscle (see also Chapter 1 on squid axons) has been obtained from isolated sarcolemmal vesicles under relatively unphysiological conditions. If these data apply to intact cells, the regulation of the resting $[Ca^{2+}]_i$ would depend on both systems. Calculations based on the measured values of intracellular $[Ca^{2+}]_i$ and $[Na]_i$ in intact muscle and the transport data from the studies on isolated sarcolemmal vesicles suggest that the two systems should carry about an equal share of the Ca-efflux, with the Ca-pump operating at half its maximum rate and the Na–Ca exchange at only 5 per cent of maximum (Chapman 1983; Philipson and Ward 1986). However, the vesicle experiments were done in the presence of an overestimated value for $[Mg^{2+}]_i$, rather than the more recent much lower estimate. As a consequence the activity of the exchange will be partially inhibited (Blatter and McGuigan 1986; Hess, Metzger and Weingart 1982). In intact cells, where the inhibition by $[Mg^{2+}]_i$ would be less, the Na–Ca exchange would contribute rather more to the Ca-efflux. What is also clear from these data is that following a perturbation of either the Ca^{2+} or Na^+ balance across the cell membrane, (which could occur either as the result of changes in the bathing solution) or during changes in intracellular $[Ca^{2+}]$ or $[Na^+]$ (which occur during the heart beat, the staircase response, or the action of cardiac glycosides), the Na–Ca exchange has the capacity to be more greatly affected. Furthermore because the exchange is electrogenic, it will have an effect upon and be affected by the membrane potential.

5.2 Experiments where tension was measured

5.2.1 Low Na contractures

Analysis of the Na–Ca exchange in intact cardiac muscle has favoured the use of the reduction of the bathing [Na], because many of the effects would seem not to be complicated by an involvement of the Ca-channels (see for review Chapman 1983). Before the development of techniques to measure $[Ca^{2+}]_i$, contraction was often used as an index of changes in $[Ca^{2+}]_i$ when induced, in intact cardiac muscle, either by depolarization of the cell membrane (by an action potential, elevated $(K)_o$ or voltage clamp) or by reduction of the bathing $[Na]_o$.

In amphibian heart the tension developed by low Na contractures shows a steep relationship to $[Na]_o$, which is shifted to higher $[Na]_o$ by depolarization (with elevated $[K]_o$ or voltage clamp), raised $[Ca]_o$, or increased $[Na]_i$ and is shifted to lower $[Na]_o$ by hyperpolarization of the membrane or manoeuvres that lower $[Na]_i$ (Chapman and Rodrigo 1986, 1987; Chapman and Tunstall 1980, 1983a,b, 1984). The effects of $[Ca]_o$ and depolarization with K-rich fluid, on the tension-$[Na]_o$ relationship were best fitted by a scheme whereby $[Ca^{2+}]_i$ is determined by a 3Na–Ca exchange (Chapman and Tunstall 1980). In the voltage clamp experiments the assumption, made by these authors about the relationship for the activation of contraction by Ca^{2+}, can be eliminated from the analysis. This is because the contracture, induced by mild reduction of the $[Na]_o$ can be abolished by imposed hyperpolarization (Fig. 5.1). When the membrane potential, which reduces the developed tension to the same value (usually zero), is plotted against the change in the Na-gradient across the cell membrane, the slope of the line yields the apparent coupling ratio for the Na–Ca exchange, of 2.7Na–Ca, while on exposure to strophanthidin the apparent coupling ratio is increased to 2.9Na/Ca (Fig. 5.2). In these

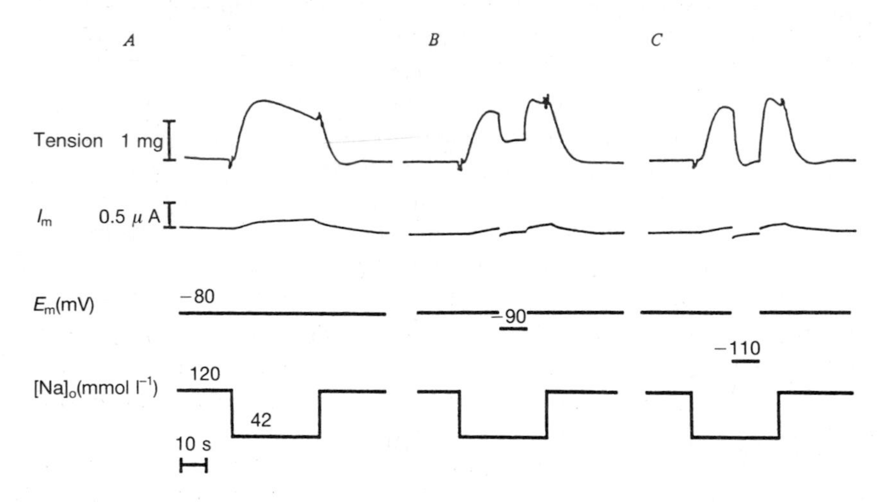

Fig. 5.1. Records of contracture tension, membrane current (I_m) and membrane potential (E_m), of a voltage-clamped frog atrial trabeculae, recorded when $[Na]_o$ was reduced from 120 to 42 mM. In *A* the membrane voltage is maintained at −80 mV. A strong contracture associated with a sustained outward current develops when the $[Na]_o$ is reduced. Both the contracture and the outward current fall to control levels when the muscle is returned to normal Ringer solution. In *B* the membrane is temporarily hyperpolarized to −90 mV during the low-Na contracture. This induces a rapid but partial relaxation, to a smaller steady tension. In *C* hyperpolarization to −110 mV relaxes the contracture tension rapidly to the pre-contracture level (from Chapman and Rodrigo 1986).

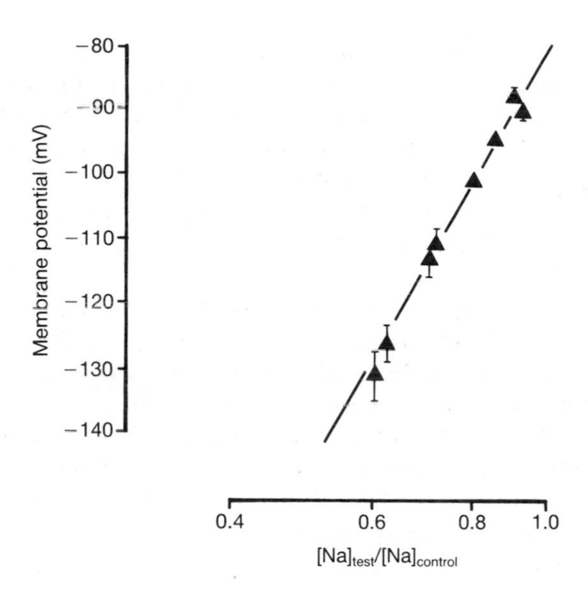

Fig. 5.2. A plot of the log Na-gradient ($[Na]_{test}/[Na]_{control}$; when $[Na]_{test}$ was the reduced $[Na]_o$ at which the contracture was evoked and $[Na]_{control}$ was the $[Na]_o$ at which an increase in tension was just detectable) against the membrane potential which relaxes tension to the pre-contracture level, in frog atrial trabeculae. The continuous line is a line of best fit to the data and has a slope of 2.7. Data +/− S.E.M. when the number of experiments exceed 6. (After Chapman and Rodrigo 1986.) This shows that changes in the log. Na-gradient have 2.7 times more effect on tension than changes in membrane potential.

experiments the resting tension is developed by the muscle in normal Ringer solution and shows a similar sensitivity to the membrane potential and $[Na]_i$, suggesting that it is regulated by the Na–Ca exchange (Chapman and Rodrigo 1986).

Caffeine contractures can be evoked in frog and mammalian heart and are inhibited by tetracaine suggesting that the Ca^{2+} that activates contraction is released from the sarcoplasmic reticulum (SR) (Chapman and Leoty 1976; Chapman and Miller 1974). If the level of $[Ca]_o$ or $[Na]_o$, or the membrane potential is altered, the strength and rate of relaxation of the caffeine contracture in frog atrial muscle is changed in a way consistent with an influence of the Na–Ca exchange (Chapman and Tunstall 1983a). These results suggest that the SR in frog atrial muscle can provide Ca^{2+} to activate tension but that Na–Ca exchange can compete with the contractile proteins for this Ca^{2+}. This is an idea consistent with the increased ^{45}Ca efflux and reduced contracture produced by suddenly raised $[Na]_o$ in frog atria exposed to caffeine (Chapman, Tunstall, and Yates 1981). The upshot would seem to be that the Na–Ca exchange in amphibian heart is the major process concerned with the long-term regulation of diastolic $[Ca^{2+}]_i$ and furthermore it competes for Ca^{2+} discharged into the sarcoplasm by other means.

Mammalian preparations develop low Na contractures that are much less than the maximum tension the muscle can generate, although the strength of these contractures and the change in $[Ca^{2+}]_i$ that accompany them is increased by raising $[Ca]_o$ or $[K]_o$, by factors that would be expected to raise $[Na]_i$, such as exposure to cardiac glycosides and by substances that reduce the Ca-uptake by the SR or the mitochondria, such as caffeine, ryanodine, and metabolic inhibitors (Allen, Eisner, and Orchard 1984; Allen *et al.* 1983; Chapman 1986; Chapman, Coray and McGuigan 1983; Hansford and Lakatta 1986). These findings have been interpreted in terms of the activity of other energy-dependent processes that remove Ca^{2+} from the sarcoplasm and, in consequence, interact more profoundly with the Na–Ca exchange than in amphibian heart to limit the rise in $[Ca^{2+}]_i$.

5.2.2 Depolarization-induced contraction

The contractile responses to prolonged depolarization often show an initial phasic and a later tonic component. In frog muscle the sensitivity of the tonic component to membrane potential (an *e*-fold increase in tension for each 25 mV of depolarization), $[Na]_o$,$[Ca]_o$ and pharmacological agents (divalent cations, tetracaine, catecholamines, cardiac glycosides etc.) are sufficient to suggest that it is evoked by activation of the Na–Ca exchange. The phasic tension, on the other hand, appears to depend on the inward Ca-current through the cell membrane and its ability to trigger Ca-release from the SR, because phasic tension (unlike tonic tension) persists in Na-free fluid when it becomes insensitive to changes in $[Ca]_o$ and is blocked by tetracaine and Ca-channel blockers. Bearing in mind the duration of the cardiac action potential, it might be expected that the amphibian heart beat is largely made up of phasic tension (Chapman and Tunstall 1981a,b). However, both the heart beat and the basic component of tension are immediately affected by changes that alter the status of the Na–Ca exchange. A substantial component of the relaxation of the phasic tension shows a sensitivity to the membrane potential, $[Na]_o$ and $[Ca]_o$ consistent with it being brought about by the Na–Ca exchange operating in the 'Ca-efflux' mode (Chapman and Rodrigo 1985; Goto *et al.* 1972; Roulet *et al.* 1979). This raises the possibility that the Na–Ca exchange may have its influence upon the heart beat by removing rather than adding to the pool of Ca^{2+} that activates contraction.

In mammalian heart, the evidence for a similar scheme has been considered to be less secure. Although a profound dependence of the strength of the heart beat on $[Na]_o$ and $[Ca]_o$ is well-established, the development of a tonic tension during sustained depolarization is only seen when $[Na]_o$ is reduced, $[Na]_i$ is elevated, or Ba^{2+} replaces Ca^{2+} (Eisner, Lederer, and Vaughan-Jones 1983; Gibbons and Fozzard 1971; Mascher 1973). In Purkinje fibres tonic tension is much slower to develop than the phasic tension but both phases show a marked dependence on $[Na^+]_i$ (Eisner, Lederer, and Vaughan-Jones 1983). At the Stowe Conference on Na–Ca exchange, two papers were presented which recognize effects due to Na–Ca exchange in isolated mammalian ventricular myocytes when the release of Ca^{2+} from the SR is compromised by ryanodine or caffeine. Barcenas-Ruiz, Beuckelmann, and Weir (1987) find that the slow but voltage-dependent changes in $[Ca^{2+}]_i$, measured with fura-2 in ryanodine-treated myocytes, are accompanied by currents through the membrane which show features consistent with an activation of Na–Ca exchange. Bridge, Spitzer, and Ershler (1988) find that, after prolonged exposure to caffeine, depolarization evokes a slowly developing tonic tension; the rate of relaxation of this tension, upon repolarization of the membrane, depends upon the membrane potential $[Na]_o$ and $[Ca]_o$, and would seem to be brought about by the Na–Ca exchange. The heart beat and the twitch tension in mammalian heart, unlike amphibian heart, is virtually abolished by ryanodine and prolonged exposure to caffeine, agents that affect the release of Ca^{2+} from the SR (Chapman and Leoty 1976; Hess and Weir 1984; Marban and Weir 1985). This suggests that the mammalian heart beat is initiated primarily by the Ca^{2+} released from the SR. The effects of the strength of the heart beat of changes in $[Na]_o$ or $[Na]_i$, or $[Ca]_o$ would therefore be due to an indirect effect of the Na–Ca exchange via its effect upon $[Ca^{2+}]_i$ rather than by a mechanism similar to that seen in amphibian heart.

5.3 Effects of Na–Ca exchange on intracellular ionic activities

The intracellular activities of Ca and Na ions have been measured and have been of interest in the attempt to establish the role of the Na–Ca exchange in intact cardiac muscle. In the resting hearts of homiotherms, $[Na]_i$, as measured with ion-sensitive micro-electrodes, shows a mean of around 7.0 mM (Lee 1981), while at lower temperatures and in poikilotherms it is higher and may reach 15 mM (Chapman 1986; Chapman and Tunstall 1984). In mammalian cardiac cells resting $[Ca^{2+}]_i$ has been measured both with ion-sensitive micro-electrodes and intracellular Ca-indicators, a value of between 60–100 nM is generally accepted (Chapman 1986; Murphy, Jacob, and Leiberman 1984; Powell, Tatham, and Twist 1984; Sheu, Sharma, and Banerjee, 1984). From these values and the membrane potential the Na–Ca exchange would be at equilibrium if the coupling ratio were between 2.5 and

2.7Na–1Ca (Bers and Ellis 1982; Sheu and Fozzard 1982). Clearly, the exchange is not at equilibrium but in a steady-state with other processes which affect $[Ca^{2+}]_i$. When $[Na^+]_i$ is elevated by reduced temperature in ferret ventricular trabeculae, or by inhibition of the Na-pump in sheep Purkinje fibres, the apparent coupling ratio approaches closer to 3Na–1Ca (Chapman 1986; Sheu and Fozzard 1982). When $[Na]_o$ is lowered the fall in $[Na^+]_i$ is much larger than would be expected from the accompanying rise in $[Ca^{2+}]_i$, if only the sarcolemmal 3Na–1Ca exchange were responsible. However, the apparent coupling ratio (as determined from the final steady values) is little changed. The rise in $[Ca^{2+}]_i$ is augmented and the fall in $[Na^+]_i$ is reduced by inhibition of metabolism (Allen *et al.* 1983, Chapman 1986). The overall conclusion from these experiments is similar to that for the contracture studies namely; that, at 37°C, the 3Na–1Ca exchange is in a steady-state with the other passive and the active processes that influence $[Na^+]_i$ and $[Ca^{2+}]_i$.

In one of the original papers on the Na–Ca exchange, Baker *et al.* (1969) proposed a central role for the effect of raised $[Na^+]_i$ on the Na–Ca exchange in the response of the heart to cardiac glycosides. A similar mechanism has been proposed to underlie the staircase reponse (e.g. Lado, Sheu and Fozzard 1982). In amphibian heart, albeit with indirect methods, it is possible to show a good correlation between the changes in a calculated value of $[Na^+]_i$ and the increase in the strength of the heart beat produced by increased heart rate, low $[K]_o$, cardiac glycosides and Na-ionophores (Chapman and Tunstall 1983b). In mammalian cardiac tissue a steep relationship between $[Na^+]_i$ and both twitch and tonic tension is seen when the Na-pump is inhibited by low $[K]_o$ or cardiac glycosides (Eisner, Lederer, and Vaughan-Jones 1983, 1984; Im and Lee 1984; Lee and Dagostino 1982). These workers fit an equation where tension $= b([Na^+]_i)^y$ and obtain values for y which vary between 3.2 and 7.0. This variability may be due to technical and species difference as Im and Lee (1984) suggest, but Eisner *et al.* (1984) and Boyett, Hart, and Levi (1986) find a temporal dissociation between $[Na^+]_i$ and tension when the Na-pump is inhibited and re-activated. These discrepancies are not necessarily destructive of the hypothesis of Baker *et al.* (1969) but merely indicate that other processes are involved. The possible processes affecting the amount of available activator-Ca^{2+}, include the intracellular Ca-buffering systems (the mitochondrial Ca-accumulation is affected by $[Na^+]_i$ Crompton, Capano, and Carafoli 1976), the loading and liberation of Ca^{2+} in the SR, and the Ca-pump in the sarcolemma (which are affected by Ca-calmodulin-dependent phosphorylation (Gopinath and Vincenzi 1977; Moore and Kraus-Freidmann 1983)), and effects upon the triggering inward Ca-currents (which have been shown to be affected by cardiac glycosides Weingart, Kass, and Tsein 1978). A possible direct effect on the contractile proteins could be mediated by a calmodulin-dependent phosphorylation of the myosin regulatory light chains. Increased stimulation rate causes the classical staircase response in cardiac muscle. Although this is associated with an increased Ca-transient (Allen and Blinks 1978), it has long been known that it can be accompanied by a progressive fall in i_{Ca} in multicellular preparations (Beeler and Reuter 1970). Recently, this relationship has been studied in ventricular myocytes isolated from the guinea-pig heart, where problems associated with the voltage clamp are much reduced as compared to multicellular preparations (Fedida, Noble, and Spindler 1988). In these myocytes, a series of long voltage clamp pulses (>100 ms) induce a progressive fall in i_{Ca} while the contractions show a positive staircase; with brief pulses (<100 ms) i_{Ca} is progressively made larger while the contraction shows a negative staircase response. These changes in i_{Ca} are greatly reduced if EGTA is introduced into the sarcoplasm. These results indicate that over a particular range the size of i_{Ca} may not be important in determining the amount of Ca^{2+} liberated from the SR. Furthermore, the increase in the Ca-transient and the heart beat during staircase response would seem to be due to an increased release of Ca^{2+} from the SR. This increased release may be associated with an increased filling of the SR. Cohen, Fozzard, and Sheu (1982) and Lado, Sheu and Fozzard (1982) find that diastolic $[Na^+]_i$ and $[Ca^{2+}]_i$ rise as the stimulation rate is increased in sheep Purkinje fibres. Comparison of the changes in $[Na^+]_i$ and $[Ca^{2+}]_i$, indicate that when the former rises by a factor of 1.2 the latter is increased by about 1.6. However, the time course of the change in both on the cessation of stimulation are very similar. This discrepancy adds further weight to the idea that the effects of the Na–Ca exchange upon the heart beat are indirect in mammalian heart.

5.4 Measurement of Ca and Na fluxes in cardiac muscle

As noted above the early experiments where the Na-dependent Ca fluxes or the Ca-dependent Na-fluxes were studied in heart tissue, favoured a coupled movement of $2Na^+$ for each Ca^{2+} (Busselen and Van Kerkhove 1978; Jundt *et al.* 1975; Reuter and Seitz 1968). Subsequently, the Na-dependent Ca-efflux from fish, frog, and mammalian heart has been shown to be affected by elevated $[K]_o$ (Busselen 1982; Chapman, Tunstall, and Yates, 1981); while the efflux of ^{22}Na from rabbit interarterial septa is affected by $[Ca]_o$ in a way consistent with a 3Na–Ca exchange (Bridge *et al.* 1981), and in Langendorff perfused heart loaded with Na^+ by exposure to cardiac glycosides, the raising $[Ca]_o$ caused a loss of Na and an accumulation of Ca such that the total Na-loss divided by the total Ca-gain came close to 3.0 (Bridge and Bassingthwaighte 1983). In frog atria, exposed to low $[Na]_o$, over 80 per cent of the uptake of ^{45}Ca occurs while the tension developed by the tissue is relaxing spontaneously, a feature that may explain that low-Na contractures are accompanied by a sustained outward current (Fig. 5.1). Agents that inhibit the Ca-influx also slow the spontaneous relaxation of tension (Chapman, Tunstall, and Yates 1983, 1984; and Fig. 5.3). These observations are strong evidence in favour of an interaction between the sarcolemmal Na–Ca exchange and an intracellular accumulation of Ca by the mitochondria. This is because quinidine reduces the accumulation of calcium in frog atria (Fig. 5.3), while in chick cultured heart cells the large uptake of Ca^{2+}, produced by raising $[Ca]_o$ when $[Na^+]_i$ is elevated, is associated with an accumulation of calcium in the mitochondria (as detected by X-ray microanalysis; Murphy *et al.* 1986). These results may explain why a large fall in $[Na^+]_i$ is accompanied by a small rise in sarcoplasmic $[Ca^{2+}]$ when $[Na]_o$ is lowered. This would be expected if the Na–Ca exchange is activated to influx Ca^{2+} into the cells, to raise $[Ca^{2+}]_i$ and reduce $[Na^+]_i$. These changes would provide a sarcoplasmic environment which would favour an accumulation of Ca^{2+} in the mitochondria by activation of the Ca-uniporter and inhibition of the efflux via the Na-dependent pathway. A positive feedback interaction between the sarcolemmal Na–Ca exchange and Ca-accumulation by the mitochondria would result, with Ca^{2+} entering the cell and being taken up by the mitochondria until $[Na^+]_i$ has fallen to a level where the Ca-influx via the Na–Ca exchange is reduced to a level balanced by the efflux via sarcolemmal Ca-pump. On restoration of the $[Na]_o$ the whole system will reverse as Ca^{2+} leaves the cell in

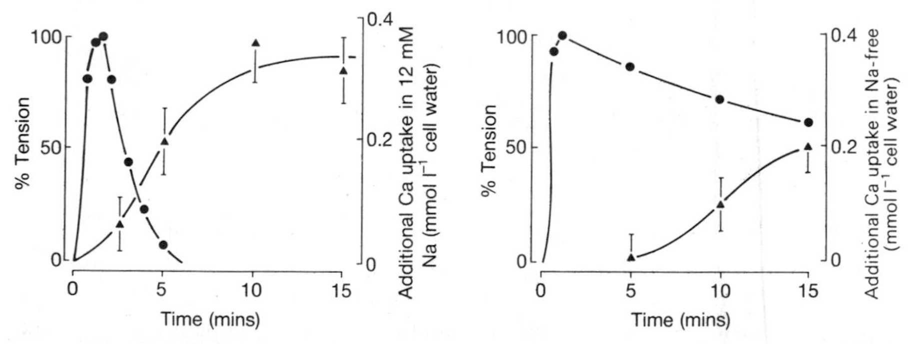

Fig. 5.3. (a) The time course of the uptake of ^{45}Ca by isolated frog half-atria (triangles) when bathed by Na-free medium is compared to the time course of the accompanying tension (circles). The Ca-uptake was calculated by subtracting the ^{45}Ca taken up by the other half of the atrium exposure to normal $[Na^+]_o$ at the same time.
(b) The time course of ^{45}Ca-uptake by isolated frog half-atria (calculated as in part a) in the presence of 1 mM quinidine, when the bathing [Na] was reduced to 12 mM (triangles), is compared to the strength of the accompanying contracture (circles). 20°C and 1 mM-Ca Ringer throughout. (From Chapman, Tunstall, and Yates 1983, 1984.)

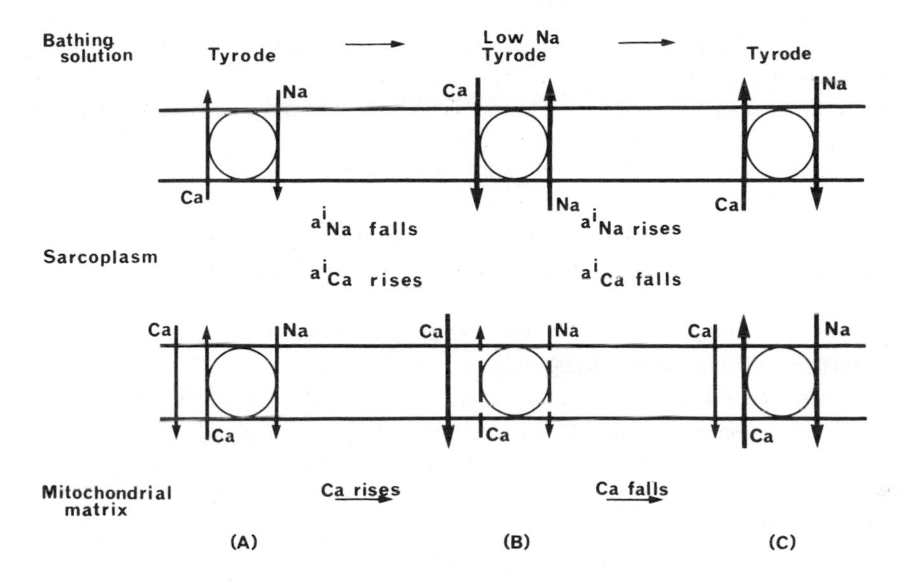

Fig. 5.4. A schematic diagram of the movements of Ca^{2+} and Na^+ across the sarcolemmal and mitochondrial membranes during exposure of a cell to changes in the bathing [Na].
A. The initial conditions in normal Tyrode solution.
B. The conditions immediately after the reduction of $[Na]_o$.
C. Immediately upon return to normal Tyrode solution. The size of the arrows is an attempt to show the magnitude of the ionic fluxes. (From Chapman 1986.)

exchange for Na^+, the rising $[Na^+]_i$ will increase the efflux of Ca^{2+} from the mitochondria and so on (Fig. 5.4). This means that the overall Na- and Ca-fluxes in intact tissue due to the Na–Ca exchange are likely to be complicated by the activity of the mitochondria.

The effects of $[Na]_o$, $[Na^+]_i$, $[Ca]_o$ and membrane potential upon the movement of ions across the cell membrane, the development of tension on the activities of intracellular Ca^{2+} or Na^+ have established the existence and nature of the Na–Ca exchange in cardiac tissue. Apart from effects upon the rate of relaxation of tension, however, it has not been possible to assess the role of the exchange during and individual heart beat, to do this it is essential to identify a current carried by the exchange and from its size and direction under voltage clamp deduce the influence of the Na–Ca exchange.

5.5 The Na–Ca exchange current, i_{NaCa}

Clearly, if an electrogenic Na–Ca exchange is active during normal cardiac activity, it will contribute a membrane current, i_{NaCa}, to that activity. Two questions then arise: can this current be identified and separated from other membrane currents, such as the calcium current, i_{Ca} and if i_{NaCa} can be identified, how does its amplitude and time course correlate with net membrane fluxes? The second of these questions raises the intriguing possibility that, given the much greater capacity of the exchange (see p. 103) to respond to changes in $[Ca]_i$ compared to that of the calcium pump, the current it generates might be sufficient to account for a net membrane calcium efflux that could balance the calcium influx through calcium channels.

Mullins (1979, 1981) proposed that a variety of known membrane currents might be carried by Na–Ca exchange. Of his suggestions the one that has been taken up is that part of the slow inward current, i_{si}, may be attributable to i_{NaCa}. This hypothesis may be further subdivided into the hypotheses: (a) that the i_{NaCa} slow inward current tails that occur following repolarization from potentials at which i_{Ca} is activated are carried by i_{NaCa}, and (b) that a slow part of i_{si} itself is carried by i_{NaCa}. It should be remembered that Ca efflux via the exchange will add to i_{si} while calcium influx will subtract current. The first hypothesis has recently become well-established experimentally, while the second is still controversial.

5.5.1 Slow tail currents

In frog heart cells, Hume and Uehara (1986) and Campbell *et al.* (1988) have shown that the inward current tails are dependent on $[Ca]_i$ and $[Na]_o$. Moreover, the current tail amplitude depends on how much calcium has entered the cell during the depolarization. Thus, while the peak i_{Ca} is reached during very short depolarizations (5–10 ms), the peak inward current tail requires several hundred milliseconds to develop. This leads to the conclusion that the exchange current will be greatest at the end of repolarization and that the resultant efflux of calcium occurs primarily during the diastolic interval.

In mammalian heart cells, the first evidence that the slow inward tail current is carried by a $[Ca]_i$-activated mechanism that requires $[Na]_o$ was obtained by Mitchell *et al.* (1984, 1987), who showed that this tail current is responsible for maintaining the late low-voltage plateau in isolated rat ventricular cells. Schouten and ter Keurs (1985) have also come to this conclusion using a multicellular rat ventricular preparation.

In guinea-pig ventricular cells, Fedida *et al.* (1987b) showed that the slow inward tail current displays kinetic and other properties that clearly distinguish it from i_{Ca}. First, at relatively negative membrane potentials, e.g. below -35 mV, i_{Ca} deactivates very rapidly (within a few ms), whereas the tail current lasts for tens or hundreds of milliseconds, the precise decay time constant depending on the temperature. Furthermore, when intracellular calcium is buffered, no slow tail current occurs and the steady-state current level is then established within 10 ms. Second, whereas i_{Ca} becomes reavailable within 100–300 ms, tail current reactivation requires several seconds, as does the reactivation of contraction and, therefore, of calcium release. Third, the current–voltage relation for the tail current displays a nearly exponential form which is characteristic of that for the inward mode of i_{NaCa} (Kimura, Miyamae, and Noma 1987), and quite unlike that for i_{Ca}.

The important difference between the results in amphibian and mammalian cells is that the mammalian tail current reaches a peak very quickly (Egan *et al.* 1987, 1989) and is very small by the time of the end of repolarization. Thus the activation occurs primarily *during* the action potential, not after it, as in the frog. This fits the idea that the amplitude of the tail current follows that of the intracellular calcium transient (Weir *et al.* 1987).

Tail currents attributable to i_{NaCa} have most frequently been studied at potentials between −40 and −80 mV, the widest range being that used by Fedida *et al.* (1987a) to obtain a current–voltage relation between −15 and −100 mV. In the case of the transient inward current, the range has been extended to +30 mV (Fedida *et al.* 1987a; Figs 5.6, 5.7, and 5.8). Since, in both cases, the voltage dependence strongly resembles that of the inward mode of i_{NaCa} (Kimura, Miyamae, and Noma 1987), the obvious conclusion is that i_{NaCa} can flow over most or even all the potential range of the action potential and that it therefore does so during depolarizations as well as manifesting itself as a tail current following repolarization.

5.5.2 Exchange current during depolarization

The reason why this hypothesis is harder to test experimentally is that i_{Ca} is usually very much larger than i_{NaCa} during depolarizations. Thus, in the DiFrancesco–Noble (1985) model which uses i_{Ca} and i_{NaCa} current–voltage relations and kinetics close to those observed experimentally, i_{NaCa} only appears a dominant component of i_{si} at relatively negative potentials. Positive to about −30 mV, i_{NaCa} merely perturbs the time course of a current that is largely attributable to i_{Ca} (see DiFrancesco and Noble 1985; Fig. 5.5). Furthermore, since $[Ca]_i$ can itself perturb the time course and amplitude of i_{Ca} by influencing its inactivation characteristics, it is not easy to discriminate between i_{Ca} and i_{NaCa} by virtue of their $[Ca]_i$ dependence.

These considerations have led to two approaches to the problem. The first is to use relatively small depolarizations that activate i_{Ca} to a small or even negligible degree but, which still induce substantial Ca release. Fedida *et al.* (1987b; see also Fig. 5.5C) found that in some guinea-pig ventricular cells it was then possible to activate a very slow inward current by depolarizing from −40 mV to just below the threshold for i_{Ca}. This current correlated well with the associated contraction. In ferret ventricular cells, Boyett, Orchard, and

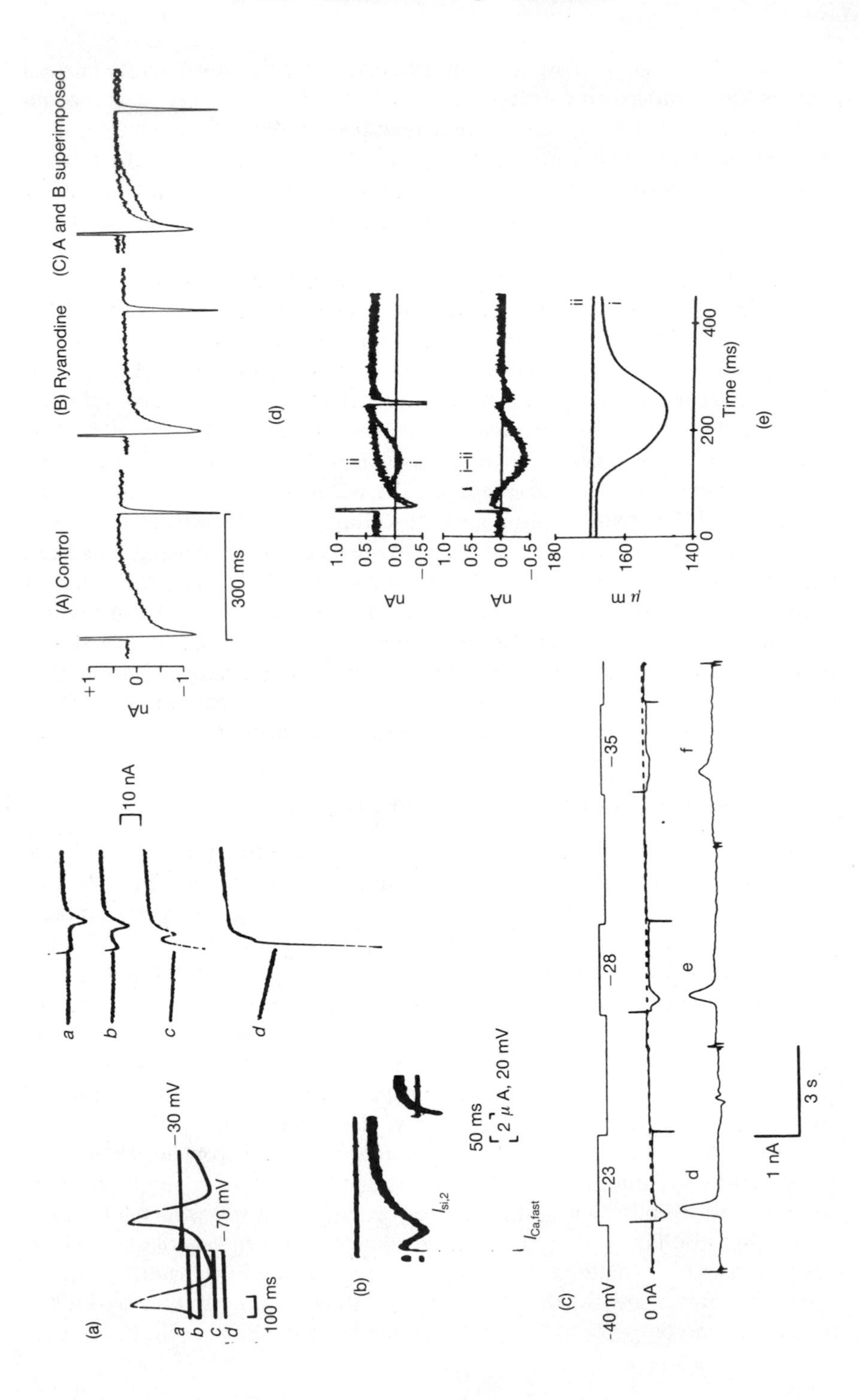
(a)
a
b
c
d
−30 mV
−70 mV
100 ms
10 nA
(b)
$I_{si,2}$
$I_{Ca,fast}$
50 ms
2 μA, 20 mV
(c)
−40 mV
0 nA
−23
−28
−35
d
e
f
1 nA
3 s
(A) Control
(B) Ryanodine
(C) A and B superimposed
nA
+1
0
−1
300 ms
(d)
ii
i
i–ii
1.0
0.5
0.0
−0.5
nA
180
160
140
μm
0
200
400
Time (ms)
(e)

Kirby (1988a) have also observed such a current and showed that it was abolished when the contraction is abolished with ryanodine (see Fig. 5.5D). This behaviour is exactly that of the slow inward current recorded at -40 mV in multicellular Purkinje fibre preparations by Eisner, Lederer, and Noble (1979). It is quite unlike the behaviour of i_{Ca} which is increased in amplitude and duration when the $[Ca]_i$ transient is removed.

The second approach is to select a species in which the relative amplitudes and time courses of i_{Ca} and i_{NaCa} favour their dissection. Thus, in ferret ventricle the slow component is sometimes sufficiently large even at potentials near zero to form a clear second phase of current decay and even to form a separate peak (Arlock and Noble 1985; Boyett, Kirby, and Orchard 1988a,b), as found in the multicellular sino-atrial node preparation (Brown *et al.* 1984). Arlock and Noble (1985) showed that this second peak, $i_{si,2}$, persists for a few pulses even when i_{Ca} has been rapidly blocked by, say, manganese ions. Recently, Boyett, Kirby and Orchard (1988b) have found similar results in some isolated ferret ventricular cells. Their experiments show that the slow component is (a) correlated with the amplitude of the slow tail current on repolarization and (b) is abolished when intracellular calcium release is blocked (e.g. with ryanodine).

Figure 5.5 shows some examples of experiments showing the slow component recorded during depolarizations. It should, however, be emphasized that

---

Fig. 5.5. Slow phases of inward current recorded during depolarization in cardiac muscle cells.

(a) Sino-atrial node of the rabbit. The clamp protocol is shown on the left superimposed on natural pacemaker activity from this tissue. Currents shown on right were recorded on clamping to -30 mV from various levels of pre-potential to generate different amplitudes of i_{Ca}. When i_{Ca} is very large (following the most negative prepotential) the slow current is evident only as a slow phase at the end of the calcium current. When i_{Ca} is small the slow phase of current forms a clearly separate peak. This occurred in some but not all preparations. In most preparations, the presence of a slow component was evident only in the change in time course following clamps from different prepotentials (Brown *et al.* 1984).

(b) Ferret ventricular strip. Voltage clamp from -40 mV to $+4$ mV generates a rapid calcium current followed by a slow phase labelled $i_{si,2}$ (Arlock and Noble 1985).

(c) Guinea-pig ventricular cell. Small depolarizations from -40 mV to -35, -28, and -23 mV generate a very slow current. This is unlikely to be attributable to clamp escape since the contraction (bottom trace) is graded (Fedida *et al.* 1987).

(d) Ferret ventricular cell. Currents recorded before and after application of ryanodine to suppress calcium release from the SR. This removes the slow phase of inward current (Boyett, Kirby and Orchard 1988a).

(e) Ferret ventricular cell. Currents recorded before (i) and after (ii) substituting sodium with lithium. This abolishes the slow peak of inward current and the contraction (bottom trace). The difference current (middle trace) gives a possible time course for the exchange current (from Boyett, Kirby and Orchard, 1988b).

records of these kinds are the exception rather than the rule. This suggests that, normally, i_{NaCa} is not large enough to form a clearly separable phase or peak of inward current decay. We will return later to the question how large we would expect i_{NaCa} to be during depolarizations to around zero potential. In order to answer this question it is necessary to consider the possible functional role of activation of i_{NaCa} during the action potential. A strong clue to this role is obtained by correlating ionic currents with net transmembrane fluxes.

5.6 Correlation with calcium fluxes

A method that enables net calcium flux to be estimated with the required time resolution (down to 2 ms) is to record extracellular calcium with a calcium-sensitive marker. Since the diffusion of calcium between the extracellular space and the bathing fluid is very slow (order of tens of seconds), fast changes occuring within milliseconds or seconds can be attributed entirely to transmembrane fluxes (Hilgemann 1986a,b). The results show that the rate of $[Ca]_o$ depletion resulting from calcium influx via i_{Ca} is maximal almost immediately, which correlates well with the fact that i_{Ca} reaches its peak within 2–5 ms. As shown in the upper panel of Fig. 6, $[Ca]_o$ continues to decrease for about 30 ms during the rabbit atrial action potential, after which re-accumulation occurs. This process is complete, with $[Ca]_o$ restored to its original level well before repolarization is complete. Thus calcium influx dominates for the first 30 ms, while calcium efflux dominates for the next 150 ms. Note that the ratio of time scales is about 1:5. This will be important when we return to the question of the expected relative amplitudes of i_{Ca} and i_{NaCa}.

The main panel in Fig. 5.6 shows a computer reconstruction of this result. The model is based on a modification of that for the rabbit sino-atrial node (Noble and Noble 1984) developed for the rabbit atrium by Hilgemann and Noble (1987). The kinetics and voltage dependence of i_{Ca} and i_{NaCa} are represented by equations that accurately reproduce the known experimental results on these two currents. It is clear that the computed curve for $[Ca]_o$ closely follows that recorded experimentally. This means, in turn, that the net calcium fluxes predicted from knowledge of i_{Ca} and i_{NaCa} are close to those recorded experimentally. The conclusion, therefore, is that calcium efflux via the Na–Ca exchange could indeed be large enough to balance calcium entry via the calcium channels and that this process occurs rapidly enough to restore calcium balance before the end of repolarization.

This result enables us to proceed to estimate the likely relative amplitudes of i_{Ca} and i_{NaCa} First we note that if the exchange ratio is 3:1 and the exchange is responsible for pumping all the calcium out then the net charge entry (i.e.

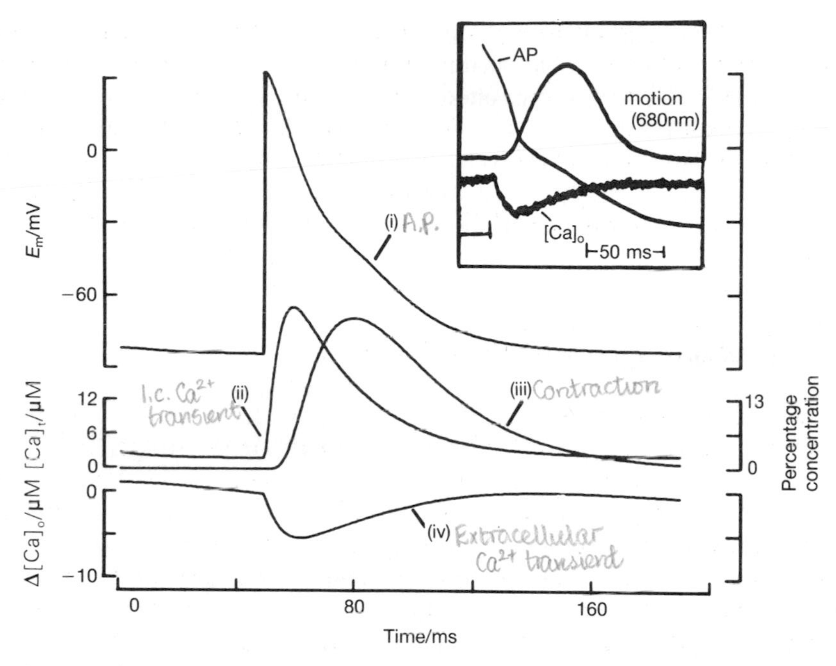

Fig. 5.6. Experimental (top right inset) and computed rabbit atrial action potentials together with contraction (iii), intracellular calcium transient (in computed case only) (ii) and extracellular calcium transient (iv). In both computed and experimental results external calcium shows rapid depletion (due to calcium entry through ionic channels) during first 30 ms followed by re-accumulation due to calcium pumping by the Na–Ca exchange during rest of action potential (from Hilgemann and Noble 1987).

the time integral of the current) represented by i_{Ca} should be twice that represented by i_{NaCa} since each calcium entering via the channels carries two charges, whereas each calcium moving via the exchange is balanced by two of the sodium ions entering, leaving one unit of charge entering for each calcium pumped out. Since the time scales over which these events occur are in the ratio 5:1, the expected ratio of average amplitudes of currents should be about 10:1 (i.e. the average value of i_{NaCa} will be only about 10 per cent of that for i_{Ca}). In most cases, therefore, this will produce only a small perturbation of the time course of inward current following i_{Ca}. If the cell is, for any reason, hypodynamic then this ratio will become even more unfavourable to i_{NaCa}. In many cases this is the case. Thus, resting guinea-pig and rat ventricular cells appear to lose most of their stored calcium, and a period of rapid repetitive stimulation is then required to generate a substantial contraction. It is significant that this also greatly enhances the slow inward tail current attributable to i_{NaCa} (Egan *et al.* 1989).

One case of particular importance in which i_{NaCa} will be virtually eliminated from the recordings of inward current is that in which experiments are performed with intracellular calcium buffered by including EGTA or BAPTA in the patch electrode, as is often done in patch clamp recording. This buffering removes the slow current tails (Fedida *et al.* 1987b) and, not surprisingly, no evidence for a contribution of i_{NaCa} to the current during depolarizing pulses can then be found.

Conversely, if for any reason the cells are hyperdynamic and the calcium stores are well loaded, then the balance will be shifted in favour of i_{NaCa}. It would only require the peak value of $[Ca]_i$ to double or treble to permit relative amplitudes of the kind shown in Fig. 5.5 to occur. Thus, in Fig. 5.5B the peak value of $i_{si,2}$ is about 30 per cent of the peak value of i_{Ca}.

5.7 Exchange current during the action potential

We now turn to the question of the influence of i_{NaCa} on the action potential and, in particular, on the speed and shape of repolarization. Hilgemann and Noble (1987) answered this question by comparing action potentials occurring with and without large $[Ca]_i$ transients. This can be achieved physiologically by using paired pulse stimulation, with the second pulse occurring before contractile repriming has had time to occur. The result is a small or negligible contraction and hardly any calcium efflux, as estimated by recording $[Ca]_o$ (see Fig. 5.7A). The Hilgemann–Noble (1987) model behaves in the same way. In both experiment and model the effects on the action potential are twofold: at positive membrane potentials, where i_{Ca} dominates, the action potential is prolonged as a result of slower i_{Ca} inactivation when $[Ca]_i$ is low; at negative potentials, where i_{NaCa} dominates, the action potential is shortened since i_{NaCa} becomes very small. This means that in rabbit atrium, the last phase of the action potential is maintained largely by the exchange current (Earm, Ho, and So 1989). Mitchell *et al.* (1984) and Schouten and ter Keurs (1985) have come to a similar conclusion as regards the low voltage slow plateau (late repolarization) in rat ventricular cells. This part of the rat ventricular action potential is abolished either when the $[Ca]_i$ transient is removed by ryanodine or when the exchange current is removed by replacing $[Na]_o$ with lithium, an ion which is not transported by the exchange.

Fedida *et al.* (1987b) have also used the two-pulse method in voltage clamp experiments. As shown in Fig. 5.7B, the inward tail current is large following the first pulse, but is small following the second pulse. The i_{Ca}, by contrast, is actually increased during the second pulse. Evidence that $[Ca]_i$ controls i_{Ca} in this way has also been obtained by Fedida, Noble, and Spindler (1988) and by Boyett, Kirby, and Orchard (1988a,b). Since these experiments were performed on guinea-pig and ferret ventricular cells, this shows that the basic mechanisms controlling i_{Ca} and i_{NaCa} are similar to those in atrial muscle and

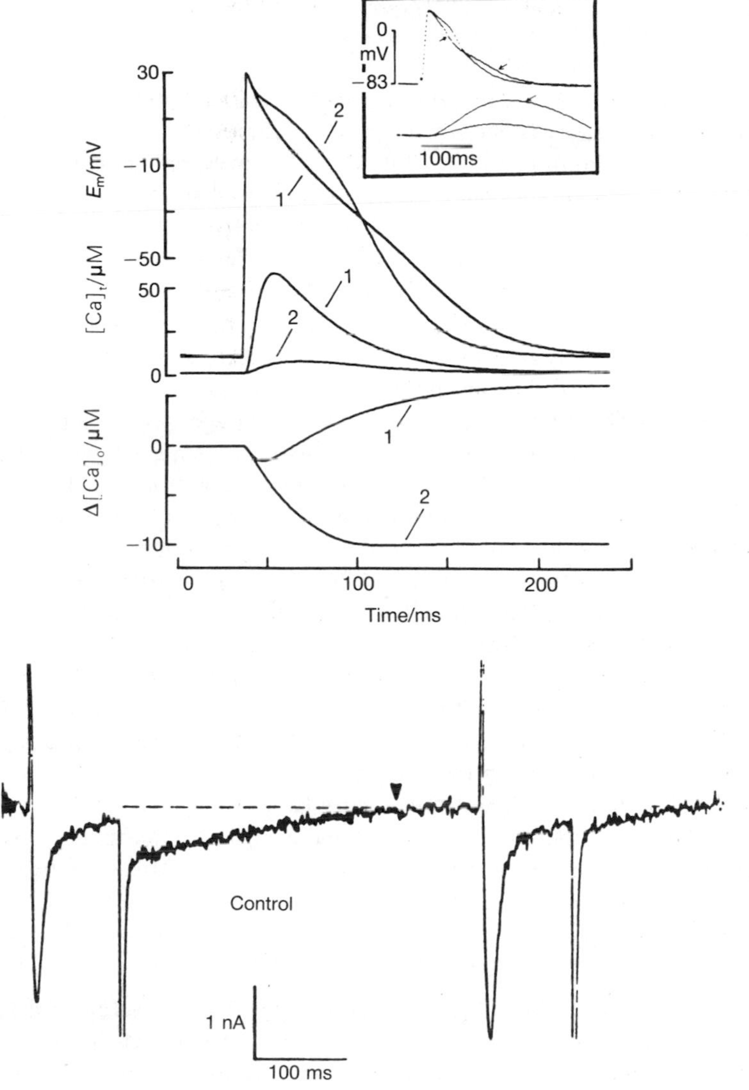

Fig. 5.7. (A) Experimental (top right inset) and computed results for paired pulse stimulation. The second pulse follows very shortly after the first so that the SR calcium release is still inactivated. The contraction accompanying the second action potential is therefore very small. There is then virtually no calcium efflux (external calcium shows depletion followed by very little re-accumulation). In the model this is attributed to much weaker activation of the Na–Ca exchange during the second response since the variation in internal calcium is too small (from Hilgemann and Noble 1987).

(B) Paired pulse voltage clamp experiment on guinea-pig ventricular cell. The membrane was held at -40 mV and pulsed to 0 mV. The slow inward current tail attributable to Na–Ca exchange is very small following second pulse (from Fedida *et al.* 1987).

in rat ventricular cells, even though in the case of guinea-pig and ferret ventricular cells, there is usually no separate phase of repolarization that can be attributed to the time course of decay of the exchange current.

Can we nevertheless estimate the time course and amplitude of i_{NaCa} during the guinea-pig ventricular action potential, despite the absence of a phase of repolarization correlated almost uniquely with the exchange current? This question is important because in the great majority of species, including man, the ventricular action potential resembles that of the guinea-pig rather than that of the rat. Clearly it is not possible to answer this question by recording i_{NaCa} directly during the action potential plateau. However, the question can be answered indirectly by making use of the known properties of i_{NaCa}. Egan *et al.* (1988a) have recently done this, and their results are shown in Fig. 5.8.

The recording labelled E_m shows the action potential itself. The curve labelled $[Ca]_i$ was obtained by measuring slow inward current tail amplitudes on suddenly interrupting the action potential at different times and clamping at the resting potential. Since these tail currents can be attributed to i_{NaCa} and because it is known that 1 μM $[Ca]_i$ generates 1 nA of exchange current (Kimura, Miyamae, and Noma 1987), the tail amplitudes can be converted to estimates of the values of $[Ca]_i$ that would be required to generate each tail current. This analysis makes the assumption that there is little or no delay in the activation of the exchange by internal calcium, but the analysis is justified by the remarkable correspondence between the curve obtained and that recorded directly using fura-2 to measure $[Ca]_i$ (Weir *et al.* 1987).

From this information and using a value of $[Na]_i$ (assumed here to be 7 mM) we can compute the reversal potential, E_{NaCa}, for the exchange process using the equation

$$E_{NaCa} = 3\ E_{Na} - 2\ E_{Ca}$$

It can be seen that the value of E_{NaCa} quickly becomes positive to that of the action potential plateau, and remains some 20–30 mV positive to E_m throughout repolarization. This means that the exchange should generate inward current and should therefore delay repolarization, as it does in the rat ventricular action potential.

Finally, using the tail current amplitudes together with the computed values of E_{NaCa}, and knowing the current–voltage relation for i_{NaCa}, we can compute the time course of i_{NaCa} during the plateau and repolarization phase. The result is shown as the lowest curve in Fig. 5.8. The exchange current is very small at the beginning of the action potential, reaches a peak inward level of about -150 pA towards the end of the plateau, and then decays quite rapidly during the final phase of repolarization. Note that the peak inward current occurs considerably later than the presumed time of peak $[Ca]_i$ The reason for this delay is that the voltage dependence of i_{NaCa} is sufficiently steep that repolarization increases the current faster than the decline of $[Ca]_i$.

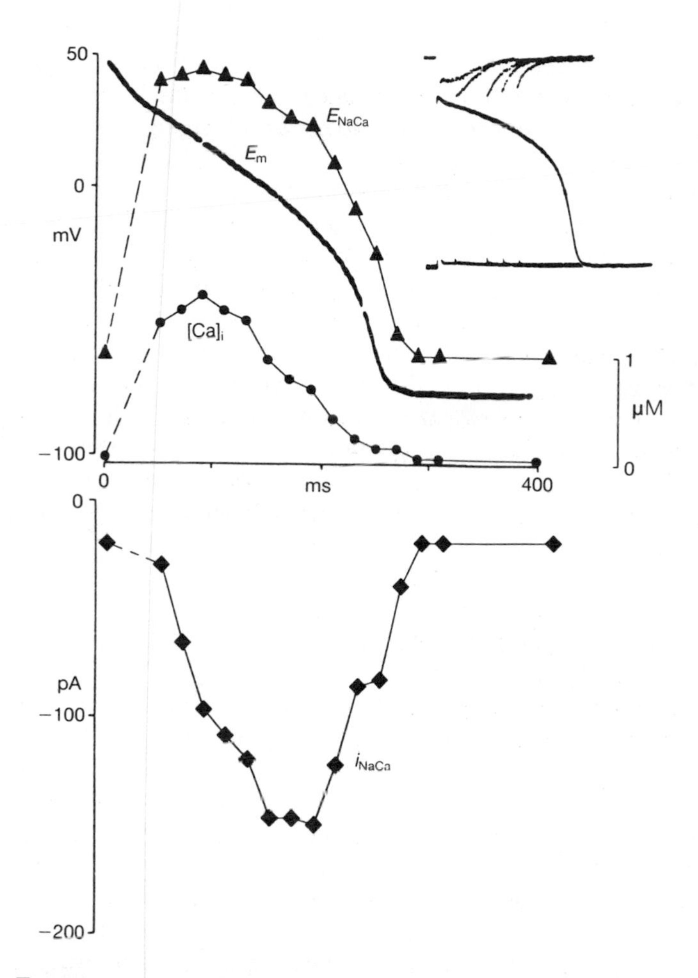

Fig. 5.8. Top inset. Result of experiment in which a guinea-pig ventricular cell is voltage clamped to the resting potential at various times during the action potential. The resulting slow inward tail currents first increase and then decrease in amplitude. Main figure. Results of another experiment similar to that shown in the inset in which the variation of tail current amplitudes has been converted to corresponding intracellular calcium levels. From this information the times courses of the Na–Ca exchange reversal potential and Na–Ca exchange current were calculated (from Egan *et al.* 1989).

How significant is this level of i_{NaCa}? The net repolarizing current during the plateau can be calculated to be about 20 pA (Egan *et al.* 1989). The estimated exchange current is large compared to this net repolarizing current, which means that the exchange current must significantly slow repolarization. Moreover, if we integrate the time course of i_{NaCa} we obtain a net charge transfer of about 20 pC (an average of 100 pA flows for about 200 ms). This corresponds to a net exit of calcium carrying 40 pC of charge, and would be quite sufficient to balance a net entry via calcium channels, since these carry about 500 pA of current for about 100 ms on clamping to the level of the plateau.

To summarize, during normal rhythmic beating in mammalian cardiac muscle the functional role of the Na–Ca exchange seems to be to pump calcium out of the cells during the process of repolarization. In cells with relatively low or short plateaux this process can generate net calcium efflux as early as 30 ms after the beginning of the action potential and can generate a second slow plateau phase. In cells with relatively high plateaux the time of onset of the exchange current and therefore of net calcium efflux is somewhat later (D. W. Hilgemann, personal communication; and see Fig. 5.8), as would be expected since the membrane potential takes a longer time to reach the range of potentials at which i_{NaCa} is larger than i_{Ca}. But, in both cases the exchange may produce an efflux of calcium sufficient to balance the calcium entry via calcium channels during each action potential.

5.8 Exchange current during abnormal rhythm

We now come to the role of the exchange in abnormal cardiac rhythm. A possible role here was proposed some time ago, when Kass *et al.* (1978) reviewed the possible mechanisms for the so-called transient inward current, i_{TI}, underlying ectopic beating in conditions of calcium overload (e.g. during sodium pump inhibition by cardiac glycosides). It is clear that this current is activated by oscillatory variations in $[Ca]_i$ resulting from calcium-induced release from the SR. Variations in $[Ca]_i$ have been recorded experimentally using aequorin (Allen *et al.* 1984). Two possible Ca-activated mechanisms were proposed by Kass *et al.*, Ca-activated cation channels showing a reversal potential near zero potential, and calcium-activation of electrogenic Na–Ca exchange. The first hypothesis became strongly favoured when Colquhoun *et al.* (1981) identified a single channel in cultured heart cells that conducts sodium and potassium ions almost equally well and which is activated by $[Ca]_i$, particularly since the work on i_{TI} in Purkinje fibres showed a reversal potential near zero potential. Indeed, in conditions in which electrogenic Na–Ca exchange has been eliminated, the reversal potential is particularly clear (Cannell and Lederer 1986).

But this result does not, of course, answer the question what generates i_{TI} in conditions when the exchange *is* active, nor whether the situation in Purkinje fibres also applies to the atrial and ventricular muscle proper. In fact, recent results strongly support the view that the majority of i_{TI} in atrial and ventricular cells is carried by i_{NaCa} (J. Kimura, personal communication, Fedida *et al.* 1987a; Lipp and Pott 1988). Thus, the current–voltage relation for i_{TI} in guinea-pig ventricular cells closely resembles that for the inward mode of the exchange: it is roughly exponential, tending towards zero at strong depolarizations but not showing a reversal potential.

The involvement of Na–Ca exchange in i_{TI} could be functionally significant since it is obviously important to activate efflux of calcium in cells that are calcium overloaded.

References

Allen, D. G. and Blinks, J. R. (1978). *Nature* **273**, 509–13.

Allen, D. G., Eisner, D. A., Lab, M. J., and Orchard, C. H. (1983). *J. Physiol.* **345**, 391–407.

Allen, D. G., Eisner, D. A., and Orchard, C. H. (1984). *J. Physiol.* **350**, 615–30.

Arlock, P. and Noble, D. (1985). *J. Physiol.* **369**, 88P.

Baker, P. F., Blaustein, M. P., Hodgkin, A. L., and Steinhardt, R. A. (1969). *J. Physiol.* **200**, 431–58.

Barcenas-Ruiz, L., Beuckelmann, D. J., and Weir, W. G. (1987). *J. Physiol.* **394**, 27P.

Beeler, G. W. and Reuter, H. (1970). *J. Physiol.* **207**, 211–29.

Bers, D. M. and Ellis, D. (1982). *Pflügers Arch.* **393**, 171–8.

Blaustein, M. P. and Hodgkin, A. L. (1969). *J. Physiol.* **200**, 497–527.

Blatter, L. A. and McGuigan, J. A. S. (1986). *Q. J. Exp. Physiol.* **71**, 467–73.

Boyett, M. R., Hart, G., and Levi, A. J. (1986). *J. Physiol.* **381**, 311–31.

Boyett, M. R., Kirby, M. S. and Orchard, C. H. (1988a). *J. Physiol.* **396**, 165P.

Boyett, M. R., Kirby, M. S., and Orchard, C. H. (1988b). *J. Physiol.* (in press).

Bridge, J. H. B., Cabeen, W. R., Langer, G. A., and Reeder, S. (1981). *J. Physiol.* **316**, 555–74.

Bridge, J. H. B. and Bassingthwaighte, J. B. (1983). *Science* **219**, 178–80.

Bridge, J. H. B., Spitzer, K. W., and Ershler, P. R. (1988). *Science* **241**, 823–5.

Brown, H. F., Kimura, J., Noble, D., Noble, S. J., and Taupignon, A. I. (1984). *Proc. Roy. Soc. B* **222**, 305–28.

Busselen, P. (1982). *J. Physiol.* **327**, 309–24.

Busselen, P. and Van Kerkhove, E. (1978). *J. Physiol.* **282**, 263–83.

Campbell, D. L., Giles, W. R., Robinson, K., and Shibata, E. F. (1988). *J. Physiol.* (in press).

Cannell, M. B. and Lederer, W. J. (1986). *J. Physiol.* **374**, 201–19.

Chapman, R. A. (1983). *Am. J. Physiol.* **245**, H535–52.

Chapman, R. A. (1986). *J. Physiol.* **373**, 163–79.

Chapman, R. A. and Miller, D. J. (1974). *J. Physiol.* **242**, 589–613.

Chapman, R. A. and Leoty, C. (1976). *J. Physiol.* **256**, 278–314.

Chapman, R. A. and Rodrigo, G. C. (1985). *Q. J. Exp. Physiol.* **70**, 447–59.

Chapman, R. A. and Rodrigo, C. G. (1986). *Q. J. Exp. Physiol.* **71**, 675–87.

Chapman, R. A. and Rodrigo, C. G. (1987). *Q. J. Exp. Physiol.* **72**, 561–70.

Chapman, R. A. and Tunstall, J. (1980). *J. Physiol.* **305**, 109–23.

Chapman, R. A. and Tunstall, J. (1981a). *J. Physiol.* **310**, 97–115.

Chapman, R. A. and Tunstall, J. (1981b). *Adv. Physiol. Sci.* **8**, 93–102.

Chapman, R. A. and Tunstall, J. (1983a). *Q. J. Exp. Physiol.* **68**, 397–412.
Chapman, R. A. and Tunstall, J. (1983b). *Q. J. Exp. Physiol.* **68**, 381–95.
Chapman, R. A. and Tunstall, J. (1984). *Q. J. Exp. Physiol.* **69**, 559–72.
Chapman, R. A., Coray, A., and McGuigan, J. A. S. (1983). *J. Physiol.* **343**, 253–76.
Chapman, R. A., Tunstall, J., and Yates, R. J. (1981). *J. Physiol.* **316**, 31–31P.
Chapman, R. A., Tunstall, J., and Yates, R. J. (1983). *J. Physiol.* **342**, 69–70P.
Chapman, R. A., Tunstall, J., and Yates, R. J. (1984). *J. Physiol.* **350**, 53P.
Cohen, C. J., Fozzard, H. A., and Sheu, S. S. (1982). *Circ. Res.* **50**, 651–62.
Colquhoun, D., Neher, E., Reuter, H., and Stevens, C. F. (1981). *Nature* **294**, 752–4.
Crompton, M., Capano, M., and Carafoli, E. (1976). *Eur. J. Biochem.* **69**, 221–41.
DiFrancesco, D. and Noble D. (1985). *Phil. Trans. Roy. Soc. B* **307**, 353–98.
Earm, Y. E., Ho, W. K., and So, I. S. (1989). *J. Physiol.* **410**, 64P.
Eisner, D. A., Lederer, W. J., and Noble, D. (1979). *J. Physiol.* **293**, 76P.
Eisner, D. A., Lederer, W. J., and Vaughan-Jones, R. D. (1983). *J. Physiol.* **335**, 723–43.
Eisner, D. A., Lederer, W. J., and Vaughan-Jones, R. D. (1984). *J. Physiol.* **355**, 251–66.
Egan, T. M., Noble, D., Noble, S. J., Powell, T., and Twist, V. W. (1987). *J. Physiol.* **381**, 94P.
Egan, T. M., Noble, D., Noble, S. J., Powell, T., Spindler, A. J., and Twist, V. (1989). *J. Physiol.* **411**, 639–61.
Fedida, D., Noble, D., Rankin, A. C., and Spindler, A. J. (1987a). *J. Physiol.* **392**, 523–42.
Fedida, D., Noble, D., Shimoni, Y., and Spindler, A. J. (1987b). *J. Physiol.* **385**, 565–89.
Fedida, D., Noble, D., and Spindler, A. J. (1988). *J. Physiol.* **405**, 439–60.
Gibbons, W. R. and Fozzard, H. A. (1971). *J. Gen. Physiol.* **58**, 483–510.
Glitsch, H. G., Reuter, H., and Scholz, H. (1970). *J. Physiol.* **209**, 25–42.
Goto, M., Kimoto, Y., Saito, M., and Wada, Y. (1972). *Jap. J. Physiol.* **22**, 637–50.
Gopinath, R. M. and Vincenzi, F. F. (1977). *Biochem. Biophys. Res. Comm.* **77**, 1203–16.
Hansford, R. G. and Lakatta, E. G. (1986). *J. Physiol.* **390**, 453–67.
Hess, P., Metzger, P., and Weingart, R. (1982). *J. Physiol.* **333**, 173–88.
Hess, P. and Weir, W. G. (1984). *J. Gen. Physiol.* **83**, 417–33.
Hilgemann, D. W. (1986a). *J. Gen. Physiol.* **87**, 675–706.
Hilgemann, D. W. (1986b). *J. Gen. Physiol.* **87**, 707–35.
Hilgemann, D. W., and Noble, D. (1987) *Proc. Roy. Soc. B.* **230**, 163–205.
Hume, J. R. and Uehara, A. (1986). *J. Gen. Physiol.* **87**, 857–84.
Im, W. B. and Lee, C. O. (1984). *Am. J. Physiol.* **247**, C478–87.
Jundt, H., Porzig, H., Reuter, H., and Stucki, J. W. (1975). *J. Physiol.* **246**, 229–53.
Kass, R. S., Lederer, W. J., Tsien, R. W., and Weingart, R. (1978). *J. Physiol.* **281**, 187–208.
Kimura, J., Miyamae, S., and Norma, A. (1987). *J. Physiol.* **384**, 199–222.
Lado, M. G., Sheu, S. S., and Fozzard, H. A. (1982). *Am. J. Physiol.* **243**, H133–7.
Lee, C. O. (1981). *Am. J. Physiol.* **241**, H4459–78.
Lee, C. O. and Dagostino, M. (1982). *Biophys. J.* 40, 185–92.
Lee, C. O., Uhm, D. Y., and Dresdner, K. (1980). *Science* **209**, 699–701.

Lipp, P. and Pott, L. (1988). *J. Physiol.* **397**, 601–30.
Lüttgau, H. C. and Niedergerke, R. (1958). *J. Physiol.* **143**, 486–505.
Marban, E. and Weir, W. G. (1985). *Circ. Res.* **56**, 133–8.
Mascher, D. (1973). *Pflügers Arch.* **3342**, 325–46.
Mitchell, M. R., Powell, T., Terrar, D. A., and Twist, V. (1984). *J. Physiol.* **351**, 40P.
Mitchell, M. R., Powell, T., Terrar, D. A., and Twist, V. (1987). *J. Physiol.* **391**, 545–600.
Moore, P. B. and Kraus-Friedmann, N. (1983). *Biochem. J.* **214**, 69–75.
Mullins, L. J. (1979) *Am J. Physiol.* **236**, C103–10.
Mullins, L. J. (1981) *Ion transport in heart.* Raven Press, New York.
Murphy, E., Wheeler, D. M., LeFurgey, A., Jacob, R., Lobaugh, L. A., and Lieberman, M. (1986). *Am. J. Physiol.* **250**, C442–52.
Murphy, E., Jacob, R., and Leiberman, M. (1984). *Fed. Proc.* **43**, 767.
Niedergerke, R. (1963). *J. Physiol.* **167**, 515–50.
Noble, D. and Noble, S. J. (1984) *Proc. Roy. Soc. B* **222**, 295–304.
Philipson, K. D. and Nishimoto, A. Y. (1981). *J. Biol. Chem.* **256**, 3698–702.
Philipson, K. D. and Ward, R. (1986). *J. Mol. Cell. Cardiol.* **18**, 943–51.
Pitts, B. J. R. (1979). *J. Biol. Chem.* **354**, 6232–5.
Powell, T., Tatham, P. E. R., and Twist, V. W. (1984). *Biochem. Biophys. Res. Comm.* **122**, 1012–20.
Reeves, J. P. and Sutko, J. L. (1980). *Science* **208**, 1461–4.
Reuter, H. and Seitz, N. (1968). *J. Physiol.* **195**, 451–70.
Roulet, M. J., Mongo, K. G., Vassort, G., and Ventura-Clapier, R. (1979). *Pflügers Arch.* **379**, 259–68.
Schouten, V. J. A. and ter Keurs, H. D. E. J. (1985). *J. Physiol.* **360**, 13–25.
Sheu, S. S. and Fozzard, H. A. (1982). *J. Gen. Physiol.* **80**, 325–51.
Sheu, S. S., Sharma, V. K., and Banerjee, S. P. (1984). *Cir. Res.* **55**, 830–4.
Weingart, R., Kass, R. S., and Tsein, R. W. (1978). *Nature* **273**, 389–92.
Weir, W. G., Cannell, M. B., Berlin, J. R., Marban, E., and Lederer, W. J. (1987). *Science*, **235**, 325–8.
von Willbrandt, W. and Koller, H. (1948). *Helv. Physiol. Pharmacol. Acta.* **6**, 208–21.

6 Numerical probes of sodium–calcium exchange

Donald W. Hilgemann

6.1 Introduction

Numerical models of Na–Ca exchange provide an important framework to consider diverse data on exchange in relation to specific molecular assumptions (Hilgemann 1988; Hodgkin and Nunn 1987; Johnson and Kootsey 1985; Laüger 1987). The simulation approach has no alternatives in this task, although precedents for caution in applying detailed hypothetical models to physiological mechanisms are strong. The attempt to distinguish between different possible models need not be the primary goal, but is probably the goal of widest interest. Therefore, the present article briefly considers simple formulations of three exchange model types: (1) the 'consecutive' exchange model, where three sodium ions and one calcium ion are moved in separate steps during each exchange cycle; (2) a simultaneous exchange model, which moves sodium and calcium in one step followed by rearrangement of the empty carrier in a second step; (3) a simultaneous exchange model with non-independent sodium and calcium binding sites on each side, which does not require empty-state reorientation. Behaviours of the simple models are summarized, which presently favour the latter type of simultaneous model.[1] Finally, the article considers the possible significance of ion binding–unbinding kinetics for current–voltage relations, current transients (or lack thereof), and estimates of exchanger density. Many issues outlined here have been treated in more detail in the articles cited above.

6.2 Three models of exchange

Schematic diagrams of the three model types are given in Fig. 6.1. Ion movements are traced through one full exchange cycle, so that the first and the last exchanger states are equivalent (E1). Open parentheses are used to designate multiple ion binding–unbinding reactions. These parentheses can

[1] Tentative support for this choice by simulations was presented by Drs M. Kootsey and E. A. Johnson at the 1987 Stowe meeting on Na^{+}–Ca^{2+} exchange.

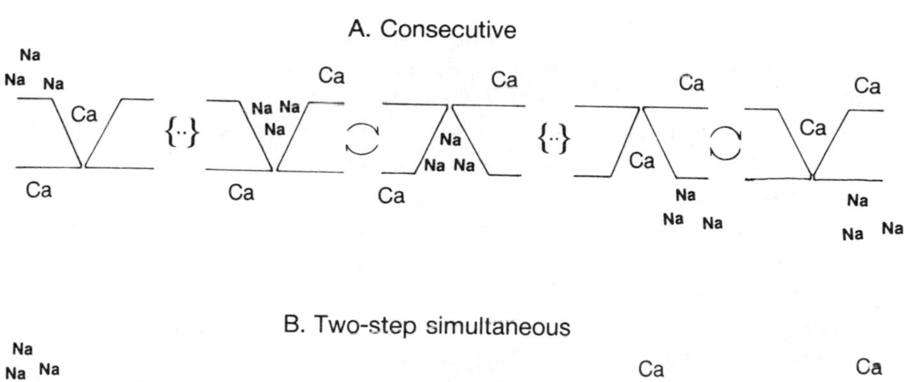

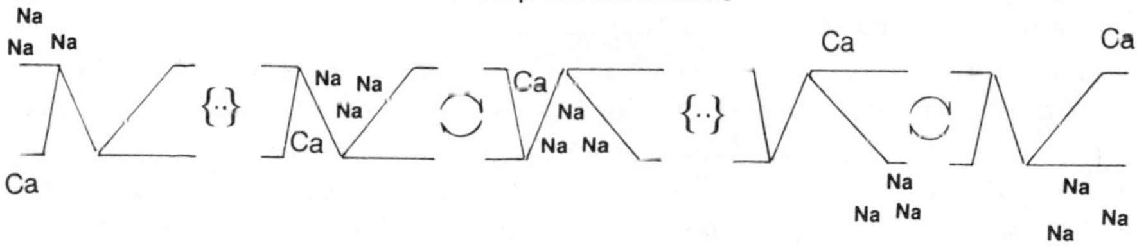

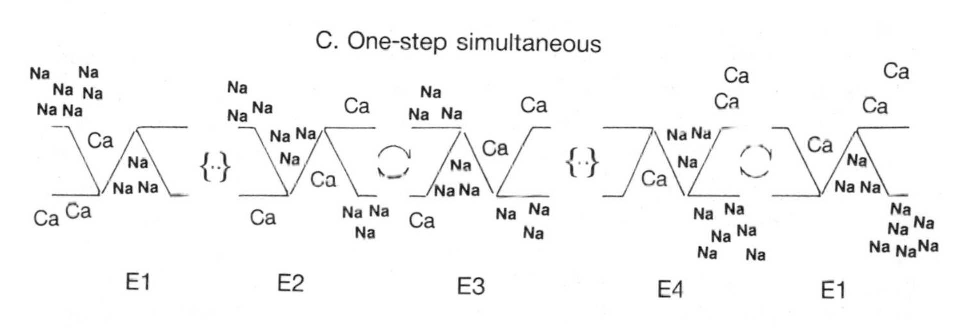

Fig. 6.1. Schematic representations of three Na–Ca exchange model types. The figure follows each model through one full exchange cycle (E1 to E2 to E3 to E4 to E1). Brackets designate multiple ion binding–unbinding steps. Circular arrows designate steps involving the movement of ions or binding sites through the membrane potential. A. Consecutive type: Multiple sodium ions and/or one calcium ion can be bound to sites, which are accessible to either the internal or external medium. Ions are reversibly translocated when three sodium ions or one calcium ion is bound. B. Two-step simultaneous type: Binding sites exclusive to three sodium ions and one calcium ion are available for binding from opposite sides of the exchanger. Reversible translocation of bound ions occurs when all binding sites are loaded. Reversible translocation of binding sites occurs when all sites are empty. C. One-step simultaneous type. Binding sites for both calcium and sodium are accessible from both internal and external media. When one calcium ion is bound on one side and exclusively three sodium ions are bound on the opposite side, reversible translocation of the bound ions occurs. A full cycle, returning the binding sites to their original orientation involves the translocation of two calcium and six sodium ions in opposite directions.

be filled in by many different binding schemes. Steps involving rearrangements across the membrane field are designated by curved arrows. The first and third reactions in each case are ion binding–unbinding reactions; the second and fourth reactions are 'translocation' reactions. The schematic representations do not imply any assumption about the physical nature of the exchanger or exchanger vestibules (see Laüger 1987) on approach of ions to binding sites. Formulations of each of the models, incorporating calcium binding–unbinding kinetics, have been used in simulations of cardiac excitation–contraction coupling (Hilgemann and Noble 1987).

Type A is referred to here as the 'consecutive' type. It is sometimes referred to as the 'ping-pong' model or 'sequential' model. The term 'sequential' is not used, because it was used to describe an exchange mechanism which simultaneously translocates ions in opposite direction (Gunn and Frölich 1979). In this scheme, exchanger binding sites are available exclusively to medium on one side or the other side of the membrane. When loaded with either three sodium ions or one calcium ion, the exchanger can translocate the ions from a site accessible to the cytosol to a site accessible to the extracellular medium and vice versa. An apparent attraction of this scheme is that Ca–Ca exchange in the absence of sodium ions and Na–Na exchange in the absence of calcium ions are inherent to its operation.

Type B is referred to as the 'two-step simultaneous' type. Three sodium ions and one calcium ion can bind to sites with exclusive sodium or calcium affinity on opposite sides of the exchanger. Fully loaded exchanger and fully empty exchanger species can reorientate so as to make binding sites accessible to the media on opposite membrane sides. For further details see Johnson and Kootsey (1985). Without additional assumptions Na–Na exchange and Ca–Ca exchange in the absence of the counter ion are not predicted by the model. It is called a 'two-step' simultaneous model because a full exchanger cycle involves two 'translocation' steps. Ions are moved in one step and the fully empty carrier undergoes a conformation change in another step, which again allows transport in the previous direction.

Type C is referred to as the 'one-step simultaneous' type. In this mechanism three sodium ions or one calcium ion can bind to either side of the exchanger. Exchanger with one calcium ion bound on one side and three sodium ions bound on the opposite side can undergo translocation. A full cycle of exchange, returning binding sites to their original orientation, translocates two calcium ions and six sodium ions. If one assumes complete symmetry of the two sets of ion binding sites, the exchange cycle is complete after only one set of ion binding reactions followed by one translocation step (i.e. 'one-step simultaneous'). Ca^{2+}–Ca^{2+} and Na^{+}–Na^{+} exchange in the absence of the counter ion could be accounted for by assuming that multiple states of the exchanger can translocate ions. From the point of view of ion binding, this mechanism is equivalent to two linked consecutive exchangers (i.e. either one calcium or three sodium ions can bind to a set of associated sites). Its function is, however, very different. Models of this type were introduced to me as an attractive possibility by both Drs K. D. Philipson and J. Reeves (personal discussions 1984–5). A scheme of this type was favoured recently by Hodgkin and co-workers (Hodgkin and Nunn 1987; Hodgkin, McNaughton, and Nunn 1987), who have presented numerical formulations of the model and demonstrated that data on ion concentration dependencies of exchange current are consistent with the model.

As reference points to other systems, the Na–K pump is with a large degree of certainty a consecutive mechanism (see Glynn 1988 for recent review), the erythrocyte anion exchanger is thought to be consecutive (Gunn and Frölich 1979), and data for the Na–H exchanger is inconclusive (Aronson 1985). A one-step simultaneous model was considered and rejected for the anion exchanger (Gunn and Frölich 1979).

6.3 Model distinction: why a one-step simultaneous model?

The distinction between models depends on establishing the unique behaviours of each model type, which can be verified or negated by experimental results. This process can (and must) be complicated by considering wide ranges of parameter settings, model variants, and different modelling approaches. At the experimental level, the distinction between Na–Ca exchange models is complicated by well-established behaviours, which are not predicted by primary assumptions in any of the models: (1) 'catalysis' of Ca–Ca exchange by monovalent cations (e.g. Allen and Baker 1986b; Slaughter, Sutko, and Reeves 1983; see Reeves 1985 for summary); (2) apparently large differences in calcium affinity on opposite membrane sides in intact systems (see Allen and Baker 1986b for recent study in absence of external sodium; see Kimura, Miyamae, and Noma 1987 for low external calcium affinity in myocytes; see Reeves 1985 for summary with references to early work on calcium fluxes); (3) secondary stimulation of exchange activity by internal calcium (and ATP) in intact systems (e.g. DiPolo 1979; see Kimura, Miyamae, and Noma 1987 for recent data from myocytes and Noda, Shepherd, and Gadsby 1988 for recent determination of concentration–response relations in myocytes).

The simplest possible model formulations are used here, because the consideration of more detailed schemes and modified schemes of each model did not lead to fundamentally new arguments in the question of model choice (Hilgemann 1988). The formulations, which are given in an appendix, assume instantaneous ion binding. Those for the consecutive model and the two-step simultaneous model are equivalent to formulations used by Laüger

(1987). Perfect 3-to-1 stoichiometry is assumed in each model. Work with the anion exchanger strongly supports remarkably tight coupling in an ion exchange mechanism; very small fluxes in the absence of counter anion are apparently due to anion movement through the exchanger in a channel-like fashion rather than carrier conformation changes involving empty binding sites (Frölich 1984). For Na–Ca exchange, purely electrophysiological measurements have provided a new line of evidence for quite tight 3-to-1 stoichiometry (Hodgkin, McNaughton, and Nunn 1987).

For the two-step simultaneous model, it is assumed here that three sodium ions can bind in random fashion on one side, and one calcium ion on the opposite side. For the consecutive model and the one-step simultaneous model, it is assumed that empty binding sites can bind either three sodium ions in random fashion or one calcium ion; occupation of a single sodium site disallows calcium binding (see Appendix and Fig. 6.9). Whether or not separate binding sites for sodium and calcium exist, they must behave in a mutually exclusive fashion to reproduce apparently competitive interactions on each membrane side (Reeves and Sutko 1983). Chemical modification of the Na^{+}–H^{+} exchanger has suggested a model of how mutual exclusivity might be accounted for in that system (Igarashi and Aronson 1987); similar proposals are not available for Na–Ca exchange.

The voltage-dependent step in each model is assumed to involve one charge movement across a single symmetrical Eyring barrier. For the consecutive model, this can be either one net positive charge translocation in the movement of three sodium ions or one net negative charge translocation in the movement of one calcium ion (see also Laüger 1987). For the two-step simultaneous model used here, voltage dependence is placed on the ion translocation step (as in Johnson and Kootsey 1985). The assumption of multiple barriers leads to less steep current–voltage relations than those found experimentally for the outward exchange current, where the prediction of e-fold change of current with about 50 mV potential change holds remarkably well (Hume and Uehara 1986; Kimura, Miyamae, and Noma 1987; Nakao *et al.* 1986; Noda *et al.* 1988). A substantial voltage dependence of ion binding (i.e. >10 per cent fall of membrane potential to both sodium and calcium binding sites) predicts biphasic current–voltage relations and voltage-dependent concentration–flux relations, which are inconsistent with data available in the cited publications (see Laüger 1987 for predictions; see Lagnado, Cervetto, and McNaughton (1988) for an interpretation of recent data in terms of the voltage dependence of an external sodium binding reaction). The assumption of multiple charge movements, one of which must be greater than a single charge (see Eisner and Lederer 1985), leads to unrealistic polyphasic current–voltage relations with steep voltage dependencies over small voltage ranges.

The consecutive model of Na–Ca exchange was challenged first by Baker and McNaughton (1978), and subsequently by multiple authors (Johnson and Kootsey 1985; see Reeves 1985 for discussion and citations up to that time). One contradiction is that the concentration-dependence of each ion's transport should shift to a lower concentration range with reduction of the concentration of the countertransported ion on the opposite side. The

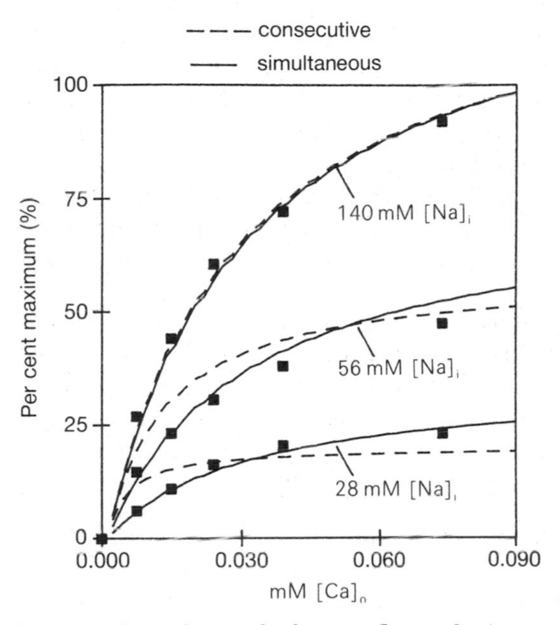

Fig. 6.2. External calcium dependence of calcium influx under 'zero-trans' conditions at the indicated internal sodium concentrations. Results of the consecutive and one-step simultaneous models are compared to experimental results from cardiac sarcolemmal vesicles given as solid squares. (Fig. 1 of Philipson and Nishitomo 1982). All results are scaled to 100 per cent at 0.09 mM external calcium. Here and in all subsequent figures, simulation parameters are given with variable definitions used in the appendix. For ionic concentrations, internal calcium is Ca_i, external calcium is Ca_o, internal sodium is Na_i and external sodium is Na_o. For the consecutive model 'K_c' and 'K_n' are the translocation rates of the calcium and three-sodium loaded carrier, respectively. For the simultaneous models, 'K_c' is the translocation rate of the one-calcium and three-sodium loaded carrier. For the two-step model, 'K_o' is the translocation rate of unloaded carrier. 'K_{mc}' and 'K_{mn}' are the equilibrium binding constants for calcium and sodium, respectively. Here $Ca_i = 0.0001$ mM, $Na_o = 0.01$ mM and $E_m = 0$ mV. For the consecutive model, $K_c = 10000\ s^{-1}$, $K_n = 5000\ s^{-1}$, $K_{mc} = 0.1$ mM, and $K_{mn} = 50$ mM. For the one-step simultaneous model, $K_c = 5000\ s^{-1}$, $K_{mc} = 0.03$ mM and $K_{mn} = 23$ mM. The calcium uptake calculations for both simultaneous models can be scaled up and down with near superimposition, as can the experimental results. In the consecutive model, half-maximal calcium uptake rates shift to lower external calcium concentrations with decreasing internal sodium concentrations.

prediction is presented in Fig. 6.2 along with a comparison of results from the simultaneous models where this is not the case. Concern about the prediction is well justified, because similar arguments provided strong evidence against a consecutive model of the Na–K pump (Glynn and Karlish 1975). Laüger (1987) recently re-examined the issue and stressed the need to study ion dependencies under 'zero-trans' conditions for confident distinction (i.e. with only sodium on one side and only calcium on the other side). Results which met these criteria were apparently overlooked in the assessment. In cardiac vesicles (Philipson and Nishimoto 1982), the external calcium dependence of calcium uptake with no external sodium does not shift with reduction of internal sodium. Results are accurately simulated by simultaneous models but not by consecutive models (see Fig. 6.2). In the meantime, experiments examining Na–Ca exchange current in salamander outer rod segments further disfavour the consecutive model for similar reasons (Hodgkin and Nunn 1987).

Might some simple modification allow the consecutive model to display the concentration–flux dependencies expected for the simultaneous models, while maintaining other experimentally established exchange behaviours? I have tried extensively to do so, but with little success (Hilgemann 1988). In addition, experimental evidence would be needed for at least two other predictions to resurrect the consecutive model:

1. 'Self-exchange' of the ion translocated in the voltage-dependent step must show bell-shaped dependence on voltage (see Fig. 6.3). Available data contradicts the prediction for Ca–Ca exchange in the absence of sodium in squid axon (Allen and Baker 1986b), and (very) tentatively challenges the possibility for Na–Na exchange in the absence of calcium in squid axon (DiPolo and Beaugé 1987). As illustrated in Fig. 6.3, further experimentation along these lines will be important because multiple factors may produce a complicating shift of the biphasic voltage dependence (e.g. asymmetrical calcium concentrations, asymmetrical surface charge, and calcium affinity changes on opposite sides).

2. In the presence of both ions on both sides, unidirectional fluxes should under many conditions correlate very poorly with each other and with changes of net ion movement (i.e. with changes of voltage or changes of concentration of the countertransported ion). This is described in Fig. 6.4 for the consecutive model configuration with voltage dependence placed on the sodium translocation step. The configuration with voltage dependence on calcium translocation results in current–voltage relationships, which saturate unrealistically in the positive potential range (Hilgemann 1988). In the configuration with voltage dependence on the sodium translocation, unidirectional calcium efflux (but not net calcium movement) is almost independent of voltage and external sodium. For the results shown, this behaviour

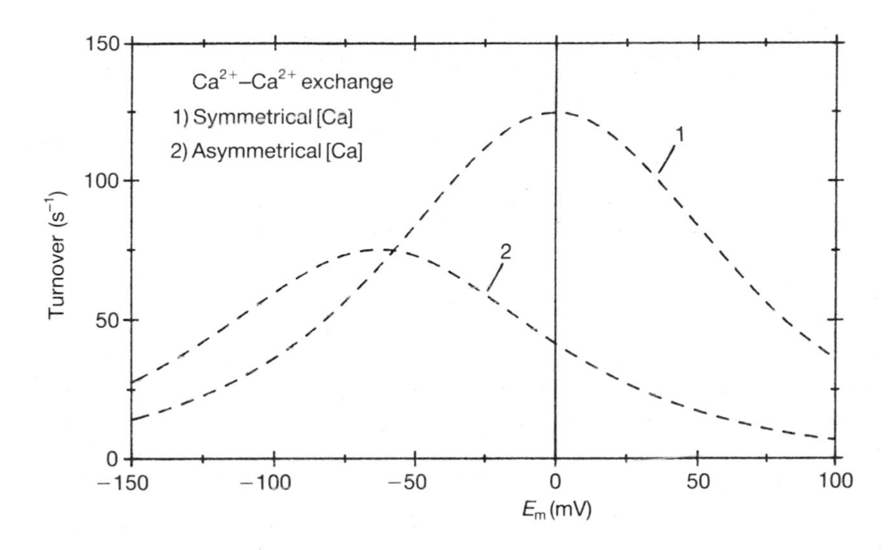

Fig. 6.3. Consecutive model. Voltage dependence of single exchanger Ca–Ca exchange with voltage dependence placed on the calcium translocation step. External and internal sodium is zero. Same parameter settings as in Fig. 6.4. For curve marked 1, $Ca_i = 0.01$ mM and $Ca_o = 0.01$ mM. For curve marked 2, $Ca_i = 0.001$ mM and $Ca_o = 10$ mM. Note the biphasic dependence of Ca–Ca exchange (=unidirectional calcium efflux or influx) on membrane potential. The relationship shifts by roughly 70 mV to more negative potentials with high external calcium and low internal calcium (curve 2). It should be remembered that Ca–Ca exchange described in this figure involves no net ion movement. It is a potential-dependent but electroneutral process.

has in fact been minimized by making the calcium translocation steps ten times slower than the sodium translocation step. It may be reminded that net ion transport has an absolute stoichiometry of 3 to 1 in the figure, and that the difference between unidirectional calcium efflux and influx rates is equal to the net calcium translocation rate. Although the complexities of ion flux in the presence of both ions on both sides have in general been avoided, evidence for the expected behaviours would have appeared with a high degree of probability in multiple studies (e.g. Allen and Baker 1986a; Blaustein 1977; Blaustein and Russell 1975).

Distinction between the two simultaneous models might first be expected on the basis of antagonism between sodium and calcium for binding sites in the one-step model, but not in the two-step model. This distinction, however, only applies to the two-step model when translocation rates of the empty-state and loaded exchanger are kept approximately equal. In any case, it must be assumed in the two-step model that the empty-state reorientation is much faster ($10^5\ s^{-1}$ or greater) than ion translocation steps. Otherwise, current–

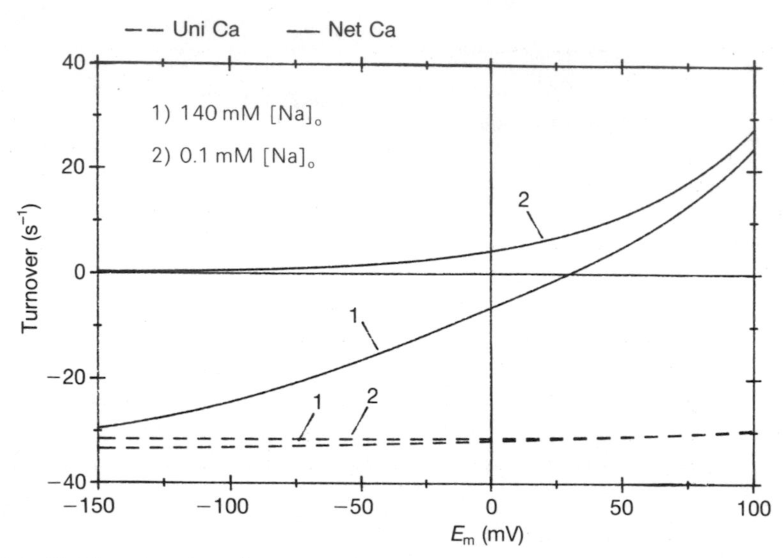

Fig. 6.4. Unidirectional and net single exchanger calcium efflux rates for the consecutive model in dependence on membrane voltage. Voltage dependence is placed on the sodium translocation step. $K_n = 5000\ s^{-1}$, $K_c = 500\ s^{-1}$, $K_{mc} = 0.1$ mM and $K_{mn} = 70$ mM. For results marked '1', (a) $Ca_i = 0.001$ mM, $Ca_o = 2$ mM, $Na_i = 8$ mM and $Na_o = 150$ mM. Note the near independence of unidirectional calcium efflex on membrane potential, while net calcium efflux changes in more expected fashion. For curves marked 2, external sodium was removed ($Na_o = 0.01$ mM). Note that unidirectional calcium efflux is nearly independent of external sodium, and that net calcium influx is nearly independent of external sodium.

voltage relations saturate too strongly in the hyperpolarizing direction, and large current transients are found on step voltage changes. With very fast empty-state reorientation ($= > 10^6\ s^{-1}$), many results interpreted as antagonism between calcium and sodium at binding sites are predicted equally well by the two-step model as by the one-step model (Hilgemann 1988). Concentration–flux relations for both sodium and calcium are shifted to higher concentrations by increasing the countertransported ion concentration on the same membrane side with little or no reduction of maximum transport rates. Maxima of concentration–flux relations are reduced when the concentration of each transported ion is increased on the membrane side to which its transport is taking place. In short, all results from studies of exchange current, interpreted in terms of a one-step model (Hodgkin and Nunn 1987), are reproduced by the two-step model. In some configurations of the two-step model, a pseudo-sodium–calcium antagonism can be more pronounced than in the models with direct sodium–calcium antagonism. Furthermore, pseudo-calcium affinity changes are possible within the two-step model over a range of more than three orders of magnitude by asymmetry of the translocation rate constants (Hilgemann 1988).

Important difficulties are encountered for the two-step model, however, in describing Ca–Ca exchange. Without further assumptions, the model catalyses no Ca–Ca exchange in the absence of sodium, but catalyses Ca–Ca exchange profusely when sodium is high on both sides. Both characteristics are contradicted by experimental data, starting with very strong recent evidence that the Na–Ca exchanger can catalyze Ca–Ca exchange rapidly in the absence of sodium. Ca–Ca exchange activity co-purifies with Na–Ca exchange activity in a variety of isolation procedures (Drs J. P. Reeves and K. D. Philipson, personal communications). Therefore, attempts were made to introduce this characteristic to the two-step model, either as an independent mode of operation, initiated when no sodium is bound, or via countertransport of the cations which catalyse Ca–Ca exchange activity (Hilgemann 1988). Contradictions arose in both cases. For example, substantial fractions of the two-step exchanger remain in the Ca–Ca exchange mode with sodium present only on one membrane side (i.e. ionic conditions approaching physiological conditions). In the case of countertransport of catalysing cations, the catalysing cation becomes a competitive inhibitor of sodium in normal exchange function, and no evidence can be cited in support. Dissociation of the catalysing cation must be made unusually slow to reproduce the finding that transmembrane flux of the catalysing cation is minimal (Slaughter, Sutko, and Reeves 1983).

Figure 6.5 summarizes the Ca–Ca exchange characteristics of the two unmodified simultaneous models. It may be stressed again that very similar results are obtained in formulations incorporating ion binding reactions. Curve 1 is for the one-step model without introduction of a specific Ca–Ca exchange step. Ca–Ca exchange rises and then falls with increasing sodium on both sides. Although the peak of Ca–Ca exchange (via Na–Ca exchange) is small, the model with these parameter settings catalyses Na–Ca exchange at fast rates with appropriate ion gradients (i.e. unequal sodium and calcium gradients). The model in this configuration reproduces accurately the results from squid axon experiments in the absence of ATP (Requena 1978); at resting potential calcium extrusion increases and then falls as sodium is increased on both sides. Curve 2 is for the one-step model after introduction of a specific Ca–Ca translocation step. Ca–Ca exchange decreases monotonically with increasing sodium on both sides, which reproduces the corresponding experiments of Slaughter, Sutko, and Reeves (1983) in cardiac vesicles. As shown in curve 3, Ca–Ca exchange increases monotonically to a very high rate in the two-step model as sodium is increased on both sides. Attempts were made to correct the behaviour of the two-step model by introducing competitive sodium binding to the calcium binding site. One striking contradiction, which arises, is a biphasic dependence of Na–Ca exchange on sodium; as sodium is increased on the side to which calcium is transported, exchange is first stimulated and then inhibited.

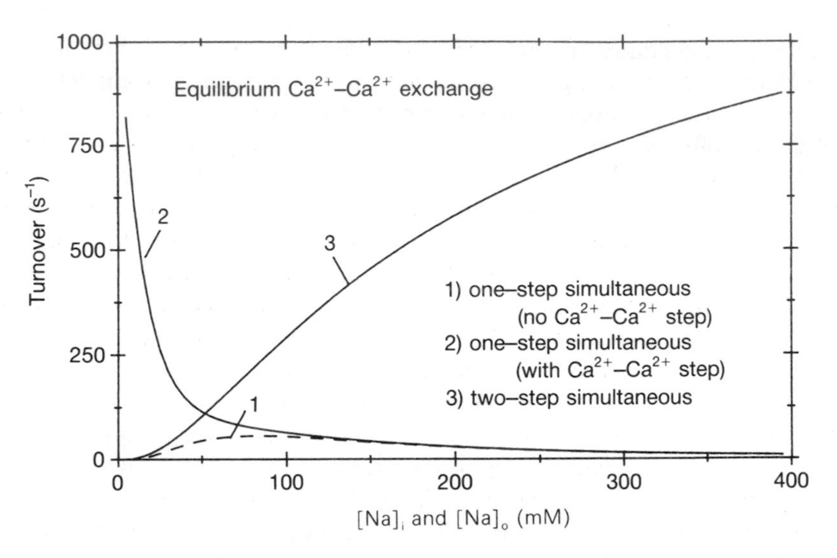

Fig. 6.5. Different equilibrium Ca–Ca exchange characteristics predicted by the one-step and two-step simultaneous models. Internal and external calcium is 0.002 mM, membrane potential is zero. Both external and internal sodium are varied, as indicated, from 0 to 400 mM. For both models, $K_c = 10\,000\ s^{-1}$, $K_{mc} = 0.005$ mM and $K_{mn} = 70$ mM. For the two-step model, the empty state translocation rate (K_o) is $10^4\ s^{-1}$. Curve 1 is for the one-step model without a specific Ca–Ca calcium translocation step ($K_{cc} = 0$), and curve 2 is after introduction of the step ($K_{cc} = 10^4\ s^{-1}$. Note that Ca–Ca exchange in curve 1 is biphasic. Curve 3 is for the two-step simultaneous model. Note that high sodium on both membrane sides causes rapid Ca–Ca exchange via the Na–Ca exchange mechanism.

6.4 Calcium binding kinetics and exchange function

Johnson and Kootsey (1985) assumed that the kinetics of ion binding reactions would be important determinants of exchange function. This may be surprising, since ion fluxes that occur via exchange mechanisms are expected to be orders of magnitude smaller than those that occur via channel mechanisms. Even in the study of channels there is apparently no established case of current–voltage relations being limited by ion diffusion to the channel mouth (Hille 1984). The issue for Na–Ca exchange is, however, not related to exceptionally high turnover rates, but to the fact that calcium movements are taking place from very low internal concentrations in the calcium efflux mode of exchange operation. The internal calcium binding reaction becomes a likely limiter of calcium extrusion. By contrast, diffusion-limited sodium binding reactions are unlikely determinants of exchange function, and the reasons for this conclusion are summarized first.

Following the approach of Johnson and Kootsey, an exchange model will contain at least as many states, which must be simulated, as there are possible combinations of different numbers of sodium and calcium ions bound to the exchanger. The one-step model contains minimally 25 states. If it is formulated with competition for calcium binding by sodium at just one of the three binding sites, for example, it contains minimally 81 states (Hilgemann 1988). Calculations of unidirectional ion fluxes greatly increase the preparatory and calculational overhead of simulating the models, and for many purposes the simulations become prohibitively slow. For these and other reasons, it appeared important to examine whether the model formulations could justifiably be simplified. A likely possibility was the treatment of sodium binding as instantaneous: sodium affinities are in the range of tens of millimolar, while calcium affinity (at least on the inside) is in the range of several micromolar. If the ion association rates are diffusion-limited, sodium dissociation must be orders of magnitude faster than internal calcium dissociation and an order of magnitude faster than external calcium dissociation. With changes of binding site availability, sodium binding–unbinding reactions will approach a steady-state much more rapidly than calcium binding–unbinding reactions.

With this in mind, four-state formulations of each model type were developed. Calcium binding reactions take place between states, and equilibrium sodium binding is assumed within each state. The simple models were compared to models with both sodium and calcium binding reactions (Hilgemann 1988). And the exercise was carried out for calculations of both net and unidirectional calcium fluxes. Using the complex models, parallel variation of sodium association coefficients and dissociation rates (i.e. keeping the affinity constant) resulted in only small changes of model behaviours as the sodium binding coefficient was increased from $5 \times 10^{7}\,M^{-1}\,s^{-1}$ to $10^{15}\,M^{-1}\,s^{-1}$. With unlimited increments of the translocation rates, sodium binding–unbinding reactions remained nearly in equilibrium because the calcium binding–unbinding rates limited exchange turnover first. The simplified four-state models reproduced accurately both net and unidirectional calcium fluxes predicted by the complex models over the full range of conditions described in the article by Johnson and Kootsey. Extenuating possibilities, resulting in slow sodium kinetics (e.g. sequential sodium binding reactions and competition of protons for sodium binding sites), are not eliminated. For the moment, however, the results focus attention on three issues, ultimately related to calcium kinetics: the relationship of calcium binding–unbinding kinetics to exchange turnover rates; exchanger site densities; lack of rapid current transients upon step voltage changes (Kimura, Miyamae, and Noma 1987; D. C. Gadsby, M. Nakao, and N. Shepherd, personal communication of work in heart cells to about 1 ms resolution; F. Bezanilla, personal communication of data from squid axon).

Cheon and Reeves (1988) have recently estimated turnover rates for exchanger reconstituted into proteoliposome vesicles at $1000\ s^{-1}$ or more per site with $100\ \mu M$ calcium in the medium. Exchange was half-saturated by calcium at $14\ \mu M$, which gives a second-order rate constant for the exchange process of $7 \times 10^7\ M^{-1}\ s^{-1}$, a number very close to the ion association coefficient used in the simulations of Johnson and Kootsey ($10^8\ M^{-1}\ s^{-1}$). In cardiac myocytes, an inward Na–Ca exchange current of about 0.5 nA is obtained with $0.5\ \mu M$ intracellular free calcium (Kimura, Miyamae, and Noma 1987; 140 mM external sodium and −70 mV). This current corresponds to extrusion of about 100 000 calcium ions per second per square micron, assuming 3-Na-to-1-Ca stoichiometry and a ventricular myocyte surface area of 25 000 square microns. The upper calcium binding coefficient measured to date appears to be $7.5 \times 10^8\ M^{-1}\ s^{-1}$ for organic calcium chelators (Bayley *et al.* 1984). A coefficient of $5 \times 10^8\ M^{-1}\ s^{-1}$ was recently obtained by indirect means for an external binding site of a calcium channel, which compared well to several organic chelators (Lansman, Hess, and Tsien 1986). Assuming $10^9\ M^{-1}\ s^{-1}$ as the internal association coefficient, the individual exchanger would have 500 chances per second to bind and extrude a calcium ion with $0.5\ \mu M$ intracellular free calcium. Assuming perfect efficiency, the lowest possible estimate of exchanger site density becomes 200 per square micron.

It is interesting that the lowest possible density estimate places the exchanger in a range two to three log units above calcium channel density (1 to 10 per square micron) and several-fold below sodium pump density in heart (1000 to 3000 per square micron[2]). At least in exchanger isolation work starting with cardiac sarcolemmal vesicles, a low density of exchanger in comparison to the Na^+–K^+ pump must be expected (J. P. Reeves and K. D. Philipson, personal communications). To stay in a low density range, a model of exchange must be required to achieve high efficacy in the calcium extrusion mode. Figures 6.6 to 6.8 describe how the one-step simultaneous model can meet the requirements. The simulations presented were made with a four-state formulation of the one-step model described in the Appendix.

The major requirement is that external calcium in the range of 1–2 mM does not inhibit exchange turnover. This is not the case with any of the models, when they are set up symmetrically with high affinity calcium binding (i.e. as expected from sarcolemmal vesicle studies and as simulated by Johnson and Kootsey 1985). While large pseudo-affinity changes can be generated in the two-step model and the consecutive model, similar possibilities have not been found for the one-step model. It must therefore be assumed

[2] Estimates can now be made from charge movement during Na^+–K^+ pump current transients (Nakao and Gadsby 1986; Bahinski, Nakao, and Gadsby 1988), as well as by glycoside binding and extrapolation of maximum sodium extrusion rates to maximum biochemically determined pump rates (see Jørgenson 1980 for estimation of maximum sodium pump rate at $167\ s^{-1}$).

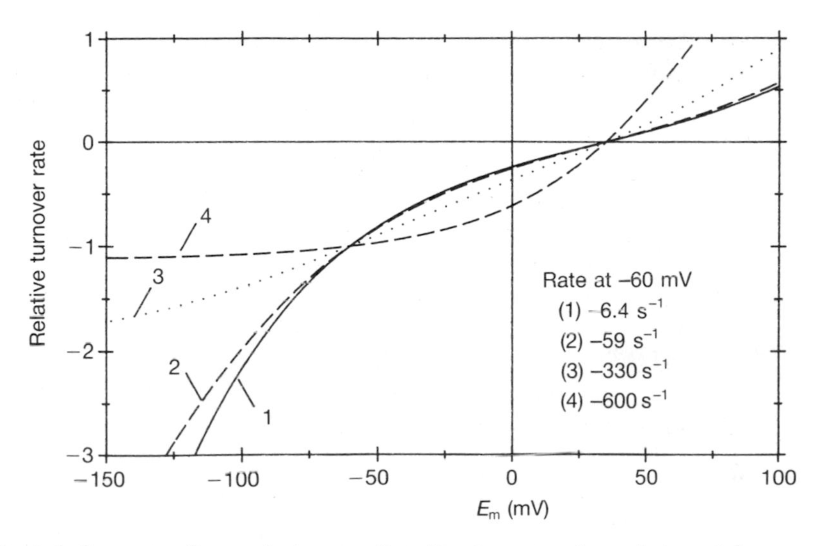

Fig. 6.6. Current-voltage relations predicted by four-state formulation of the one-step simultaneous model with different intrinsic translocation rates (K_c). For all results, the calcium association coefficient (K_{con}) is 10^9 M^{-1} s^{-1}, internal calcium dissociation (K_{cioff}) is 3×10^3 s^{-1}, and external calcium dissociation (K_{cooff}) is 3×10^6 s^{-1}. $K_{mn} = 60$ mM, $Ca_i = 0.001$ mM, $Ca_o = 2$ mM, $Na_i = 8$ mM, and $Na_o = 150$ mM. For curve 1, $K_c = 10^3$ s^{-1}; for curve 2, $K_c = 10^4$ s^{-1}; for curve 3, $K_c = 10^5$ s^{-1}; and for curve 4, $K_c = 10^6$ s^{-1}. The results have been scaled at -60 mV, and turnover rates achieved are given at right. As the translocation rate is increased, the 'hyperbolic sine' form of current–voltage relations is distorted first for the inward current. At very high translocation rates voltage dependence is lost in the negative potential range.

that calcium dissociation at the external side takes place three orders of magnitude faster than at the internal side. A kinetic compensation must be introduced to maintain the 'detailed balances' of reactions (Onsager 1931). This is carried out as if the affinity change involves asymmetrical fixed charges, which are part of the calcium binding site on one side only (i.e. they are not moved upon translocation of ions). The voltage-dependent step is logically stimulated in one direction and inhibited in the other direction by introduction of asymmetry. This is similar to effects of generalized asymmetrical surface charge on the basis of classical surface charge theory (Hilgemann 1988). No suggestion can be made as to why the asymmetry is lost in preparation of vesicles (Philipson and Nishimoto 1982) or why sodium affinity is not similarly asymmetrical in the one-step model.

Figure 6.6 shows how current–voltage relations change as translocation rates are increased to push turnover rates into the range of the internal calcium binding reaction. Steady-state current–voltage relations with increasing translocation rates have been scaled at -60 mV, and the corresponding turnover rates are given at right. Ionic conditions are: 2 mM

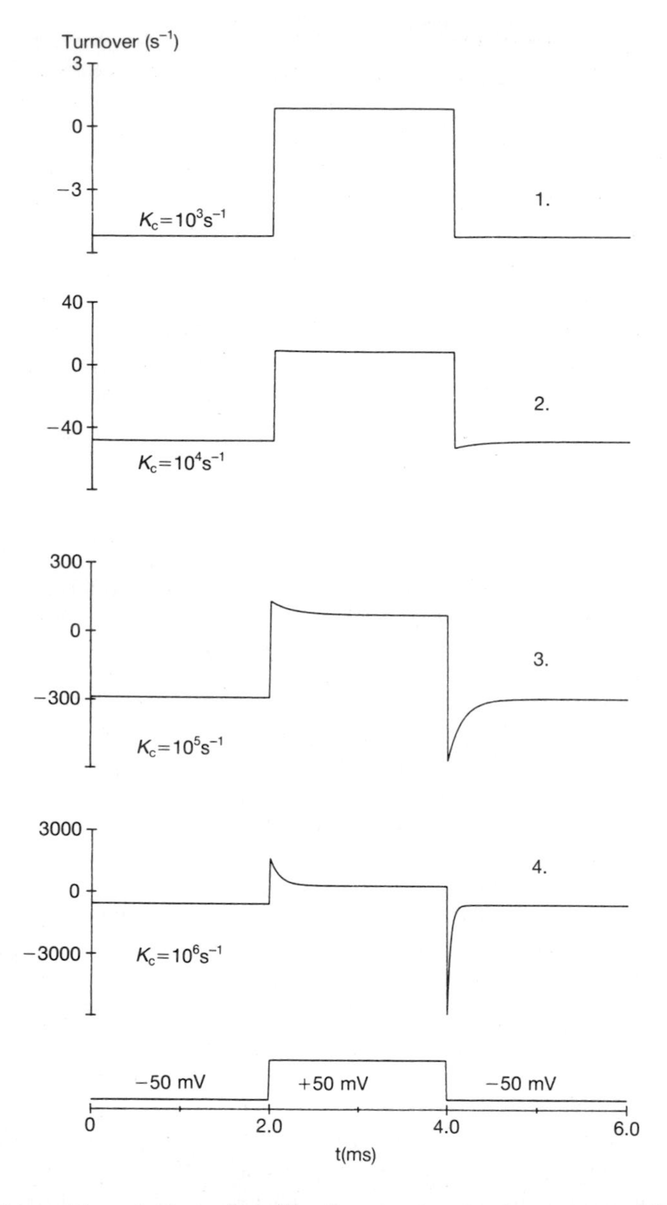

Fig. 6.7. Current transients predicted by the one-step simultaneous model under the conditions of Fig. 6.6. As indicated, membrane potential was steped from − 50 mV to 50 mM for 2 ms and back to − 50 mV. The designation of results is the same as in Fig. 6.6. With very fast translocation rates (curve 4) current transients are larger than steady-state currents. Note that the time course of currents transients speeds up as the translocation rate is increased. The time courses do not reflect calcium binding–unbinding kinetics in this configuration of the one-step model.

external calcium, 1 μM internal calcium, 150 mM external sodium, and 8 mM internal sodium. The 'intrinsic' translocation rate is increased in log units from $10^3\ s^{-1}$ to $10^6\ s^{-1}$ in curves 1 to 4, respectively. In curve 1, the current–voltage relation is a simple hyperbolic sine function. As the translocation rate is increased (curves 2 to 4), the inward current saturates with hyperpolarization at a single exchanger turnover rate of about $600\ s^{-1}$. Maximum possible turnover in the calcium efflux mode would be $1\,000\ s^{-1}$, determined by the assumed $10^9\ M^{-1}\ s^{-1}$ calcium binding coefficient with 1 μM internal calcium.

Figure 6.7 shows predicted current transients upon step voltage change from −50 to +50 mV under the conditions of Fig. 6.6. The results were made by first-order implicit integration with accuracy checks to 0.01 per

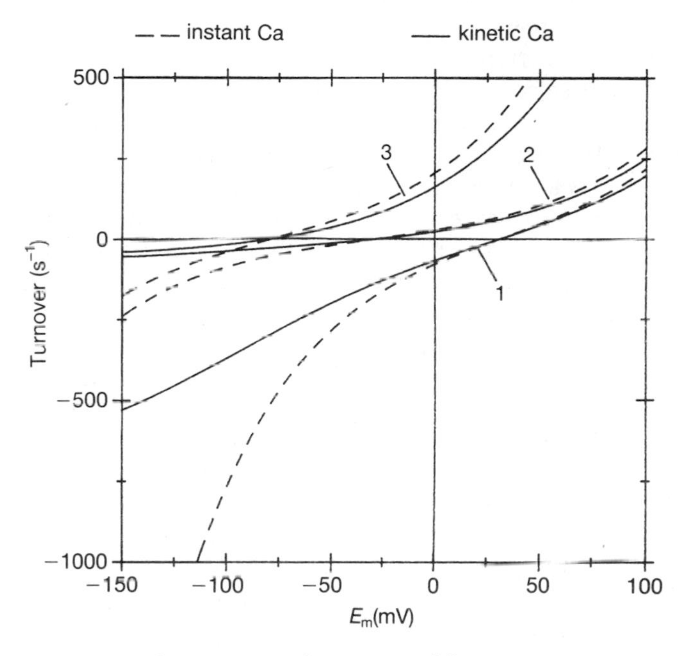

Fig. 6.8. Best estimate of one-step simultaneous model parameter settings to reproduce myocyte current–voltage relations. Dotted results assume instantaneous calcium binding–unbinding. Solid results are with binding parameters of Figs 6.6 and 6.7; $K_c = 6 \times 10^4\ s^{-1}$. Curves marked 1 are for the same ionic conditions as previous figures. Note that the kinetics of calcium binding limit the current–voltage relation almost exclusively in the negative potential range (dotted versus solid results). Curves marked 2 are with 0.0001 mM internal calcium. Note the relative insensitivity of outward current to internal calcium. Curves 3 are with 0.0001 mM internal calcium and 15 mM internal sodium. Note that the outward current–voltage relation is essentially exponential with a factor 2.3 increase of current over 50 mV.

cent. At low translocation rates (examples 1 and 2), instantaneous current jumps are expected; ion binding–unbinding reactions remain in steady-state. It should be stressed that in the consecutive and two-step simultaneous models, the voltage-independent translocation steps must be made very fast to avoid substantial current transients with step voltage changes in similar simulations (Hilgemann 1988). At a translocation rate resulting in a turnover of $300\,s^{-1}$ at -50 mV (curve 3), small current transients are generated. The transient upon hyperpolarization is largest and decays with a time constant of about 200 μs. At the highest translocation rate (curve 4), which results in a strongly saturating current–voltage relation (see Fig. 6.6), transient currents are much larger than the steady-state current and are much faster than in curve 3. The time courses of current transients are limited here by the translocation rates; in other configurations of the one-step model, a 'slow' phase of current transients can be determined by the internal calcium dissociation reaction.

Figure 6.8 shows current–voltage relations for the one-step model, which are the present best guess about physiological exchange function. Dotted results are for the model with instantaneous calcium binding (i.e. with hyperbolic sine I–V relations). Solid results are with calcium binding reactions. Curves 1 are for high internal calcium: 2 mM external calcium, 1 μM internal calcium, 150 mM external sodium, and 8 mM internal sodium. Curves 2 are for resting internal calcium: 0.1 μM internal calcium and no other changes. Curves 3 are for resting internal calcium and high internal sodium: 15 mM internal sodium and no other changes. Outward current increases essentially exponentially with depolarization, both at normal internal sodium and at high internal sodium. The current shows e-fold increase with about 50 mV change. The parameter settings predict somewhat smaller current transients than those of curve 3 in Fig. 6.7. For application of the model to a cardiac action potential, for example, steady-state simulation of the model would remain justified.

The chosen translocation rate ($60\,000\,s^{-1}$) with the chosen internal calcium kinetics ($10^9\,M^{-1}\,s^{-1}$ association coefficient; $3\,000\,s^{-1}$ dissociation rate) give a current–voltage relation for the inward current consistent with available current measurements in cardiac myocytes (Kimura *et al.* 1987) and calcium flux measurements in squid axon (Allen and Baker 1986b; DiPolo *et al.* 1985). The inward current–voltage relation is 'non-exponential', can be nearly linear over substantial potential ranges, and tends to saturate at negative potentials. As the model is set up, high internal calcium can return the inward current–voltage relation to an exponential shape (not shown). It should be possible to test this prediction in systems where very high internal calcium is tolerated (i.e. squid axon or outer rod segments). Note that at -150 mV, the exchanger extrudes 500 of the 1000 calcium ions striking the exchanger per second in curve 1 (1 μM internal calcium and other concentrations roughly physiological). An exchanger density of approximately 500 per square micron must be assumed to generate current densities found in cardiac myocytes.

6.5 Discussion

The tentative first choice of model type for Na–Ca exchange has fallen to the simultaneous one-step model (Hilgemann 1988; Hodgkin and Nunn 1987; presentation of E. A. Johnson and M. Kootsey at Stowe meeting). For the time being, this choice appears to be unique among ionic mechanisms moving two ion species. The choice therefore strengthens the idea that the Na–Ca exchange cycle may be fundamentally different from that of an ATP-dependent ion pump and (presumably) the erythrocyte anion exchanger. It will be of great interest to see whether similar studies of Na–H exchange yield similar conclusions. An easily basic attraction of the one-step model is that it accounts for simple hyperbolic sine-shaped current–voltage relations and an apparent lack of current transients upon step voltage changes. The one-step model thus explains why a very simple 'empirical' formulation (DiFrancesco and Noble 1985) turned out to be a good predictor of these two aspects of exchange function. In strong contrast to the sodium pump (Nakao and Gadsby 1986), the voltage-dependent step of the exchange cycle must be the slowest step under most conditions.

Multiple concerns about the choice remain to be resolved. While ion self-exchange in the absence of counter ion is inherent to consecutive models, it must be introduced as a secondary process in the simultaneous models. An alternative, which deserves attention, is that ion translocation in 'simultaneous' exchangers is not in fact simultaneous, but takes place in rapid sequential steps (i.e. first one ion is translocated and then the counter ion). The complexity of corresponding models is impressive, but there is good reason to think that such complexity is essential. How electrochemically can calcium affinity, but not sodium affinity, change by three orders of magnitude on opposite membrane sides, although sodium binds to the same sites (or closely associated sites)? The addition of internal ATP in the squid axon fundamentally changes calcium flux characteristics. Available data is still not sufficient to determine whether the one-step model can account for calcium fluxes via exchange under these more physiological conditions. The effect of ATP has been described as a 'suspension' of exchange inhibition by internal sodium, whereby calcium efflux is inhibited little if at all by increasing internal sodium up to 70 mM (DiPolo 1974; Requena 1978). Since the equilibrium of exchange must be approached (unless extracellular calcium is zero), experimental results suggest that rapid Ca–Ca exchange might be taking place with high sodium on both membrane sides (see curve 3 of Fig. 6.5).

Independent of model choice, the internal calcium binding reaction demands attention. In cardiac vesicles, calcium uptake by exchange is stimulated by buffering calcium with several calcium chelators up to chelator concentrations of about 50 μM (Trosper and Philipson 1984); the apparent effect is an increase of calcium affinity. If exchange turnover rates can be limited by calcium binding rates in the calcium efflux mode (i.e. operating with micromolar free-calcium concentrations), it might well be expected that calcium gradients to the exchanger mouth would exist and limit exchange activity in unbuffered, unstirred media. Even if the exchanger is highly effective in the calcium efflux mode (e.g. current voltage relation 3 in Fig. 6.6), the minimum exchanger density in cardiac cells must be placed at about 500 per square micron, which by the standard of the sodium pump is not a low density. Two possibilities are given: substantial portions of the total exchanger population might be lost in isolation procedures, which indicate up to now a low density system in the cardiac cell membrane; or, calcium binding at the internal site might be optimized under physiological conditions to substantially exceed binding rates expected from random calcium diffusion in aqueous media (i.e. $10^9\,M^{-1}\,s^{-1}$; Eigen and Mass 1966). Theoretical work suggests that some degree of optimization might be possible by 'funnelling' of field lines to the exchanger binding sites in a relatively wide vestibule with fixed negative charges (Dani 1986). Largest effects are expected for divalent cations. Recent experimental data suggest that faster ion binding rates are indeed possible in ionic mechanisms at cell surfaces than by small organic chelators (Prod'hom, Pietrobon, and Hess 1987). A starting point would be to test rigorously for internal charge screening by small cations on calcium efflux in the presence of ATP.

Although the one-step model reproduces a range of experimental results more readily than the other models, a simple application of the model to cellular simulations raises contradictions to available experimental data. In recent simulations of cardiac electrical activity and E–C coupling, action potential changes and extracellular calcium transient changes associated with calcium efflux could be accounted for excellently by Na–Ca exchange as underlying mechanism (Hilgemann and Noble 1987). The magnitudes of exchange currents predicted during the falling action potential correlated well with currents measured in ion-controlled myocytes. Nevertheless, two important contradictions arose. First, free internal calcium at rest was predicted to be very close to the exchange equilibrium (i.e. 10^{-8} M, not 10^{-7} M), in spite of substantial resting calcium influx. Second, a rapid pulse of calcium influx by exchange was predicted upon action potential upstroke, for which experimental evidence is lacking (Hilgemann and Noble 1987; Pizarro, Cleeman, and Morad 1985; see Bers 1987 for indications of a late component of sodium-dependent calcium influx under special conditions, consistent also with optical resolution of influx under similar conditions; Hilgemann, unpublished). The contradictions are overcome to only a small extent by using detailed formulations of exchange. Introduction of rapid secondary exchange modulation by internal calcium appears essential (unpublished results). As free calcium falls below a contractile threshold of around 10^{-7} M, extrusion must be secondarily inhibited. And if extrusion by exchange is inhibited during diastole (e.g. to limit unnecessary calcium extrusion), it will presumably be inhibited as a calcium influx mechanism upon the subsequent action potential upstroke.

In summary, the full range of available data on Na–Ca does not allow simple unifying simulations of exchange function at either the molecular or cellular levels. A one-step simultaneous model of exchange must be favoured at present, primarily because the other models display behaviours which contradict available experimental data. The model choice must remain tentative until further methodological advances (e.g. development of specific inhibitors) allow resolution of partial reactions of the exchanger cycle. Recent estimates of exchange turnover rates suggest that 'non-exponentiality' of current–voltage relations for inward current may be explained by the physically limited rate at which internal calcium ions can bind to the exchanger in relation to intrinsic rates of ion translocation. As long as a relatively high exchanger density is not verified, the possibility that internal calcium binding coefficients might exceed $10^9\ M^{-1}\ s^{-1}$ must be addressed. At present, a better definition of physiological exchange function demands kinetic resolution of secondary exchange modulation by internal calcium.

Appendix: simple formulations of Na–Ca exchange

I. Models with instantaneous ion binding

A. *Definitions and preliminary calculations*

Fractional steady-state binding to single, independent sodium and calcium binding sites is calculated for internal and external media.

F_{ci} = fractional occupation by internal calcium
F_{co} = fractional occupation by external calcium
F_{ni} = fractional occupation by internal sodium
F_{no} = fractional occupation by external sodium
K_{mc} = calcium equilibrium binding constant (mM)
K_{mn} = sodium equilibrium binding constant (mM)
Ca_i = internal calcium concentration (mM)
Ca_o = external calcium concentration (mM)
Na_i = internal sodium concentration (mM)

Na_o = external sodium concentration (mM), whereby:

$$F_{ci} = Ca_i/(K_{mc} + Ca_i)$$
$$F_{co} = Ca_o/(K_{mc} + Ca_o)$$
$$F_{ni} = Na_i/(K_{mn} + Na_i)$$
$$F_{no} = Na_o/(K_{mn} + Na_o).$$

For the case of three independent sodium binding sites

F_{3ni} = fractional occupation of all three internal sites
F_{3no} = fractional occupation of all three external sites
F_{3zi} = fractional occupation of no internal sites
F_{3zo} = fraction occupation of no external sites, whereby:

$$F_{3ni} = F_{ni}^{3}$$
$$F_{3no} = F_{no}^{3}$$
$$F_{3zi} = (1 - F_{ni})^3$$
$$F_{3zo} = (1 - F_{no})^3.$$

For the case that empty exchanger can bind one calcium ion or three sodium ions in random fashion (see scheme A in Fig. 6.9)

F_{xco} = fractional occupation of external sites by calcium
F_{xci} = fractional occupation of internal sites by calcium
F_{x3no} = fractional occupation of external sites by three sodium
F_{x3ni} = fractional occupation of internal sites by three sodium, whereby:

$$F_{xco} = [Ca_o \times F_{3zo}]/[Ca_o \times F_{3zo} + K_{mc}]$$
$$F_{xci} = [Ca_i \times F_{3zi}]/[Ca_i \times F_{3zi} + K_{mc}]$$
$$F_{3no} = (1 - F_{xco}) \times F_{3no}$$
$$F_{3ni} = (1 - F_{xci}) \times F_{3ni}).$$

For a single charge movement over a single energy barrier, the appropriate reaction steps are modified by membrane voltage with the constant, K_{em}, where:

$$K_{em} = \exp[E_m \times F/(2 \times R \times T)], \text{ with}$$
T = 306 K for all results presented.

B. *Routines for models with instantaneous ion binding*

General definitions:

K_1 to K_{10}, and D are temporary variables
R_{io} = unidirectional single exchanger calcium efflux rate
R_{oi} = unidirectional single exchanger calcium influx rate
R_{net} = net single exchanger turnover rate.

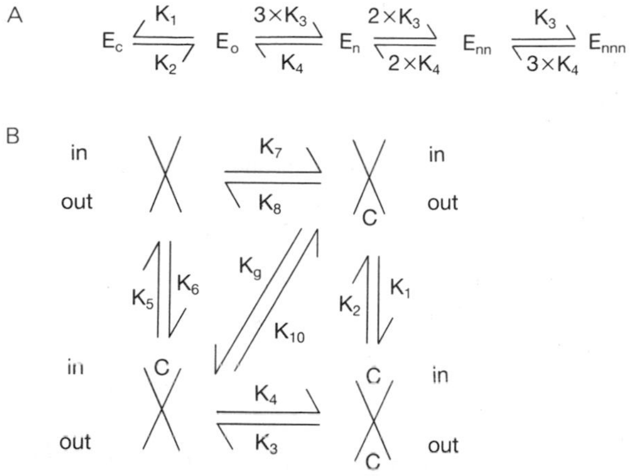

A

$$E_c \underset{K_2}{\overset{K_1}{\rightleftharpoons}} E_o \underset{K_4}{\overset{3\times K_3}{\rightleftharpoons}} E_n \underset{2\times K_4}{\overset{2\times K_3}{\rightleftharpoons}} E_{nn} \underset{3\times K_4}{\overset{K_3}{\rightleftharpoons}} E_{nnn}$$

B

in / out — K_7 / K_8 — in / out, C

K_5, K_6 — K_9, K_{10} — K_2, K_1

in / out, C — K_4 / K_3 — in / out, C, C

Fig. 6.9. Part A. The ion binding scheme used for simulations of the consecutive and the one-step simultaneous models. Empty binding sites (E_o) can bind either one calcium ion to form E_c or three sodium ions in random fashion, forming E_n, E_{nn}, and E_{nnn}. The association rates for calcium and sodium, respectively, are K_1 and K_3 (i.e. association coefficients × ion concentrations). Dissociation rates for calcium and sodium, respectively, are K_2 and K_4, respectively. The steady state solutions for E_c and E_{nnn} may be written in many different ways. The chosen form (see text) uses the equilibrium binding constants. Part B. State diagram for the one-step simultaneous model with calcium binding reactions and equilibrium sodium binding. The rates designated by Ks are unrelated to those in part A. The four states are defined as the four possible exchanger species with different calcium binding configurations. Sodium binding is instantaneous within each state. Only those binding sites with no bound sodium can bind calcium (K_1, K_4, K_6 and K_7 in the state diagram; see routine for simulations in part C of appendix). The internal calcium dissociation reactions are K_2 and K_5; the external calcium dissociation reactions are K_3 and K_8. The exchanger translocates ions only when one calcium ion is bound on one side and three sodium ions on the opposite side (K_9 and K_{10} in the state diagram).

1. *Consecutive model*

Specific definitions

K_c = intrinsic translocation rate of calcium-loaded exchanger

K_n = intrinsic translocation rate of sodium-loaded exchanger.

Calculation routines

For voltage dependence on calcium translocation:

$$K_1 = F_{xci} \times K_c/K_{em}$$
$$K_2 = F_{xco} \times K_c \times K_{em}$$
$$K_3 = F_{x3ni} \times K_n$$
$$K_4 = F_{x3no} \times K_n.$$

For voltage dependence on sodium translocation:

$K_1 = F_{xci} \times K_c$
$K_2 = F_{xco} \times K_c$
$K_3 = F_{x3ni} \times K_n \times K_{em}$
$K_4 = F_{x3no} \times K_n \times K_{em}$.

For both cases:

$D = K_1 + K_2 + K_3 + K_4$
$R_{io} = K_1 \times (K_2 + K_4)/D$
$R_{oi} = K_2 \times (K_1 + K_3)/D$
$R_{net} = R_{oi} - R_{io}$.

2. *Two-step simultaneous model*
Specific definitions

K_c = intrinsic translocation rate of ion-loaded exchanger
K_o = translocation rate of empty carrier.

Calculation routine

$K_1 = F_{3no} \times F_{ci} \times K_c/K_{em}$
$K_2 = F_{3ni} \times F_{co} \times K_{em}$
$K_3 = (1 - F_{ci}) \times F_{3zo} \times K_o$
$K_4 = (1 - F_{co}) \times F_{3zi} \times K_o$
$D = K_1 + K_2 + K_3 + K_4$
$R_{io} = K_1 \times (K_2 + K_4)/D$
$R_{oi} = K_2 \times (K_1 + K_3)/D$
$R_{net} = R_{oi} - R_{io}$.

3. *One-step simultaneous*
Specific definitions

K_c = 'intrinsic' translocation rate of Na–Ca loaded exchanger
K_{cc} = translocation rate of Ca–Ca loaded exchanger.

Calculation routine

$R_{oi} = F_{x3ni} \times F_{xco} \times K_c \times K_{em} + F_{xco} \times F_{xci} \times K_{cc}$
$R_{io} = F_{x3no} \times F_{xci} \times K_c/K_{em} + F_{xco} \times F_{xci} \times K_{cc}$
$R_{net} = R_{oi} - R_{io}$.

C. One-step simultaneous with calcium binding reactions

The state diagram for the model is given in part B of Fig. 6.9.

Definitions

K_{con} = calcium association coefficient on both exchanger sides

K_{cioff} = internal calcium dissociation rate
K_{cooff} = external calcium dissociation rate
X_1 to X_4, and F_{asym} are additional temporary variables.

Routine for net single exchanger turnover

$$
\begin{aligned}
F_{asym} &= (K_{cioff}/K_{cooff}) \\
K_1 &= C_o \times K_{con} \times F_{3zo} \\
K_2 &= K_{cooff} \\
K_3 &= K_{cioff} \\
K_4 &= \mathrm{Ca}_i \times K_{con} \times F_{3zi} \\
K_5 &= K_{cioff} \\
K_6 &= \mathrm{Ca}_o \times K_{con} \times F_{3zo} \\
K_7 &= \mathrm{Ca}_i \times K_{con} \times F_{3zi} \\
K_8 &= K_{cooff} \\
K_9 &= K_c \times F_{3no}/(K_{em} \times F_{asym}) \\
K_{10} &= K_c \times F_{3ni} \times K_{em} \times F_{asym}
\end{aligned}
$$

$$
\begin{aligned}
X_1 &= K_2 \times K_4 \times (K_7 + K_6) + K_5 \times K_7 \times (K_2 + K_3) + K_{10} \times (K_2 + K_3) \times (K_7 + K_6) \\
X_2 &= K_1 \times K_7 \times (K_4 + K_5) + K_4 \times K_6 \times (K_1 + K_8) + (K_{10} \times K_1 + K_9 \times K_4) \times (K_7 + K_6) \\
X_3 &= K_1 \times K_3 \times (K_7 + K_6) + K_8 \times K_6 \times (K_2 + K_3) + K_9 \times (K_6 + K_7) \times (K_3 + K_2) \\
X_4 &= K_2 \times K_8 \times (K_4 + K_5) + K_3 \times K_5 \times (K_1 + K_8) + (K_9 \times K_5 + K_{10} \times K_8) \times (K_3 + K_2) \\
D &= X_1 + X_2 + X_3 + X_4; \\
R_{net} &= (X_3 \times K_{10} - X_1 \times K_9)/D.
\end{aligned}
$$

Acknowledgements

Sincere gratitude is expressed to Dr Denis Noble for encouragement and discussions over the last 3 years. I am indebted to Drs John Reeves, David Gadsby, and Ken Philipson for discussion and comments, and to many others for their willingness to exchange published and unpublished data and respond to my queries. Completion of the work was made possible by members of the Laboratoire de Neurobiologie Cellulaire et Moléculaire, Centre National de la Recherche Scientifique, Gif sur Yvette. For their great generosity, I express thanks to Drs Ray Kado, Jordi Molgo', and Jacques Stinnakre. Special gratitude is expressed to Dr Christiane Baud for support and assistance. This work was supported by the American Heart Association, National Center, and by the American Heart Association, Greater Los Angeles Affiliate.

References

Allen, T. J. A. and Baker, P. F. (1986a). Comparison of the effects of potassium and membrane potential on the calcium-dependent sodium efflux in squid axons. *J. Physiol.* **378**, 53–76.

Allen, T. J. A. and Baker, P. F. (1986b). Influence of membrane potential on calcium efflux from giant axons of *Loligo*. *J. Physiol.* **378**, 77–96.

Aronson, P. S. (1985). Kinetic properties of the plasma membrane Na^+–H^+ exchanger. *Ann. Rev. Physiol.* **47**, 545–60.

Bahinski, A., Nakao, M., and Gadsby, D. C. (1988). Potassium translocation by the Na–K pump is voltage insensitive. *Proc. Natl. Acad. Sci. USA* (in press).

Baker, P. F. and McNaughton, P. A. (1978). The influence of extracellular calcium binding on the calcium efflux from squid axons. *J. Physiol.* **276**, 127–50.

Bayley, P., Ahlstrom, P., Martin, S. R., and Forsen, S. (1984). The kinetics of calcium binding to calmodulin: Puin 2 and ANS stopped-flow fluorescence studies. *Biochem. Biophys. Res. Comm.* **120**, 185–91.

Bers, D. M. (1987). Mechanisms contributing to the cardiac inotropic effect of Na pump inhibition and reduction of extracellular Na. *J. Gen. Physiol.* **90**, 479–504.

Blaustein, M. P. (1977). Effects of internal and external cations and of ATP on sodium–calcium and calcium–calcium exchange in internally dialyzed squid axons. *Biophys. J.* **20**, 79–111.

Blaustein, M. P. and Russell, J. M. (1975). Sodium–calcium exchange and calcium–calcium exchange in internally dialyzed squid giant axons. *J. Membr. Biol.* **22**, 285–312.

Cheon, J. and Reeves, J. P. (1988). Site density of the sodium–calcium exchange carrier in reconstituted vesicles from bovine cardiac sarcolemma. *J. Biol. Chem.* (in press).

Dani, J. (1986). Ion-channel entrances influence permeation: Net charge, size, shape and binding considerations. *Biophys. J.* **41**, 607–18.

DiFrancesco, D. and Noble, D. (1985). A model of cardiac electrical activity incorporating ionic pumps and concentration changes. *Phil. Trans. Roy. Soc. B* **307**, 353–98.

DiPolo, R. (1974). The influence of nucleotides on calcium fluxes. *Fed. Proc. Biol.* **35**, 2579–82.

DiPolo, R. (1979). Calcium influx in internally dialyzed squid giant axons. *J. Gen. Physiol.* **73**, 91–113.

DiPolo, R. and Beaugé, L. (1987). Characterization of the reverse Na–Ca exchange in squid axons and its modulation by Ca_i and ATP. Ca_i-dependent Na_i–Ca_o and Na_i–Na_o exchange modes. *J. Gen. Physiol.* **90**, 505–25.

DiPolo, R., Bezanilla, F., Caputo, C., and Rojas, H. (1985). Voltage dependence of the Na–Ca exchange in voltage-clamped dialysed squid axons. *J. Gen. Physiol.* **86**, 457–78.

Eigen, M. and Mass, G. (1966). Über die Kinetic der Metallkomplexibildung der Alkali- und Erdalkaliionen in wässrigen Lösungen. *Z. Physik. Chem.* **49**, 163–77.

Eisner, D. A. and Lederer, W. J. (1985). Na–Ca exchange: stoichiometry and electrogenicity. *Am. J. Physiol.* **248**, C189–202.

Frölich, O. (1984). Relative contributions of the slippage and tunneling mechanisms to anion net efflux from human erythrocytes. *J. Gen. Physiol.* **84**, 877–93.

Glynn, I. M. (1988). The coupling of enzymatic steps to the translocation of sodium and potassium. In: *Proceedings of the 5th International Conference on Na,K-ATPase.* Alan Liss (in press).

Glynn, I. M. and Karlish, S. (1975). The sodium pump. *Ann. Rev. Physiol.* **37**, 13–15.

Gunn, R. B. and Frölich, O. (1979). Asymmetry in the mechanism for anion exchange in human red blood cell membranes. Evidence for reciprocating sites that react with one transported anion at a time. *J. Gen. Physiol.* **74**, 351–74.

Hilgemann, D. W. (1988). Numerical approximations of sodium–calcium exchange. *Prog. in Biophys.* **51**, 1–45.

Hilgemann, D. W. and Noble, D. (1987). Excitation–contraction coupling and extracellular calcium transients in rabbit atrium: Reconstruction of basic cellular mechanisms. *Proc. Roy. Soc. Lond. B* **230**, 163–205.

Hille, B. (1984). *Ionic channels of excitable membranes.* Sinauer Associates Inc. Sunderland, Mass.

Hodgkin, A. L. and Nunn, B. J. (1987). The effect of ions on sodium–calcium exchange in salamander rods. *J. Physiol.* **391**, 371–98.

Hodgkin, A. L., McNaughton, P. A., and Nunn, J. (1987). Measurement of sodium–calcium exchange in salamander rods. *J. Physiol.* **391**, 347–70.

Hume, J. R. and Uehara, A. (1986). 'Creep currents' in single frog atrial cells may be generated by electrogenic Na–Ca-exchange. *J. Gen. Physiol.* **87**, 857–84.

Igarashi, P. and Aronson, P. S. (1987). Covalent modification of the renal Na^+–H^+ exchanger by N,N'-dicyclohexylcarbodiimide. *J. Biol. Chem.* **262**, 860–8.

Johnson, E. A. and Kootsey, J. M. (1985). A minimum mechanism for Na^+–Ca^{++} exchange: Net and unidirectional Ca^{++} fluxes as functions of ion composition and membrane potential. *J. Membr. Physiol.* **86**, 167–87.

Jorgensen, J. (1980). Sodium and potassium ion pump in kidney tubules. *Physiol. Rev.* **60**, 864–917.

Kimura, J., Miyamae, S., and Noma, A. (1987). Identification of sodium–calcium exchange current in single ventricular cells of guinea-pig. *J. Physiol.* **384**, 199–222.

Laüger, P. (1987). Voltage dependence of sodium–calcium exchange: Predictions from kinetic models. *J. Membr. Biol.* **99**, 1–12.

Lagnado, L., Cervetto, L., and McNaughton, P. A. (1988). Ion transport by the Na–Ca exchange in isolated rod outer segments. *Proc. Natl Acad. Sci. USA* **85**, 4548–52.

Lansman, J. B., Hess, P., and Tsien, R. W. (1986). Blockade of current through single calcium channels by Cd^{2+}, Mg^{2+} and Ca^{2+}: Voltage and concentration dependence of calcium entry into the pore. *J. Gen. Physiol.* **88**, 321–47.

Nakao, M. and Gadsby, D. C. (1986). Voltage dependence of Na translocation by the Na^+–K^+ pump. *Nature*, **323**, 628–30.

Nakao, M., Shepherd, N., Bridge, J. H. B., and Gadsby, D. C. (1986). Membrane current generated by Na–Ca exchange in isolated heart cells. *Fed. Proc.* **45**, 770 (abstr.).

Noda, M., Shepherd, R. N., and Gadsby, D. C. (1988). Activation by [Ca]i, and block by 3',4'-dichlorobenzamil, of outward Na–Ca exchange current in guinea-pig ventricular myocytes. *Biophys. J.* **53**, 342a.

Onsager, L. (1931). Reciprocal relations in irreversible processes. I. *Phys. Rev. Ser. 2*, **37**, 495–526.

Pizarro, G., Cleeman, L., and Morad, M. (1985). Optical measurement of voltage-dependent Ca^{2+} influx in frog heart. *Proc. Natl. Acad. Sci. USA*, **82**, 1864–8.

Philipson, K. D. and Nishimoto, A. Y. (1982). Na^{+}–Ca^{++} exchange in inside-out cardiac sarcolemmal vesicles. *J. Biol. Chem.* **256**, 5111–17.

Prod'hom, B., Pietrobon, D., and Hess, P. (1987). Direct measurement of proton transfer rates to a group controlling the dihydropyridine-sensitive Ca^{2+} channel. *Nature*, **329**, 243–6.

Reeves, J. P. (1985). The sodium–calcium exchange system. *Curr. Top. Membr. Trans.* **25**, 77–127.

Reeves, J. P. and Sutko, J. L. (1983). Competitive interactions of sodium and calcium with the sodium–calcium exchange system of cardiac sarcolemmal vesicles. *J. Biol. Chem.* **258**, 3178–82.

Requena, J. (1978). Calcium efflux from squid axons under constant sodium electrochemical gradient. *J. Gen. Physiol.* **72**, 443–70.

Slaughter, R. S., Sutko, J. L., and Reeves, J. P. (1983). Equilibrium calcium–calcium exchange in cardiac sarcolemmal vesicles. *J. Biol. Chem.* **258**, 3183–90.

Trosper, T. L. and Philipson, K. D. (1984). Stimulatory effects of calcium chelators on Na^{+}–Ca^{2+} exchange in cardiac sarcolemmal vesicles. *Cell Calcium*, **5**, 211–22.

7 The molecular biochemistry of the sodium–calcium exchanger

Hannah Rahamimoff

7.1 Introduction

The ultimate goal in studying the molecular properties of a transport protein is to understand the rules by which its structure determines its function. In the last few years, this goal was approached successfully by the combination of available recombinant DNA technology (see, for example, Davis, Dibher, and Batty 1986; Maniatis, Fritsch, and Sambrook 1982) with automated protein sequencing and oligonucleotide synthesis, and resulted in the successful cloning of several transport proteins (Noda *et al.* 1984; Numa *et al.* 1983). The activity of the cloned products could be evaluated in appropriate expression systems such as *Xenopus* oocytes (Goldin *et al.* 1986) and compared to the activity obtained either following specific site-directed mutagenesis (Mishina *et al.* 1985) or following protein engineering of chimaeric products (Imoto *et al.* 1986).

While with some transport proteins (see, for example, Noda *et al.* 1984, Numa *et al.* 1983) this approach is not a biochemist's 'wild dream', it is still a goal to be achieved with the Na–Ca exchanger.

In this chapter the strategies used in the elucidation of the molecular properties of the Na–Ca exchanger will be summarized, the difficulties encountered along the way will be discussed, and the progress made so far will be reviewed.

The existence of the Na–Ca exchanger was established about 20 years ago when a counter-current-type mechanism that was responsible for the exchange of Ca^{2+} and Na^{+} across the plasma membrane was proposed in the heart (Reuter and Seitz 1968) and in the squid giant axon (Baker *et al.* 1969). Since then, a considerable amount of information has been obtained regarding the exchanger's mode of action (Johnson and Kootsey 1985; Mullins 1979, 1981), role in cellular Ca^{2+} homeostasis (Baker 1986; Barry *et al.* 1986; Mullins 1979, 1981; Sheu, Sharma, and Uglesity 1986), its regulation by ATP (Blaustein 1977; Caroni and Carafoli 1983; DiPolo and Beaugé

1987), and its universal distribution not only in excitable and secretory cells, but also in a large number of non-excitatory and non-secreting cells (see, for example, Bernstein and Santacana 1985, Brass 1984, Grubb and Bentley 1985, Ortiz and Sjodin 1984, Parker 1978, Rengasmy, Soura, and Feinberg 1987). Progress towards elucidation of the Na–Ca exchangers' molecular biochemistry, however, has started to evolve only in the last few years. Two major difficulties were responsible for this:

1. The Na–Ca exchanger is a minor membrane component in terms of abundancy. In the experimental preparations that are suitable for biochemical characterization its quantity is very low. Calculated estimates of site density in cardiac sarcolemma and determination of the amount of purified protein obtained in synaptic plasma membranes (Barzilai 1987; Barzilai, Spanier, and Rahamimoff 1984; J. P. Reeves, personal communication) showed that the Na–Ca exchanger comprises probably less than 0.1 per cent of the total membrane proteins. Therefore, in the absence of a rich source for the protein, large amounts of membrane have to be processed in order to obtain a small amount of purified protein suitable for analytical purposes.

2. The second difficulty is even more serious, and is the lack of a specific inhibitor, toxin, or affinity label that can be used to identify the purified protein, by binding. Several pharmacological agents were found to inhibit the Na–Ca exchanger. Amiloride and its derivatives (Kaczorowski *et al.* 1985; Schellenberg *et al.* 1985; Schellenberg and Swanson 1982), doxorubicin (Caroni, Villani, and Carafoli 1981), and verapamil (Erdreich and Rahamimoff 1984; Erdreich, Spanier, and Rahamimoff 1983; Miyamoto and Racker 1980; Van Amsterdam and Zaagsma 1986) were among them. However, none of these is specific and therefore cannot be used as an exclusive probe to label the exchanger protein. Thus, demonstration of Na–Ca antiport activity is the sole index for detection of its presence. As a consequence, purification of the Na–Ca exchanger protein requires solubilization of the appropriate plasma membrane, followed by functional reconstitution of the isolated putative protein(s) into a phospholipid membrane to demonstrate transport activity and identification of the specific protein entity responsible for that activity. In view of these difficulties, development of native and reconstituted membrane vesicle preparations preceded purification attempts of the Na–Ca exchanger.

7.2 Na–Ca exchange activity in native membrane vesicle preparations

Resealed membrane vesicles have been extensively used to study transport of a large number of solutes. Their major advantage is that choice ionic gradients can be created at will for limited periods of time and used to drive the transport of substances having appropriate protein carriers within the membrane vesicles. Meaningful results can be obtained if the transport process to be measured has a significantly higher reaction rate than the passive dissipation of the ionic gradient driving force. Usage of membrane vesicles also permits the separation between membrane-associated processes and those that are cytosol dependent. While this can be an advantage for understanding the mechanistic aspects of some transport reactions, it should be remembered that cytosol-derived regulatory processes can be lost.

Cardiac sarcolemma and synaptic plasma membranes are the most commonly used sources for membrane vesicle preparations containing Na–Ca antiport activity and the most extensively characterized ones (see, for example, Gill, Grollman, and Kohn 1981, Michaelis and Michaelis 1981, Philipson and Nishimoto 1981, Pitts 1979, Rahamimoff and Spanier 1979, Reeves and Sutko 1979, Schellenberg and Swanson 1981, Wakabayshi and Goshima 1981). Other preparations such as membrane vesicles from skeletal muscle (Gilbert and Meissner 1982), pituitary (Kaczorowski *et al.* 1984), renal (Jayakumar *et al.* 1984), pancreatic acinar (Bayerdorffer, Haase, and Schultz 1985), neurohypophysis (Saermark and Gratzl 1986), and platelet (Rengasmy, Soura, and Feinberg 1987) were also successfully used.

The basic methodology used for preparing plasma membrane vesicles involves tissue disruption by different means and separation of the appropriate membrane fraction by differential centrifugation. Tissue mincing and disruption by Polytron (Jones *et al.* 1979; Reeves and Sutko 1979) or Waring blender (Philipson and Nishimoto 1981), has been extensively used to prepare canine, chick, or bovine heart ventricular-tissue-derived sarcolemmal vesicles, while gentler means such as Dounce or Potter–Elvhejm teflon homogenizers were used with rat or guinea-pig brains (Gill, Grollman, and Kohn 1981; Rahamimoff and Spanier 1979, Schellenberg and Swanson 1981). Although all the different experimental protocols led to successful preparation of membrane vesicles in which Na–Ca exchange activity could be demonstrated and the general characteristics of the different systems were similar, a considerable diversity in the kinetic parameters measured has been reported. The K_m values for Ca^{2+}, when high inside Na^+ gradient-dependent Ca^{2+} influx at different external Ca^{2+} concentrations was measured, ranged between 0.5 μM and 50 μM, or even higher values (Caroni and Carafoli 1983; Gill and Chueh 1984; Philipson and Nishimoto 1984; Rahamimoff and Spanier 1984; Reeves and Sutko 1983; Wakabayshi and Goshima 1981). Moreover, using the same methodology and biological preparation yielded a wide range of K_m values at different times (Gill and Chueh 1984; Philipson and Nishimoto 1984; Reeves and Sutko 1983). The reason for this is as yet unclear, and could be related to a large number of factors known to modulate Na–Ca exchange activity such as the possible presence of endogenous proteolytic activity in the preparation (Philipson and Nishimoto 1982), the degree of membrane phosphorylation (Caroni and Carafoli 1983; DiPolo and Beaugé 1987), the amount of Ca^{2+} bound to unspecified surface charges (Scott, Akerman, and Nicholls 1980), or the proportion of inside-out to right-side-out vesicles in the preparation (Bers, Philipson, and Nishimoto 1980; Caroni and Carafoli 1983; Gill 1982; Gill *et al.* 1986), and its degree of homogeneity (Tomlins *et al.* 1986). The sideness of the membrane vesicles' preparation has been estimated by measuring release of sialic acid, and activities of specific plasma membrane enzyme markers such as Na^{+}–K^{+} ATPase or Ca^{2+} ATPase in the presence or absence of permeabilizing agents such as triton X100 SDS or deoxycholate. Effects of side-oriented drugs such as ouabain, tetrodotoxin, and veratridine on Na^{+} fluxes were measured (Bers, Philipson, and Nishimoto 1980; Gill 1982; Gill *et al.* 1986). However, the ranges that were reported for the amount of inside-out vesicles in the right-side-out preparation varied between 35 per cent and 80 per cent (Bers, Philipson, and Nishimoto 1980; Caroni and Carafoli 1983; Gill 1982; Gill *et al.* 1986). Although the Na–Ca exchanger exhibits an apparent reversible action, it also possesses some asymmetric properties such as modulation by ATP (Caroni and Carafoli 1983; DiPolo and Beaúge 1987), inhibition by La^{3+} (Rahamimoff and Spanier 1984), and pH dependence of the antiport process (Rahamimoff *et al.* 1984). The presence of variable amounts of inside-out vesicles in the right-side-out population could determine the magnitude of the kinetic parameter measured. In addition, the degree of homogeneity of the preparation could be variable. In most plasma membrane vesicles preparations employed for studying Na–Ca exchange activity, the presence of significant amounts of intracellular membranes such as sarcoplasmic reticulum in sarcolemma or mitochondrial membranes in synaptic plasma membranes was ruled out by measuring marker enzyme activities such as oxalate and ATP-dependent Ca^{2+} transport (Jones *et al.* 1979) or α-ketogluterate dehydrogenase activity (Erdreich, Spanier, and Rahamimoff 1983). However, the presence of contaminating plasma membranes from other sources (such as endothelial membranes in cardiac preparations or glial membranes in synaptic plasma membrane preparation) has not been excluded. Recent estimates claim (Tomlins *et al.* 1986) that substantial amounts of endothelial plasma membrane co-purifies with sarcolemmal preparations. Until the absence or presence of Na–Ca exchange activity in these membranes is established this could be a partial explanation for the variability of the kinetic parameters. The same applies to brain plasma membranes.

7.3 Reconstitution of the Na–Ca exchanger

The Na–Ca exchanger in native plasma membrane vesicles exhibits a remarkable degree of stability. Native vesicles can be kept frozen at $-80°$

without loss of activity for prolonged periods of time (6–12 months at least). Thawing small aliquots of plasma membrane vesicles at 37°C and refreezing them rapidly in liquid nitrogen does not lead to any loss of activity. This high degree of stability of the exchanger protein is manifested not only when the protein is embedded in its native membrane, but also following its solubilization by different detergents or proteolytic enzyme treatment, and subsequent reconstitution. The first successful reconstitution of the Na–Ca exchanger (Miyamoto and Racker 1980) was carried out by solubilization of bovine sarcolemmal vesicles in 2 per cent cholate in the presence of 2.4 per cent soybean phospholipids (asolectin). The membrane protein to phospholipid ratio during solubilization was maintained at 1:12 (by weight). Reconstitution was achieved by diluting 5–12-fold the solubile protein:phospholipid mixture with a high Na^+-containing buffer followed by harvesting of the proteoliposomes by centrifugation. Ca^{2+} transport was initiated by diluting an aliquot of the Na^+-loaded proteoliposomes into a KCl (or NaCl) and Ca^{2+}-containing medium and determination of the Na^+ gradient-dependent fraction of the Ca^{2+} influx. In these studies (Miyamoto and Racker 1980), solubilization of the exchanger protein and its subsequent successful reconstitution depended on the percentage of cholate used, the presence of added soybean phospholipids, and the extent of the dilution of the solubile mixture. Lowering the cholate concentration to 1.5 per cent did not solubilize the exchanger protein and the optimal reconstitution in terms of the resulting Na–Ca exchange activity was obtained following a 6–7-fold dilution. Under these conditions, a 3–5-fold increase in specific transport activity was obtained. The reconstituted preparation exhibited Na^+ gradient dependency of the Ca^{2+} uptake process and was inhibited by externally added Na^+ and La^{3+}. The transport was stimulated by valinomycin as expected for an electrogenic process in which more Na^+-derived charges are transported across the membrane than Ca^{2+}-derived charges. Similar methodology was used to reconstitute the rat brain synaptic plasma membrane Na–Ca exchanger (Schellenberg and Swanson 1982). In this preparation the transport properties of the Na–Ca exchanger were also conserved in the reconstituted preparation. Following this, a large number of experimental protocols have been used to reconstitute the Na–Ca exchanger from various sources. Plasma membranes from calf (Hale *et al.* 1984; Porzig and Blochlinger 1987; Soldati, Longoni, and Carafoli 1985), dog (Luciani 1984; Philipson *et al.* 1987), and chick heart (Wakabayashi and Goshima 1982) were used to reconstitute the cardiac Na–Ca exchanger. Adult and young rat brains (Barzilai and Rahamimoff 1983; Barzilai, Spanier, and Rahamimoff 1984; Michaelis, Nunley, and Michaelis 1987; Schellenberg and Swanson 1982) were the source for preparation of the reconstituted synaptic plasma membrane Na–Ca exchanger and bovine kidney cortex was used for reconstitution of renal basolateral Na–Ca exchanger (Talor and Arruda 1986). Table 7.1 summarizes and compares the different protocols that were used for reconsti-

Table 7.1 Reconstitution of the Na–Ca exchanger: the experimental approaches used

Plasma membrane source	Detergent used	Phospholipid added* mg per mg membrane protein	Method used for detergent removal	Na–Ca exchange activity**	Reference
Bovine heart	Cholate 2%	Asolectin 12	Dilution	20 nmol mg min^{-1}	Miyamoto and Racker (1980)
Adult rat cerebral cortex	Cholate 2.5%	Asolectin 10	Dilution	2.5 nmol mg min^{-1}	Schellenberg and Swanson (1982)
Chick heart	Cholate 2%	Asolectin 12.5	Dilution	3 nmol mg s^{-1}	Wakabayashi and Goshima (1982)
14-day-old rat brains	Cholate 2%	Brain phospholipids 28–80	Hollow fibre dialysis or Sephadex G50	20–40 nmol mg min^{-1}	Barzilai *et al.* (1984, 1987) Barzilai (1987)
		Asolectin 25		26.5 nmol mg 5 min^{-1}	
		PC		14.5 nmol mg 5 min^{-1}	
		PC:PE 25		12.2 nmol mg 5 min^{-1}	

Dog heart	Cholate 2%	Asolectin PC+PS PC+PE 20	Hollow fibre dialysis	16 nmol mg 5 min^{-1} 12.5 nmol mg 5 min^{-1} 4 nmol mg 5 min^{-1}	Luciani (1984)
Bovine heart	Cholate 2%	Asolectin 12.5	Dilution	42.5 nmol mg s^{-1}	Hale *et al.* (1984)
Calf heart	Triton X100 1.4%	Asolectin PC+PS PC+PI	Amberlite XAD-2	relative values	Soldati *et al.* (1985)
Bovine kidney	Cholate 1%	Asolectin 8.3	Dilution	1.7 nmol mg min^{-1}	Talor and Arruda (1986)
Dog heart	Cholate 2%	Asolectin 25	Dilution	30 nmol mg s^{-1}	Philipson *et al.* (1987)

* The concentration of membrane protein and phospholipid added varies in different protocols.
** Na–Ca exchange activity refers to the crude reconstituted preparation without further purification.

tution of the Na–Ca exchanger. Cholate was the choice detergent in most of the procedures used (Barzilai, Spanier, and Rahamimoff 1984; Hale *et al.* 1984; Luciani 1984; Miyamoto and Racker 1980; Schellenberg and Swanson 1982; Wakabayashi and Goshima 1982), although triton X100 (Soldati, Longoni, and Carafoli 1985) and CHAPS (Michaelis, Nunley, and Michaelis 1987; Porzig and Blochlinger 1987) were used as well. Reconstitution has been achieved by diluting the detergent soluble protein–phospholipid mixture (Hale *et al.* 1984; Miyamoto and Racker 1980; Philipson *et al.* 1987; Schellenberg and Swanson 1982; Wakabayashi and Goshima 1982) or detergent removal by dialysis (Barzilai, Spanier, and Rahamimoff 1984; 1987), by gel filtration (Barzilai and Rahamimoff 1987; Barzilai, Spanier, and Rahamimoff 1987; Luciani 1984), or by binding it to Amberlite XAD-2 beads (Soldati 1985). The function of the Na–Ca exchanger was retained in all the different detergents used, although the activity obtained varied and depended on several parameters. One of these was the choice of the phospholipid environment used for reconstitution. This seems to play a considerable role in at least two respects: the extent of overall specific transport activity of the reconstituted Na–Ca exchanger obtained; and the amount of passive Ca^{2+} flux measured across the reconstituted membrane in the absence of a Na^{+} gradient. Both the composition of the phospholipid head groups and the fatty acyl chains contribute to the resulting Na–Ca exchange activity. Asolectin (soybean phospholipids) was extensively used for reconstitution of the Na–Ca exchanger (see Table 7.1). The Na^{+} gradient-dependent Ca^{2+} transport activity with cardiac membrane vesicle preparation was quite substantial. In the different procedures used for reconstitution in the absence of any prior purification, up to 30–42.5 nmol per mg of membrane protein per second of Na^{+} gradient-dependent Ca^{2+} uptake activity has been measured (Hale *et al.* 1984; Philipson *et al.* 1987). Following reconstitution, the specific Na^{+} gradient-dependent Ca^{2+} transport activity of the reconstituted preparation was consistently higher than the basal activity in the native vesicles (Hale *et al.* 1984; Philipson *et al.* 1987; Soldati, Longoni, and Carafoli 1985). The reason for this apparent activation is not known. Reconstitution of brain synaptic plasma membrane protein into an asolectin-containing membrane (Barzilai 1987; Barzilai, Spanier, and Rahamimoff 1984; Schellenberg and Swanson 1982), resulted in lower Na^{+} gradient-dependent Ca^{2+} transport activity and higher Na^{+} gradient-independent Ca^{2+} uptake activity than when purified brain phospholipids were used for reconstitution (Barzilai 1987; Barzilai, Spanier, and Rahamimoff 1984). Phosphatidylcholine (PC) as the sole lipid used for reconstitution was inept in supporting significant Na–Ca antiport activity (Barzilai, Spanier, and Rahamimoff 1984; Luciani 1984; Soldati, Longoni, and Carafoli 1985) with both synaptic plasma membranes or sarcolemmal preparations. Nor was a mixture of PC and PE (phosphatidylethanolamine) adequate to elicit optimal activity (Barzilai, Spanier, and

Rahamimoff 1984; Luciani 1984; Soldati, Longoni, and Carafoli 1985). On the other hand, acidic phospholipids such as phosphatidylinositol (PI) and especially phosphatidylserine (PS) had a pronounced stimulatory effect on Na–Ca exchange activity (Luciani 1984; Soldati, Longoni, and Carafoli 1985). This finding can also explain why brain phospholipids, which have a relatively high content of PS are the preferential phospholipid environment for reconstitution of the synaptic plasma membrane Na–Ca exchanger (Barzilai, Spanier, and Rahamimoff 1984). The addition of phospholipase D to native sarcolemmal vesicle led to an almost 400-fold stimulation in Na–Ca exchange activity and a parallel 10-fold increase in phosphatidic acid content (Philipson and Nishimoto 1984). Stimulation of Na–Ca exchange activity by anionic (negatively charged) fatty acids (Philipson and Ward 1985) and other anionic ampiphiles has also been shown (Philipson 1984; Philipson and Ward 1987).

7.4 Purification of the Na–Ca exchange protein(s)

Two different experimental strategies were adopted in order to purify the Na–Ca exchanger. The 'classical' approach was based on advantageous exploration of protein fractionation techniques of the solubilized plasma membrane preparation and reconstitution of the partially purified protein(s) into a phospholipid membrane in order to detect the presence of the Na–Ca exchanger by its activity. The relationship between the increased specific Na–Ca antiport activity and the protein profile, obtained following electrophoresis on SDS-containing polyacrylamide gels, served as an index for identification of the transporter (Hale *et al.* 1984; Michaelis, Nunley, and Michaelis 1987; Philipson *et al.* 1987; Soldati, Longoni, and Carafoli 1985; Wakabayashi and Goshima 1982). The second approach adopted the 'transport specific fractionation technique' (Goldin and Rhoden 1978). This technique was developed originally to purify the red blood cell glucose transporter (Goldin and Rhoden 1978) and was successfully employed to purify the synaptosomal vesicles Ca^{2+} transport ATPase (Papazian, Rahamimoff, and Goldin 1979). In principle it can be used with every transport protein. It is based on exploiting a density difference created between the reconstituted vesicles containing the transporter to be purified and the bulk of reconstituted vesicles. The reconstitution is carried out in the presence of a very large excess (50–80-fold by weight) of external phospholipids respective to the solubilized protein. This results in the distribution of the original native membrane proteins among different liposomes. A density difference between the various reconstituted vesicles is created by transporting a specific solute across the vesicles' membrane. This can be done, for example, by ATP-dependent Ca^{2+} loading of the vesicles containing the ATP-dependent Ca^{2+} pump (Papazian *et al.* 1979) or by Na^+ gradient-dependent Ca^{2+} loading of the vesicles containing the Na–Ca exchanger. An intravesicular anion that yields a low solubility product such as oxalate or phosphate for Ca^{2+} transport is used. As a result, separation of the vesicles containing that transporter by density–gradient centrifugation can be obtained. Analysis of the protein profile of the purified and reconstituted vesicles by SDS-containing polyacrylamide gel electrophoresis can indicate the protein entity specifically associated with the transport activity purified.

7.4.1 Purification of the Na–Ca exchanger by selective degradation or extraction of unrelated proteins

It has been shown that proteinase treatment of sarcolemmal membranes leads to an increase in Na–Ca exchange activity (Philipson and Nishimoto 1982). Trypsin, chymotrypsin, or papain treatment of native vesicles resulted in a considerable reduction in the apparent K_m to Ca^{2+} while the V_{max} increased. This result, which reflects probably the considerable resistance of the exchanger protein (or at least its catalytic moiety) to proteolytic digestion in comparison to other membrane proteins that are degraded, has been explored in an independent fashion for purification purposes. Pronase digestion of the solubilized chick cardiac plasma membrane proteins in the presence of added asolectin (Wakabayashi and Goshima 1982) led to a 10-fold increase in specific initial transport rate of Na^+ gradient-dependent Ca^{2+} influx following reconstitution by dilution. The 10-fold purification obtained could not be associated with any specific protein band in particular, but five major bands (as opposed to more than 50 in the native membrane) were detected. Those corresponded to 20 000, 21 000, 30 000, 32 000, and 35 000 M_r proteins. Pronase digestion of the solubilized membrane proteins, followed by purification of bovine heart sarcolemmal vesicles (Hale *et al.* 1984) using a Sephacryl column, resulted in a 30–100-fold increase in specific transport activity following reconstitution. The 82 000 M_r band, however (which correlated with the increased transport activity), was found later (J. P. Reeves, personal communication) to be retained by a concanavalin A column while Na–Ca exchange activity was excluded from the gel.

Specific Na^+ gradient-dependent Ca^{2+} transport activity of cardiac sarcolemma can be increased also by alkaline treatment of the native membranes. This treatment leads to the removal of about 33 per cent of peripheral membrane proteins and thereby to an enrichment of the preparation in the exchanger protein (Philipson *et al.* 1987). The number of proteins remaining after alkaline treatment is still too high to permit any tentative identification.

7.4.2 Separation of solubilized sarcolemmal proteins by rate zonal centrifugation

Triton X100 solubilization of calf sarcolemmal membranes (Soldati, Longoni, and Carafoli 1985) in the presence of added asolectin, followed by rate zonal centrifugation of the solubilized membrane and reconstitution, led to identification of a 33 000 M_r protein as the cardiac Na–Ca exchanger candidate. Although several other proteins (of 77 000, 68 000, and 63 000 M_r) were present in the fractions containing the enriched transport activity, the 33 000 M_r protein corresponded best to the amount of protein and specific transport activity present in each purified fraction (Soldati, Longoni, and Carafoli 1985). Measuring the initial velocity showed that as a consequence of the purification the Na^+ gradient-dependent Ca^{2+} transport activity increased 128-fold from 7.1 nmol per mg protein per second in the native sarcolemma to 908 nmol per mg protein per second following purification.

7.4.3 Purification of synaptic plasma membrane Na–Ca exchanger by 'transport specific fractionation'

Purification of the synaptic plasma membrane Na–Ca exchanger (Barzilai and Rahamimoff 1983; Barzilai, Spanier, and Rahamimoff 1984) by 'transport specific fractionation' showed that increased specific transport activity correlated with the enriched appearance of a 70 000 M_r protein. Figure 7.1 shows a typical two-step preparative purification procedure of the synaptic plasma membrane Na–Ca exchanger. Native synaptic plasma membranes were made soluble in 2 per cent cholate in the presence of a 50-fold excess by weight of purified brain phospholipids (Barzilai 1987; Barzilai, Spanier, and Rahamimoff 1984, 1987). Reconstitution was performed by dialysis through hollow fibres against an Na^+-phosphate buffer. The reconstituted vesicles were harvested by centrifugation and exposed to Ca^{2+} (unlabelled) in the absence of a Na^+ gradient. Under these conditions Ca^{2+} is associated only with vesicles that are permeable to Ca^{2+} or bind Ca^{2+} in an unspecific manner. As a consequence, due to a density difference between Na^+-phosphate (NaPi) and Ca^{2+}-phosphate (CaPi), these vesicles are denser. Centrifugation of the entire vesicle population through a linear sucrose density gradient (Fig. 7.1A) leads to partitioning of the reconstituted vesicles into two peaks. The majority of the reconstituted vesicles are present in the lower density sucrose as a sharp peak (fractions 21–30 marked by bar), a minority of the vesicles is present in the higher density sucrose region (fractions 10–20). These high-density vesicles contain about 30 per cent of the total phospholipid layered on the sucrose gradient and variable amounts of protein—which can sometimes be as high as 80 per cent. Most of the Na–Ca exchange activity is retained within the major phospholipid peak (fractions marked by bar 21–30). Only

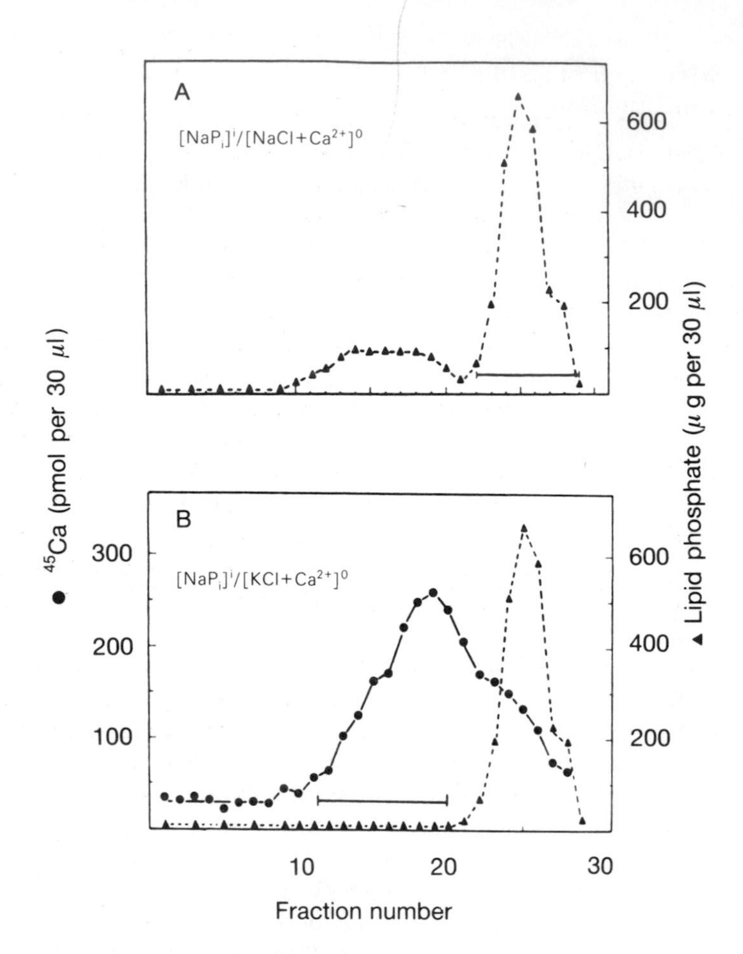

Fig. 7.1. Purification of the Na–Ca antiporter by sucrose density gradient.
A. Reconstituted NaPi preloaded vesicles were exposed to 50 μM $^{40}CaCl_2$ in an iso-osmotic NaCl solution buffered with 10 mM tris HCl pH 7.4 and containing 5 mM β-mercaptoethanol for 20 min at 23°C. The vesicles were layered over a linear 0–0.5 M sucrose gradient in an identical NaCl solution, except that it did not contain calcium. The Beckman SW–28 or Kontron TST 28 rotor was used at 100 000 g for 15 h at 4°. The vesicles marked by bar were collected, pooled and harvested by centrifugation at 161 000 g for 4 h in the Beckman Ti60 rotor.
B. Pooled partially purified reconstituted vesicles (from Fig. 7.1A – marked by bar) were diluted into an iso-osmotic KCl solution buffered with 10 mM tris HCl pH 7.4 and containing 50 μM $^{45}CaCl_2$. Following an incubation for 20 min at 23°C the $^{45}Ca^{2+}$ loaded vesicles were layered over an identical sucrose gradient as in Fig. 7.1A, except that the NaCl solution was replaced by KCl and centrifuged as in Fig. 7.1A. (from Barzilai 1987; Barzilai, Spanier and Rahamimoff 1987.)

traces of activity are detected in the minor high-density peak. The reconstituted vesicles present in the major phospholipid peak are pooled, concentrated by centrifugation, and exposed to $^{45}Ca^{2+}$ in the presence of an Na^{+} gradient (external KCl). Under these conditions, reconstituted vesicles containing the Na–Ca exchanger in their membrane will take up Ca^{2+}. Although the Na–Ca exchanger is electrogenic and 3–5 Na^{+} ions (Barzilai and Rahamimoff 1987; Mullins 1979, 1981) leave the vesicles for each Ca^{2+} that enters a density shift following centrifugation on a linear sucrose gradient (Barzilai 1987) occurs. Figure 7.1B shows the separation between the major phospholipid peak and the $^{45}Ca^{2+}$-containing fractions following centrifugation. From the distribution of $^{45}Ca^{2+}$ and phospholipid phosphate along the sucrose density gradient the specific $^{45}Ca^{2+}$ content in each fraction can be calculated. Figure 7.2 shows the distribution of the calculated specific Ca^{2+} content in fractions 9–21 of the sucrose gradient presented in Fig. 7.1B. The major phospholipid peak is also shown. The overall yield of the purified exchanger depends on the degree of separation obtained between the phospholipid peak and the Ca^{2+} peak. The better the separation between these two peaks and the smaller the overlap, the more $^{45}Ca^{2+}$-containing fractions can be pooled.

The degree of purification obtained in such an experiment is presented in Table 7.2. The sucrose gradient fractions 11–21, 22–23, and 24–28, shown in Fig. 7.1B were pooled; their $^{45}Ca^{2+}$ and phospholipid content was

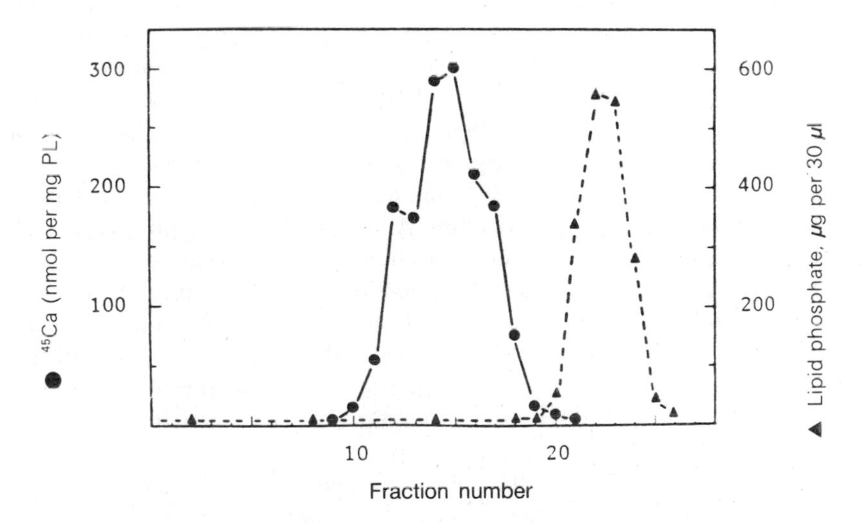

Fig. 7.2. Distribution of specific $^{45}Ca^{2+}$ content in different fractions of the sucrose density gradient from Fig. 7.1B. The calculated specific $^{45}Ca^{2+}$ content (nmol mg phospholipid) in fractions 9–21 of the sucrose gradient shown in Fig. 7.1B (●——●). The phospholipid peak (▲——▲) is also shown (from Barzilai 1987).

Table 7.2 Distribution of the specific $^{45}Ca^{2+}$ content in the different fractions of the second sucrose density gradient

Sucrose gradient fractions pooled	$^{45}Ca^{2+}$ nmol	Phospholipid mg	$^{45}Ca^{2+}$ content nmol per mg PL*
11–21	43.2	0.256	168
22–23	30.5	2.23	13.7
24–28	21.9	52.0	0.41

Sucrose density gradient fractions as specified from the experiment presented in Fig. 7.1B were pooled, their $^{45}Ca^{2+}$ and phospholipid content determined, and the specific $^{45}Ca^{2+}$ content per mg phospholipid calculated.
* PL = phospholipid.

measured. The high-density peak (fractions 11–21) contains about 0.5 per cent of the total phospholipid content (and about 0.1 per cent of the protein—not shown). The major advantage of the two-step density gradient (Barzilai 1987; Barzilai, Spanier, and Rahamimoff 1987) relative to a one-step sucrose density gradient (Barzilai 1987; Barzilai, Spanier, and Rahamimoff 1984) is in the degree of purity of the Na–Ca exchanger obtained. The first density gradient (Fig. 7.1A) removes from the reconstituted vesicle preparation those vesicles that take up calcium in an unspecific manner. A substantial amount of protein is also removed in this step without apparent loss of Na^+ gradient-dependent Ca^{2+} transport activity. After this step, the degree of purification obtained as determined by the specific Na^+ gradient-dependent Ca^{2+} transport activity in the major phospholipid peak is between 3- and 5-fold. After the second sucrose density gradient, which is carried out following Na^+ gradient-dependent Ca^{2+} loading of the vesicles, the increase in specific transport activity can be 300–500-fold as compared to the native synaptic plasma membrane vesicle preparation. A typical SDS-containing polyacrylamide gel electrophoresis profile of the proteins present in fractions 11–21 (Fig. 7.1B) is shown in Fig. 7.3. Native synaptic plasma membranes are shown in lane A and the purified 70 000 M_r band obtained is shown in lane B.

This band is consistently present in an enriched form in all the purified preparations of the reconstituted synaptic plasma membrane vesicles in conjunction with increased specific Na–Ca exchange activity. Sometimes, especially when large quantities of purified material are analysed, a 33 000 M_r protein (Barzilai 1987; Barzilai, Spanier, and Rahamimoff 1987) can also be detected. The presence of this protein is inconsistent and its amount (when present) relative to the major 70 000 M_r protein variable. Therefore, and in view of the tentative identification of a 33 000 M_r protein in cardiac sarcolemma (Soldati, Longoni, and Carafoli 1985) as the most likely candidate to be the Na–Ca exchanger, its direct identification was required.

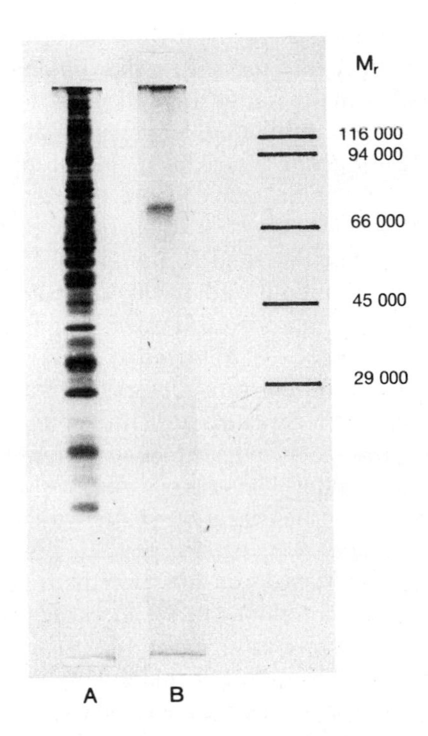

Fig. 7.3. SDS-polyacrylamide gel electrophoresis pattern of purified Na–Ca exchanger from synaptic plasma membranes. Fractions containing the highest specific $^{45}Ca^{2+}$ content (such as 11–20 shown in Fig. 7.1B) were pooled, and separated by SDS polyacrylamide gel electrophoresis (68, 74). The gel was silver stained. A. Native synaptic plasma membranes; B. Purified fractions containing the enriched specific Na–Ca transport activity.

7.5 Immunological identification of the Na–Ca exchanger

Taking the increased specific transport activity and the enriched appearance of a protein band as a sole index for identification of the Na–Ca exchanger can lead to misleading conclusions, since it cannot be ruled out that the Na–Ca exchanger is one of the minor, barely detectable proteins present in the purified preparation.

In the absence of a chemical or pharmacological agent that could be used to identify the exchanger protein, the immunological approach seems to be the sole method available. The experimental approach is based on the preparation of specific antibodies against the protein entity identified as the potential Na–Ca exchanger and using them for binding studies designed to identify, on immunoblots of the native membrane proteins, the specific entity recognized by the antibody, inhibition of transport activity in native membranes containing the antigen in active form, and/or immunoprecipitation of the transport activity by the antibodies from the native membrane. These strategies were adopted to identify both the synaptic plasma membrane Na–Ca exchanger (Barzilai, Spanier, and Rahamimoff 1987) and the cardiac sarcolemmal exchanger (Longoni and Carafoli 1987). In principle, similar methodology was used. The Na–Ca exchanger was purified by triton X100 solubilization of cardiac membranes followed by zonal centrifugation (Soldati, Longoni, and Carafoli 1985) and transport specific fractionation of synaptic plasma membranes (Barzilai, Spanier, and Rahamimoff 1984, 1987). The proteins that were present in purified preparations were separated by polyacrylamide gel electrophoresis, stained, electro-eluted, and used for the immunization of rabbits. For identification of the synaptic plasma membrane Na–Ca exchanger, each one of the stained proteins that could be detected on the gel was sliced out separately and used for the immunization of rabbits (Barzilai, Spanier, and Rahamimoff 1987). In cardiac sarcolemma, a 1 cm region around two specific proteins of about 30 000 and 140 000 M_r was cut out and used for the immunization of rabbits (Longoni and Carafoli 1987). The proteins in the 30 000 M_r region were treated with a reducing mercapto reagent—DTT (dithiothreitol)—prior to immunization, while the 140 000 M_r protein was obtained in the absence of any reducing mercapto agent. Binding studies of these antibodies to immunoblots of cardiac sarcolemma revealed that both antibodies bound to proteins of 33 000, 70 000, and 140 000 M_r. However, the relative amounts of these proteins in the purified membrane preparation depended on the amount of reducing mercapto agent present. Under non-reducing conditions, the 140 000 M_r form of the protein prevailed over the other forms. Under reducing conditions the 140 000 M_r protein disappeared entirely and the relative amounts of the 70 000 and 33 000 M_r proteins increased (Longoni and Carafoli 1987).

In the preparative gels of the purified synaptic plasma membrane Na–Ca exchanger five proteins could be detected. These corresponded to about 33 000, 50 000, 60 000, 70 000, and 120 000 M_r. Each protein was electro-eluted separately and injected into different rabbits. All the antigens were prepared in the presence of mercaptoethanol. Two antibodies that bound specifically to synaptic plasma membranes were obtained. One antibody was obtained from the serum of the rabbit immunized with the 33 000 M_r protein and the second antibody from the rabbit immunized with the 70 000 M_r protein. Binding studies of these antisera to immunoblots of native synaptic membranes (Barzilai, Spanier, and Rahamimoff 1987) showed that both antibodies bound to a single protein of 70 000 M_r. When the purified 33 000 and 70 000 M_r were used as antigens, both antibodies cross-reacted and bound to both purified proteins (Barzilai, Spanier, and Rahamimoff 1987).

The antibodies prepared against the cardiac Na–Ca exchanger inhibited Na–Ca exchange activity. Up to 50 per cent inhibition was obtained (Longoni and Carafoli 1987).

The antibodies prepared against the synaptic plasma membrane Na–Ca exchanger did not inhibit Na^+ gradient-dependent Ca^{2+} uptake activity when added to native synaptic plasma membrane vesicles (Barzilai, Spanier, and Rahamimoff 1987). This was to be expected, since the rabbits were immunized with the reduced, denaturated, gel-eluted protein (Barzilai, Spanier, and Rahamimoff 1987) rather than with its native conformation-retaining form. Therefore, immunoprecipitation of Na–Ca exchange activity was carried out by using solubilized synaptic plasma membranes. Treatment of the cholate-solubilized membranes with IgG prepared from either the anti-70 000 M_r protein antiserum or the anti-33 000 M_r protein antiserum conjugated to protein A–Sepharose beads (Barzilai, Spanier, and Rahamimoff 1987) resulted in a considerable decrease in Na–Ca exchange activity following reconstitution of the membrane proteins excluded from the beads. The extent of inhibition depended on the amount of IgG bound to the beads and could reach 82 per cent. The specificity of the immunoprecipitation reaction was tested by determination in parallel to the Na^+ gradient-dependent Ca^{2+} uptake of the ATP-dependent Ca^{2+} transport activity which is present in the same membrane (Rahamimoff and Spanier 1984). Employing conditions that led to 82 per cent inhibition of the Na^+ gradient-dependent Ca^{2+} transport activity, the ATP-dependent Ca^{2+} transport did not decrease.

Attempts to raise monoclonal antibodies against the partially purified reconstituted vesicles (marked by bar in Fig. 7.1B) containing the enriched Na–Ca exchange activity were unsuccessful both with synaptic plasma membranes (Barzilai, Spanier, and Rahamimoff 1983) and cardiac sarcolemma (Porzig and Blochlinger 1987). A monoclonal antibody against the non-T, non-B, acute lymphoblastic leukaemia cell line HOON has been shown to inhibit the cardiac Na–Ca exchanger (Michalak, Quackenbush, and Letarte 1986). It bound to proteins of 125 000 M_r in the absence of DTT and 38 000 and 95 000 M_r in the presence of DTT.

The immunological evidence available to date and the previous purification experiments strongly link several protein entities to the Na–Ca exchange activity. Antibodies that bind specifically to proteins of 33 000, 70 000, and 140 000 M_r in synaptic plasma membranes or in cardiac sarcolemma are capable also either to inhibit Na–Ca exchange activity (Longoni and Carafoli 1987) or to immunoprecipitate it (Barzilai, Spanier, and Rahamimoff 1987). Which of these proteins, however, comprises the Na–Ca exchanger and what is the molecular relationship between them?

Two lines of experimental evidence could be helpful in finding the possible relationship between the different molecular entities linked to the Na–Ca exchange activity: the discovery that disulphide bonds are involved in modulating Na–Ca exchange activity (Longoni and Carafoli 1987; Orlicky *et al.* 1987); and the finding that ATP-dependent phosphorylation plays a role in regulating the Na–Ca exchanger.

7.6 The molecular form of the Na–Ca exchanger: what is the relationship between the proteins?

At least three molecular entities can be directly linked to the Na–Ca exchanger: the 33 000, 70 000, and 140 000 M_r proteins in cardiac sarcolemma (Longoni and Carafoli 1987; Soldati, Longoni, and Carafoli 1985) and the 33 000 and 70 000 M_r proteins in synaptic plasma membranes (Barzilai, Spanier, and Rahamimoff 1984, 1987). One possible explanation that could account for the heterogeneity in molecular entities and the apparent immunological cross-reactivity as demonstrated with specific polyclonal antibodies prepared against each one of them (Barzilai, Spanier, and Rahamimoff 1987; Longoni and Carafoli 1987), was proposed by Longoni and Carafoli (1987). Based on the shift obtained between the 140 000 M_r protein and the 70 000 and 33 000 M_r proteins as a function of increasing DTT concentrations, they proposed a monomer–dimer–tetramer-type relationship between these molecular entities. The interaction between the monomers and the other mers is based on disulphide bonds. As well as the shift in molecular ratio following treatment with mercapto-reagents, the relevance of disulphide bonds to Na–Ca exchange was demonstrated by inhibition of the antiport with increasing DTT concentrations. Involvement of thiol reagents in Na^+ gradient-dependent Ca^{2+} uptake has been demonstrated also in brain microsomal fraction (Orlicky *et al.* 1987).

Mercaptoethanol was included during the purification of the synaptic plasma membrane Na–Ca exchanger (Barzilai, Spanier, and Rahamimoff 1984, 1987). The polyclonal antibodies obtained against both the 70 000 and 33 000 M_r proteins bound to native synaptic membranes only following membrane reduction with mercaptoethanol. Therefore, the presence of a 140 000 M_r tetramer under non-reducing conditions could not be tested, nor was its presence in purified preparations of the synaptic plasma membrane Na–Ca exchanger obtained in the absence of added mercaptoethanol ever detected. The presence of the 33 000 M_r protein in purified preparations of the synaptic plasma membrane Na–Ca exchanger was inconsistent (Barzilai, Spanier, and Rahamimoff 1987). Its spontaneous shift to the 70 000 M_r form occurred in the presence of mercaptoethanol. Thus, no direct evidence is available to support the monomer–dimer hypothesis in synaptic membranes.

Recent evidence (Philipson and Ward 1988) suggests that in canine cardiac sarcolemma following alkaline extraction and purification on DEAE–Sepharose and WGA–agarose columns, two protein species are revealed on SDS-containing polyacrylamide gels—a 70 000 M_r and a 120 000 M_r protein. Mild chymotrypsin treatment of the purified preparation leads to disappearance of the latter form with the concomitant increase in the 70 000 form. Under non-reducing conditions only a 160 000 M_r protein has been detected on the gels (Philipson and Ward 1988).

7.7 The modulation of Na–Ca exchange by ATP

The involvement of ATP in regulating the Na^+–Ca^{2+} antiport has been demonstrated in squid giant axon (Blaustein 1977; DiPolo and Beaugé 1987) and cardiac sarcolemma (Caroni and Carafoli 1983). Although the mechanism of ATP modulation is not yet fully understood, evidence suggests that: ATP (or its hydrolysable analogues) stimulates both external Na^+-dependent Ca^{2+} efflux and internal Na^+-dependent Ca^{2+} influx; Mg^{2+} is required for ATP stimulation of Na–Ca exchange; the affinity of the exchanger to Ca^{2+} and Na^+ is increased in the presence of ATP or adenosine-5′-γ-thio-triphosphate; kinase-mediated phosphorylation is most probably involved in this modulation; in sarcolemma, a decrease in Na–Ca exchange activity can be elicited by treatment either with phosphorylase phosphatase or Mg^{2+}, Ca^{2+}, and calmodulin (Caroni and Carafoli 1983).

There is no evidence to indicate which molecular entities are phosphorylated. Is it the Na–Ca exchange molecule itself, one of its subunits, or a related regulatory protein that undergoes phosphorylation? All these questions are still to be answered. Any of these proteins might shed some light on the relationship between the different molecular proteins found in the purified preparation of the Na–Ca exchanger.

7.8 Conclusions

Different experimental methods developed in the last few years led to considerable purification of the Na–Ca exchanger and identification of the purified protein(s). The conclusions reached in cardiac sarcolemma and synaptic plasma membranes are quite compatible (Barzilai, Spanier, and Rahamimoff 1984, 1987; Longoni and Carafoli 1987; Philipson and Ward 1988; Soldati, Longoni, and Carafoli 1985). It is clear, though, that the identification of the complete molecular entity of the Na–Ca exchanger present in native cardiac or synaptic plasma membranes or its organization within the membrane requires further investigation. Immunological evidence suggests that most of the multiple molecular entities detected in purified preparations of the Na–Ca exchanger contain partially overlapping epitophs and that they are related to the catalytic Na–Ca exchange activity. The question of whether the Na–Ca exchanger contains only a catalytic protein or whether its activity involves participation of a phosphorylatable regulatory subunit is yet to be elucidated. Moreover, whether the appearance of multiple molecular entities in purified preparations containing highly enriched Na–Ca exchange activity is due to endogenous proteolysis, is an artefact of lengthy handling, or is due to partially or completely reduced disulphide bonds, remains to be discovered. It is hoped that the progress made in the last few years will continue in forthcoming years and that studies will lead to an understanding of how the Na–Ca exchanger's structure determines its function.

The foundation for the studies on the molecular biochemistry of the Na–Ca exchanger was laid in the lifetime work of Peter Baker, who knew how to combine physiology with biochemistry, and to whose memory this chapter is dedicated.

Acknowledgements

I would like to thank Professor Rami Rahamimoff for reading this manuscript and his many helpful comments.

The results presented here, originating from the author's laboratory, were supported by the Muscular Dystrophy Association, New York; the Israel Academy of Sciences Basic Research Division and the United States–Israel Binational Science Foundation.

Editors' note

In two recent papers (Sigel *et al.* 1988; Longoni *et al.* 1988) expression of Na–Ca exchange activity in *Xenopus* oocytes has been reported. Oocytes were injected with total mRNA isolated from chick hearts (Sigel *et al.* 1988) or rabbit hearts (Longoni *et al.* 1988). Na^+-gradient-dependent ^{45}Ca uptake was observed in both sets of experiments. Fractionation of poly$(A)^+$ RNA led to the identification of a 25 S fraction (Sigel *et al.* 1988) or 3–5 Kb message size (Longoni *et al.* 1988) competent for expression of Na–Ca exchange activity.

References

Baker, P. F. (1986). The sodium–calcium exchange system. *Ciba Found. Symp.* **122**, 73–92.

Baker, P. E., Blaustein, M. P., Hodgkin, A. L., and Steinhardt, R. A. (1969). The influence of calcium on sodium efflux in squid axons. *J. Physiol.* **200**, 431–58.

Barzilai, A. (1987). PhD thesis. Approved by the Senate of the Hebrew University of Jerusalem, Israel.

Barzilai, A. and Rahamimoff, H. (1983). Isolation, purification and reconstitution of the Na^+ gradient dependent Ca^{2+} transporter from rat brain synaptic plasma membranes. *Proc. 1st Joint Meeting of the Israeli Societies for Life Sciences* (Abstr. C207).

Barzilai, A. and Rahamimoff, H. (1987). Stoichiometry of the sodium–calcium exchanger in nerve terminals. *Biochemistry* **26**, 6113–18.

Barzilai, A., Spanier, R., and Rahamimoff, H. (1984). Isolation, purification and reconstitution of the Na^+ gradient dependent Ca^{2+} transporter (Na^+–Ca^{2+} exchanger) from brain synaptic plasma membranes. *Proc. Natl. Acad. Sci. USA* **81**, 6521–5.

Barzilai, A., Spanier, R., and Rahamimoff, H. (1987). Immunological identification of the synaptic plasma membrane Na^+–Ca^{2+} exchanger. *J. Biol. Chem.* **262**, 10315–20.

Barry, W. H., Rasmussen, C. A. Jr., Ishida, H., and Bridge, J. H. (1986). External Na-independent Ca extrusion in cultured ventricular cells. Magnitude and functional significance. *J. Gen. Physiol.* **88**, 393–411.

Bayerdorffer, E., Haase, W., and Schultz, J. (1985). Na^+–Ca^{2+} countertransport in plasma membrane of rat pancreatic ascinar cells. *J. Membr. Biol.* **87**, 107–19.

Bernstein, J. and Santacana, G. (1985). The Na^+–Ca^{2+} exchange system of the liver cell. *Res. Comm. Chem. Pathol. Pharmacol.* **47**, 3–34.

Bers, D. M., Philipson, K. D., and Nishimoto, A. Y. (1980). Na^+–Ca^{2+} exchange and sideness of isolated cardiac sarcolemmal vesicles. *Biochem. Biophys. Acta* **601**, 358–71.

Blaustein, M. P. (1977). Effects of internal and external cations and of ATP on sodium–calcium and calcium–calcium exchange in squid axons. *Biophys. J.* **20**, 79–111.

Brass, L. F. (1984). The effect of Na^+ on Ca^{2+} homeostasis in unstimulated platelets. *J. Biol. Chem.* **259**, 12571–5.

Caroni, P. and Carafoli, E. (1983). The regulation of the Na–Ca exchanger of heart sarcolemma. *Eur. J. Biochem.* **132**, 451–60.

Caroni, P., Villani, F., and Carafoli, E. (1981). The cardiotoxic antibiotic doxorubicin inhibits the Na^+–Ca^{2+} exchange of dog heart sarcolemmal vesicles. *FEBS Lett.* **130**, 184–6.

Davis, L. G., Dibher, M. D., and Battey, J. F. (1986). *Basic methods in molecular biology.* Elsevier, New York, Amsterdam, London.

DiPolo, R. and Beaugé, L. (1987). In squid axons, ATP modulates Na^+–Ca^{2+} exchange by a Ca^{2+}-dependent phosphorylation. *Biochem. Biophys. Acta* **897**, 347–54.

Erdreich, A. and Rahamimoff, H. (1984). The inhibition of calcium uptake in cardiac membrane vesicles by verapamil. *Biochem. Pharmacol.* **33**, 2315–23.

Erdreich, A., Spanier, R., and Rahamimoff, H. (1983). The inhibition of Na-dependent Ca-uptake by verapamil in synaptic plasma membrane vesicles. *Eur. J. Pharmacol.* **90**, 193–202.

Gilbert, J. R. and Meissner, G. (1982). Sodium–calcium ion exchange in skeletal muscle sarcolemmal vesicles. *J. Membr. Biol.* **69**, 77–84.

Gill, D. L. (1982). Na^+ channel, Na^+ pump and Na^+–Ca^{2+} exchange activities in synaptosomal plasma membrane vesicles. *J. Biol. Chem.* **257**, 10986–90.

Gill, D. L. and Chueh, S. H. (1984). Functional importance of the synaptic plasma membrane calcium pump and sodium–calcium exchanger. *J. Biol. Chem.* **259**, 10807–13.

Gill, D. L., Grollman, E. F., and Kohn, D. L. (1981). Calcium transport mechanisms in membrane vesicles from guinea-pig brain synaptosomes. *J. Biol. Chem.* **256**, 184–92.

Gill, D. L., Chueh, S. H., Noel, M., and Ueda, T. (1986). Orientation of synaptic plasma membrane vesicles containing calcium pump and sodium–calcium exchange activities. *Biochem. Biophys. Acta* **856**, 856–73.

Goldin, A. L., Snutch, T., Lubbert, M., Dowsett, A., Marshall, J., Auld, J., Downey, W., Fritz, L. C., Lester, H. A., Dunn, R., Catterall, W. A., and Davidson, N. (1986). Messenger RNA coding for only the α-subunit of the rat brain Na^+ channel is sufficient for expression of functional channels in *Xenopus* oocytes. *Proc. Natl. Acad. Sci. USA*, **83**, 7503–7.

Goldin, S. M. and Rhoden, V. (1978). Reconstitution and 'transport specific fractionation' of the human erythrocyte glucose transport system. *J. Biol. Chem.* **253**, 2575–83.

Grubb, B. R. and Bentley, P. J. (1985). Calcium exchange in isolated cutaneous epithelium of toad *Bufo marinus*. *Am. J. Physiol.* **249**, 172–8.

Hale, C. C., Slaughter, R. S., Ahrens, D. C., and Reeves, J. P. (1984). Identification and partial purification of the cardiac sodium–calcium. *Proc. Natl. Acad. Sci. USA*, **81**, 6569–73.

Imoto, K., Methfessel, C., Sakmann, B., Mishina, M., Mori, Y., Konno, T., Fukuda, K., Kurasaki, M., Bujo, H., Fujita, Y., and Numa, S. (1986). Location of a δ-subunit region determining ion transport through the acetylcholine receptor channel. *Nature* **324**, 670–4.

Jayakumar, A., Cheng, L., Liang, T. C., and Sactor, B. (1984). Sodium dependent calcium uptake in renal basolateral membrane vesicles. *J. Biol. Chem.* **259**, 10827–33.

Johnson, E. A. and Kootsey, J. M. (1985). A minimum mechanism for Na^+–Ca^{2+} exchange: net and undirectional Ca^{++} fluxes as functions of ion composition and membrane potential. *J. Membr. Biol.* **86**(2), 167–87.

Jones, L. R., Besch, H. R., Fleming, J. W., McConnaughty, M. M., and Watanabe, A. M. (1979). Separation of vesicles of cardiac sarcolemma from vesicles of cardiac sarcoplasmic reticulum. *J. Biol. Chem.* **254**, 530–9.

Kaczorowski, G. J., Barros, F., Dethmers, J. K., Trumble, M. J., and Cargoe, E. J. Jr. (1985). Inhibition of Na^+–Ca^{2+} exchange in pituitary plasma membrane vesicles by analogues of amiloride. *Biochemistry* **24**, 1394–403.

Kaczorowski, G. J., Costello, L., Dethmers, J., Trumble, M. J., and Vandlen, R. L. (1984). Mechanism of Ca^{2+} transport in plasma membrane vesicles prepared from cultured pituitary cells: 1. Characterization of the Na^{+}–Ca^{2+} exchange activity. *J. Biol. Chem.* **259**, 9395–403.

Longoni, S. and Carafoli, E. (1987). Identification of the Na^{+}–Ca^{2+} exchanger of calf heart sarcolemma with the help of specific antibodies. *Biochem. Biophys. Res. Comm.* **145**, 1059–63.

Longoni, S., Loady, M. J., Ikeda, T., and Philipson, K. D. (1988). Expression of cardiac sarcolemmal Na^{+}–Ca^{2+} exchange activity in *Xenopus laevis* oocytes. *Am. J. Physiol.* **255**, C870–3.

Luciani, S. (1984). Reconstitution of the sodium–calcium exchanger from cardiac sarcolemmal vesicles. *Biochem. Biophys. Acta* **772**, 127–31.

Maniatis, T., Fritsch, E. F., and Sambrook, J. (1982). *Molecular cloning: a laboratory manual.* Cold Spring Harbor Laboratory.

Michaelis, M. L. and Michaelis, E. K. (1981). Ca^{2+} fluxes in resealed synaptic plasma membrane vesicles. *Life Sci.* **28**, 37–45.

Michaelis, M. L., Nunley, E. W., and Michaelis, E. K. (1987). Metal chelate affinity chromatography in the isolation of the synaptic membrane Na^{+}–Ca^{2+} exchanger. *First International Meeting on Sodium–Calcium Exchange* (Abstr. 25).

Michalak, M., Quackenbush, E. J., and Letarte, M. (1986). Inhibition of Na^{+}–Ca^{2+} exchanger activity in cardiac and skeletal muscle sarcolemmal vesicles by monoclonal antibody 44D7. *J. Biol. Chem.* **261**, 92–5.

Mishina, M., Tobimatsu, T., Imoto, K., Tanaka, K., Fugita, Y., Fukuda, K., Kurasaki, M., Takahashi, H., Morimoto, Y., Hirose, T., Inayama, S., Takahashi, T., Kuno, M., and Numa, S. (1985). Location of functional regions of acetylcholine receptor α-subunit by site-directed mutagenesis. *Nature* **313**, 364–9.

Miyamoto, H. and Racker, E. (1980). Solubilization and partial purification of the Ca^{2+}–Na^{+} antiporter from the plasma membrane of bovine heart. *J. Biol. Chem.* **255**, 2656–8.

Mullins, L. J. (1979). The generation of electric currents in cardiac fibers by Na/Ca exchange. *Am. J. Physiol.* **236**, 103–10.

Mullins, L. J. (1981). *Ion transport in heart.* Raven Press, New York.

Noda, M., Shimizu, S., Tanabe, T., Takai, T., Kayano, T., Ikeda, T., Talenhashi, H., Nakayama, M., Kanaoka, Y., Minamiro, N., Kangawa, K., Matsuo, H., Raftery, M. A., Hirose, T., Inayama, S., Mayashida, M., Miyata, T., and Numa, S. (1984). Primary structure of *Electrophorus electricus* sodium channel deduced from cDNA sequence. *Nature* **312**, 121–7.

Numa, S., Noda, M., Takahashi, H., Tanabe, T., Toyasoto, M., Furutani, Y., and Kikyotami, S. (1983). Molecular structure of the nicotinic acetylcholine receptor. *Cold Spring Harbor Symp. Quant. Biol.* **48**, 57–9.

Orlicky, J., Ruscak, M., Juhasz, O., and Zachar, J. (1987). Effects of sulfhydryl reagents on Na^{+}–Ca^{2+} exchange in rat brain microsomal membranes. *Gen. Physiol. Biophys.* **6**(2), 155–62.

Ortiz, O. E. and Sjodin, R. A. (1984). Sodium and adenosine-triphosphate-dependent calcium movements in membrane vesicles prepared from dog erythrocytes. *J. Physiol.* **354**, 287–301.

Papazian, D., Rahamimoff, H., and Goldin, S. M. (1979). Reconstitution and purification by 'transport specificity fractionation' of an ATP-dependent calcium transport component from synaptosomes-derived vesicles. *Proc. Natl. Acad. Sci. USA* **76**, 3708–12.

Parker, J. C. (1978). Sodium and calcium movements in dog red blood cells. *J. Gen. Physiol.* **71**, 1–17.

Philipson, K. D. (1984). Interaction of charged amphipiles with Na^+–Ca^{2+} exchange in cardiac sarcolemmal vesicles. *J. Biol. Chem.* **259**, 13999–4002.

Philipson, K. D. and Nishimoto, A. Y. (1981). Efflux of Ca^{2+} from sarcolemmal vesicles: Influence of external Ca^{2+} and Na^+. *J. Biol. Chem.* **256**, 3698–702.

Philipson, K. D. and Nishimoto, A. Y. (1982). Stimulation of Na^+–Ca^{2+} exchange in cardiac sarcolemmal vesicles by proteinase pretreatment. *Am. J. Physiol.* **243**, C191–5.

Philipson, K. D. and Nishimoto, A. Y. (1984). Stimulation of Na^+–Ca^{2+} exchange in cardiac sarcolemmal vesicles by phospholipase D. *J. Biol. Chem.* **259**, 16–19.

Philipson, K. D. and Ward, R. (1985). Effects of fatty acids on Na^+–Ca^{2+} exchange and Ca^{2+} permeability of cardiac sarcolemmal vesicles. *J. Biol. Chem.* **260**, 9666–71.

Philipson, K. D. and Ward, R. (1987). Modulation of Na^+–Ca^{2+} exchange and Ca^{2+} permeability by doxystearic acids. *Biochem. Biophys. Acta* **897**, 152–8.

Philipson, K. D. and Ward, R. (1988). Purification of the cardiac Na^+–Ca^{2+} exchange protein (submitted).

Philipson, K. D., McDonough, A. A., Frank, J. S., and Ward, R. (1987). Enrichment of Na^+–Ca^{2+} exchange in cardiac sarcolemmal vesicles by alkaline extraction. *Biochem. Biophys. Acta* **899**, 59–66.

Pitts, B. J. R. (1979). Stoichiometry of sodium calcium exchange in cardiac sarcolemmal vesicles. *J. Biol. Chem.* **254**, 6232–5.

Porzig, H. and Blochlinger, B. (1987). Partial purification of the Na–Ca exchange protein from bovine heart. *1st International Meeting on Na^+–Ca^{2+} Exchange* (Abstr. 8).

Rahamimoff, H. and Spanier, R. (1979). Sodium-dependent calcium uptake in membrane vesicles derived from rat brain synaptosomes. *FEBS Lett.* **104**, 111–14.

Rahamimoff, H. and Spanier, R. (1984). The asymmetric effect of lanthanides on Na^+-gradient dependent Ca^{2+} uptake in synaptic plasma membrane vesicles. *Biochem. Biophys. Acta* **773**, 279–89.

Rahamimoff, H., Barzilai, A., Erdreich, A., and Spanier, R. (1984). Molecular properties of isolated Ca^{2+} transport systems from nerve terminals. In: *Calcium, neuronal function, and transmitter release* (eds R. Rahamimoff and B. Katz), pp. 47–63. Martinus Nijhoff Press, Boston, Mass.

Reeves, J. P. and Sutko, J. L. (1979). Sodium–calcium ion exchange in cardiac membrane vesicles. *Proc. Natl. Acad. Sci. USA* **76**, 590–4.

Reeves, J. P. and Sutko, J. L. (1983). Competitive interactions of Na^+ and Ca^{2+} with the Na^+–Ca^{2+} exchange system of cardiac sarcolemmal vesicles. *J. Biol. Chem.* **258**, 3178–82.

Rengasmy, A., Soura, S., and Feinberg, H. (1987). Platelet Ca^{2+} homeostasis: Na^+–Ca^{2+} exchange in plasma membrane vesicles. *Thromb. Haemost.* **57**, 337–40.

Reuter, H. and Seitz, N. (1968). The dependence of calcium efflux from cardiac muscle on temperature and external ion composition. *J. Physiol.* **195**, 451–70.

Saermark, T. and Gratzl, M. (1986). Na^{+}–Ca^{2+} exchange in coated microvesicles. *Biochem. J.* **233**, 643–8.

Schellenberg, G. D. and Swanson, P. D. (1981). Sodium-dependent and calcium-dependent calcium transport by rat brain microsomes. *Biochem. Biophys. Acta* **648**, 13–27.

Schellenberg, G. D. and Swanson, P. D. (1982). Solubilization and reconstitution of membranes containing Na^{+}–Ca^{2+} exchange carrier from rat brain. *Biochem. Biophys. Acta* **690**, 133–44.

Schellenberg, G. D., Anderson, L., Cargoe, E. J. Jr., and Swanson, P. D. (1985). Inhibition of synaptosomal membrane Na^{+}–Ca^{2+} exchange transport by amiloride and amiloride analogues. *Mol. Pharmacol.* **27**, 537–43.

Scott, J. D., Akerman, E. O., and Nicholls, D. (1980). Calcium ion transport by intact synaptosomes. *Biochem. J.* **192**, 873–80.

Sheu, S. S., Sharma, V. K., and Uglesity, A. (1986). Na^{+}–Ca^{2+} exchange contributes to increase of cytosolic Ca^{2+} concentration during depolarization in heart muscle. *Am. J. Physiol.* **250**, 651–6.

Sigel, E., Baur, R., Porzig, H., and Reuter, H. (1988). mRNA-induced expression of the cardiac Na^{+}–Ca^{2+} exchanger in *Xenopus* oocytes. *J. Biol. Chem.* **263**, 14614–16.

Soldati, L., Longoni, S., and Carafoli, E. (1985). Solubilization and reconstitution of the Na^{+}–Ca^{2+} exchanger of cardiac sarcolemma. *J. Biol. Chem.* **260**, 13321–7.

Talor, Z. and Arruda, J. A. (1986). Na–Ca exchange in renal tubular basolateral membranes. *Min. Electrol. Metab.* **22**(4), 239–45.

Tomlins, B., Harding, S. E., Kirby, M. S., Pool-Wilson, P. A., and Williams, A. Y. (1986). Contamination of cardiac sarcolemmal preparation with endothelial plasma membrane. *Biochem. Biophys. Acta* **856**, 137–43.

Van Amsterdam, F. T. and Zaagsma, J. (1986). Modulation of ATP-dependent calcium extrusion and Na^{+}–Ca^{2+} exchange across rat cardiac sarcolemma by calcium antagonists. *Eur. J. Pharmacol.* **123**, 441–9.

Wakabayshi, S. and Goshima, K. (1981). Comparison of kinetic characteristics of Na^{+}–Ca^{2+} exchange in sarcolemma vesicles and cultured cells from chick heart. *Biochem. Biophys. Acta* **645**, 311–17.

Wakabayashi, S. and Goshima, K. (1982). Partial purification of Na^{+}–Ca^{2+} antiporter from plasma membrane of chick heart. *Biochem. Biophys. Acta* **693**, 125–33.

8 The electrogenic sodium–calcium exchange

D. A. Eisner and W. J. Lederer

8.1 Introduction

Since the discovery of the Na–Ca exchange (Baker *et al.* 1969; Reuter and Seitz 1968) there has been speculation that it may transport more than 2 Na ions in exchange for each Ca ion. Such a stoichiometry has two consequences:

1. The exchange will generate a membrane current, the activity of which will be a function of Na and Ca transport.

2. The activity of the exchange will be voltage dependent. It has been suggested that both the electrogenicity and the voltage dependence will have important effects on excitation and contraction in the heart (Mullins 1979, 1981; Noble 1986).

When we last reviewed this area (Eisner and Lederer 1985) there was very little experimental data on the physiological effects of electrogenic Na–Ca exchange. The explosion of work in this area in the last few years demands a reappraisal of the field. The work reviewed here leaves little doubt that Na–Ca exchange is voltage dependent and can generate an electric current. There is still, however, considerable uncertainty concerning the functional importance of this voltage dependence and electrogenicity.

8.2 Stoichiometry of the exchange

Although it has become generally accepted that the Na–Ca exchange transports about 3 Na ions in exchange for each Ca ion, much of the evidence for this conclusion is indirect. In this section we will review the various experimental approaches which have provided data.

8.2.1 Direct flux measurements

The simplest and most direct method one could use to determine the stoichiometry would be to measure the isotopic fluxes of Na and Ca through the Na–Ca exchange. There are, however, two problems with this approach:

1. It is difficult to identify what fraction of the normal fluxes of Na and Ca is through the Na–Ca exchange. This is largely because there is no specific inhibitor of the exchange. Therefore the Na–Ca exchange fluxes are usually defined in an operational way as being either the transmembrane fluxes of Na which are affected by the Ca gradient or the fluxes of Ca which are affected by the Na gradient. The problem with this approach is that Ca can affect transmembrane Na movements via mechanisms other than Na–Ca exchange. For example, an increase of intracellular Ca activates a non-specific cation conductance which will increase Na entry into the cell (Cannell and Lederer 1986; Colquhoun *et al.* 1981; Hill, Coranado, and Strauss 1988; Yellen 1982). A further complication is that many of the ionic manipulations used in experiments alter the membrane potential and may thereby affect ionic fluxes.

2. Another problem with isotopic measurements concerns the presence of Ca–Ca (Blaustein and Russell 1975; Slaughter, Sutko, and Reeves 1983) and Na–Na exchanges (DiPolo and Beaugé 1987). The presence of these exchanges (which do not affect the net movement of ions) will interfere with measurements of overall stoichiometry.

These problems are not unique to studies on Na–Ca exchange and can be illustrated by analogy with the Na–K pump. This system has a specific inhibitor (ouabain) and there is therefore no problem in identifying which of the fluxes of Na and K are via the Na–K pump. However, the problem of exchange diffusion (in this case K–K exchange) has confounded attempts to use isotopic fluxes to provide accurate estimates of the stoichiometry (Garrahan and Glynn 1967).

The first experiments which gave data on the stoichiometry of the Na–Ca exchange were performed on squid giant axons (Baker *et al.* 1969). The so-called 'reverse' fluxes (i.e. Na efflux for Ca influx) were measured giving results of 2–5 Na^+ exchanging with each Ca^{2+}. Similar experiments in *Myxicola* found a mean value of 2.8 Na^+ but, again, with a sufficiently large error to preclude accurate determination of the stoichiometry (Sjodin and Abercrombie 1978). Measurements in the squid axon of the 'forward' fluxes (Na influx and Ca efflux) gave results consistent with 2–3 Na^+ per Ca^{2+} (Blaustein and Russell 1975). Recent experiments on internally perfused barnacle muscle fibres (Rasgado-Flores and Blaustein 1987) found a very precise stoichiometry of $3Na^+:Ca^{2+}$.

The studies cited above have all used whole cells to measure the stoichiometry of the Na–Ca exchange. One potential problem with experiments on intact cells is that intracellular organelles may sequester (and release) an appreciable amount of Ca. Much recent work has, therefore, used plasma membrane vesicles which, in principle, should be free of organelles. Set against this, however, is the small size and high surface to volume ratio making it difficult to obtain measurements of the initial rates of fluxes. Studies using such vesicles have found ratios of about $3Na^+:Ca^{2+}$ (Pitts 1979; Wakabayashi and Goshima 1981).

The problem of contamination from Na–Na and Ca–Ca exchange can be avoided by measuring the net fluxes of Na and Ca. It is worth noting that the most precise measurements for the stoichiometry of the Na–K pump come from this sort of experiment (Post and Jolly 1957). This approach was used by Bridge and Bassingthwaighte (1983) who first inhibited the Na–K pump with ouabain to elevate $[Na^+]_i$. Subsequent increase of $[Ca^{2+}]_o$ increased $[Ca^{2+}]_i$ and decreased $[Na^+]_i$ (presumably via Na–Ca exchange). The ratio of the net loss of Na to the gain of Ca gave an apparent stoichiometry of 3. This sort of experiment does, however, suffer from the problem that changes of $[Ca^{2+}]_o$ or $[Ca^{2+}]_i$ can affect fluxes of Na by mechanisms other than Na–Ca exchange (Deitmer and Ellis 1978).

8.2.2 Measurements of stoichiometry from Na–Ca exchange currents

The existence of membrane currents produced by Na–Ca exchange (see later) implies that at least 3 Na ions exchange per Ca ion. Theoretically, one could compare the magnitude of this current with either the measured Ca or Na flux and thus calculate the stoichiometry. However, attempts to do this have not met with unqualified success (Lederer and Nelson 1983). A different approach has been employed on experiments on photoreceptors (Yau and Nakatani 1984). As mentioned later (Section 8.3.2 and Fig. 8.3) the rod outer segment can be loaded with Ca via a light-sensitive channel. The amount of Ca entering can be estimated from the magnitude of the current. When the Ca is subsequently pumped out of the cell $(n-2)$ charges will enter the cell per Ca expelled. From a comparison of the charges moved a stoichiometry of 3Na:Ca was obtained.

8.2.3 Thermodynamic measurements of stoichiometry

The problems outlined above in trying to obtain direct measurements of the stoichiometry of Na–Ca exchange have led to the adoption of a thermodynamic approach. This is best illustrated by considering Na–Ca exchange as a reversible reaction.

$$n\ Na_o^+ + Ca_i^{2+} \rightleftharpoons n\ Na_i^+ + Ca_o^{2+} \qquad (1)$$

It can be shown (Blaustein and Hodgkin 1969) that this reaction will produce net Ca efflux if:

$$[Ca^{2+}]_i/[Ca^{2+}]_o > ([Na^+]_i/[Na^+]_o)^n \exp[(n-2)VF/RT] \quad (2)$$

or

$$n > 2(E_{Ca} - V)/(E_{Na} - V) \quad (2a)$$

Similarly, for there to be a net Ca influx on the exchange then, from equation 2:

$$n < 2(E_{Ca} - V)/(E_{Na} - V) \quad (2b)$$

Finally, for there to be no net Ca flux (i.e. the reaction is at equilibrium):

$$n = 2(E_{Ca} - V)/(E_{Na} - V) \quad (2c)$$

If one has independent information as to the direction of the *net* flux through the exchange and can control or measure the intracellular and extracellular concentrations of Na and Ca as well as the membrane potential, then the appropriate version of equation 2 can be used to provide one of the following estimates of n: a minimum value (equation 2a); a maximum value (equation 2b); or the exact value (equation 2c). We will consider these cases in turn.

8.2.3.1 *Minimum value for* n

It has been appreciated since the original work of Blaustein and Hodgkin (1969) that, with physiological ionic conditions, ($[Ca^{2+}]_i$, 100 nM; $[Na^+]_i$, 10 nM; $[Ca^{2+}]_o$, 1 mM; $[Na^+]_o$, 140 mM; V, −80 mV) the value of n must be greater than 2 for the exchange to produce a net Ca efflux. The greatest uncertainty in using this approach is the question of whether the Na–Ca exchange does, indeed, produce net Ca efflux. There is evidence in many cell types of an 'uncoupled' Ca pump using the energy of ATP hydrolysis directly to extrude Ca ions (Schatzmann 1966; Caroni and Carafoli 1980). In many cells including the squid axon (DiPolo and Beaugé 1979) and cardiac muscle (Philipson and Ward 1986) at rest the rate of the uncoupled Ca pump may be comparable to that of the Na–Ca exchange. Furthermore, at least under experimental conditions, mitochondria and other organelles may accumulate considerable amounts of Ca. Given the existence of a significant Ca–ATPase and intracellular sequestering mechanisms (e.g. Carafoli 1985), there is no need to assume that the Na–Ca exchange does produce net Ca efflux.

8.2.3.2 *Maximum value for* n

When the sodium pump is completely inhibited, $[Na^+]_i$ rises to a concentration of about 35 mM (Deitmer and Ellis 1978). This is well below thermodynamic equilibrium. Deitmer and Ellis (1978) presented evidence suggesting that this low value resulted because the Na–Ca exchange extruded *Na* in exchange for net *Ca* influx. Under such conditions Sheu and Fozzard (1982) have measured $[Na^+]_i$ and $[Ca^{2+}]_i$ to be 33 mM and 1.5 μM, respectively, with a membrane potential of −69 mV. $[Na^+]_o$ and $[Ca^{2+}]_o$ were, respectively, 151 mM and 1.8 mM. This result suggests that $n <= 3$. One important consequence of this experiment should be noted. If the Na–Ca exchange is extruding Na from the cell then the consequent Ca influx must be accommodated either by pumping Ca out of the cell on, say, a Ca–ATPase or by sequestering it inside intracellular organelles. There must therefore be mechanisms other than Na–Ca exchange for removing Ca from the cytoplasm.

8.2.3.3 *Determination of the exact value of* n

Much work using this method has assumed that the Na–Ca exchange is at equilibrium. This may be approximately true if the maximum rate of the exchange is much greater than that of other systems for Na and Ca transport. In this case only a small disturbance from equilibrium will be required for the exchange to produce significant net fluxes. Sheu and Fozzard (1982) and Bers and Ellis (1982) measured $[Na^+]_i$, $[Ca^{2+}]_i$, and E_m in mammalian cardiac muscle under a variety of conditions. On the whole the data could be fit by assuming a value of n of around 3. Related analyses have been performed by Ledvora and Hegyvary (1983), Chapman (1986), and Axelsen and Bridge (1985).

A somewhat different approach was used by Eisner, Lederer, and Vaughan-Jones (1983) who used tonic tension in sheep Purkinje fibres as a measure of $[Ca^{2+}]_i$. This is the component of tension which is maintained throughout a depolarizing pulse lasting several seconds (as opposed to the rapid twitch). There is evidence that the Ca ions responsible for this component of tension enter the cell on Na–Ca exchange (Chapman and Tunstall 1980, 1981; Eisner, Lederer, and Vaughan-Jones 1983, 1984). This tonic tension is increased by either increasing the magnitude of the depolarization or by elevating intracellular Na (Fig. 8.1). Specifically, the effects of increasing $[Na^+]_i$ can be cancelled out by decreasing the magnitude of the depolarization. From the quantitative relationship between the effects of $[Na^+]_i$ and membrane potential it is possible to calculate n. A value of about 3 was found.

A similar, but more direct, approach was used by Reeves and Hale (1984). They used cardiac sarcolemmal vesicles in which there was initially no gradient of Na or Ca (Fig. 8.2). The imposition of a membrane potential (using valinomycin) resulted in a net Ca flux. This net flux could be removed by imposing a Na gradient. Again, the magnitude of the Na gradient required to compensate for a given change of membrane potential gives a value of 3 for n. The elegance of this experiment lies in the fact that conditions are found in which there is no net flux of Ca and therefore it is certain that the exchange is at equilibrium. A similar approach has been used by Talor and Arruda (1985) in renal epithelium.

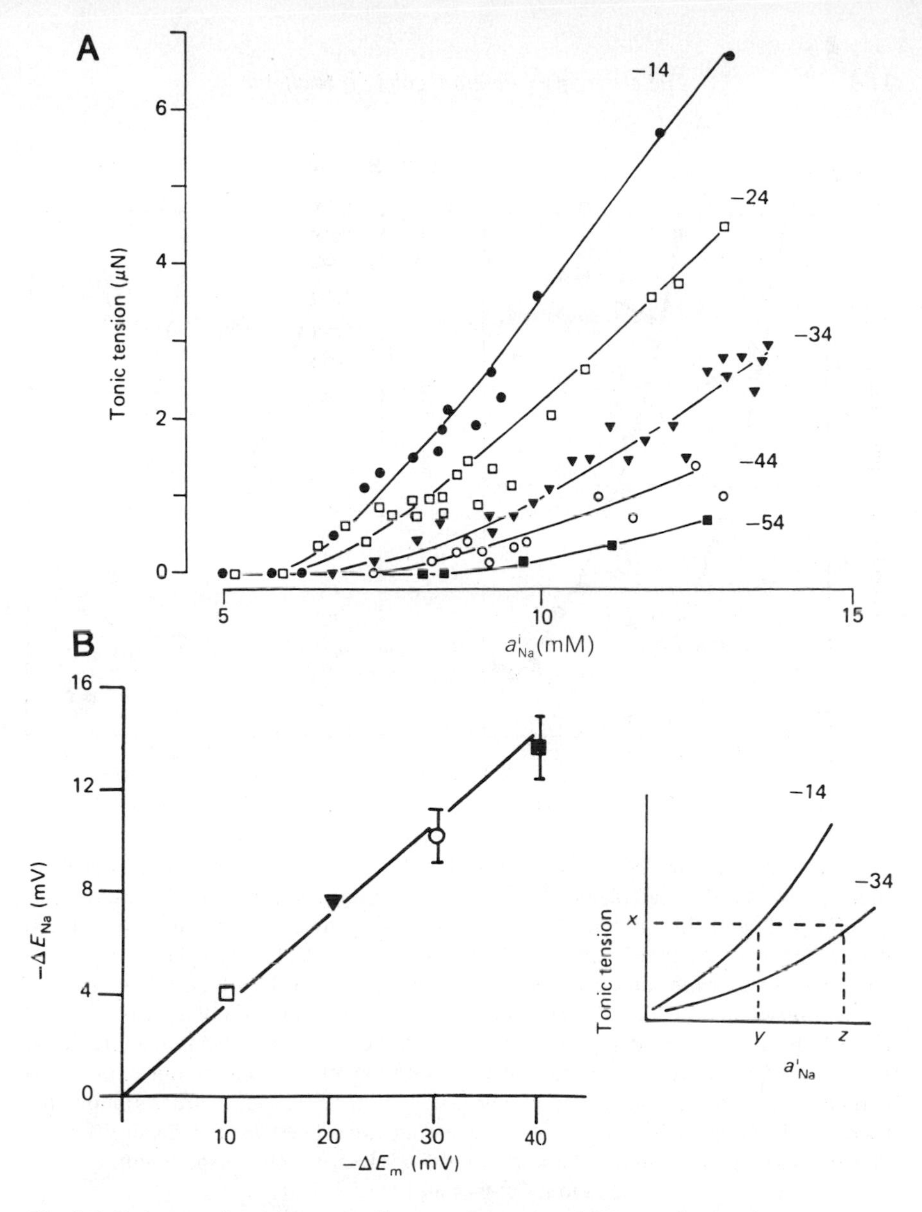

Fig. 8.1. Estimate of stoichiometry from tonic tension. (A) Dependence of tonic tension on membrane potential and intracellular Na activity (a^i_{Na}). Tonic tension was measured at the end of 2 s depolarizing pulses to the potentials indicated. a^i_{Na} was varied by inhibiting the Na^+–K^+ pump by removing external K. (B) Comparison of the effects of membrane potential (E_m) and E_{Na} on tonic tension. As shown in the inset all data are referred to − 14 mV. A hyperpolarization to another voltage would decrease tension but this can be compensated for by elevating a^i_{Na}. Here a ΔE_{Na} of RT/F in y/z would be equivalent to a ΔE_m of 20 mV. This comparison has been made with respect to − 14 mV for each of the other potentials of A. at various levels of a^i_{Na}. The symbols are as in A. The slope of the line corresponds to a stoichiometry of 3.3 Na:Ca. See Eisner, Lederer, and Vaughan-Jones (1983) for further details.

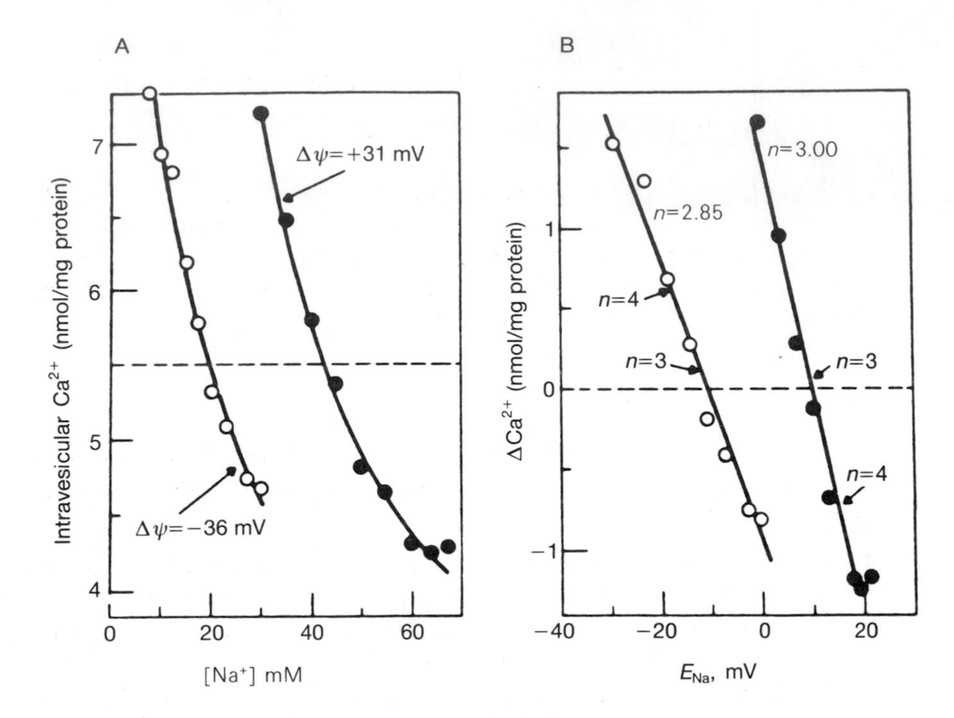

Fig. 8.2. Determination of the stoichiometry of Na–Ca exchange in cardiac sarcolemmal vesicles. Cardiac sarcolemmal vesicles containing 0.1 mM $^{45}CaCl_2$, 30 mM NaCl, 20 mM KCl, and 110 mM LiCl treated with valinomycin were put into solution containing 0.1 mM $^{45}CaCl_2$ and the [NaC1] indicated. $[K^+]_o$ was adjusted to give a calculated membrane potential of either − 36 mV (○) or + 31 mV (●). (A) Measurement of vesicular Ca content as a function of $[Na^+]_o$. The points show the Ca content of the vesicles 2 sec. after adding them to the incubation solution. The dotted line shows the Ca content of vesicles suspended in a solution of identical composition to its contents. (B) Change of Ca content as a function of E_{Na}. The arrows show the theoretical values of E_{Na} at which there would have been no net Ca flux for the stoichiometry indicated. It can be seen that $n = 3$ is nearest to the experimental values. Taken from Reeves and Hale (1984).

Recent measurements of the reversal potential of the electrogenic Na–Ca exchange current (see Section 8.3) can also be used to estimate *n*. Kimura, Miyamae, and Noma (1987) could control both intracellular and extracellular Na and Ca concentrations in guinea-pig ventricular myocytes. The use of Ni as an inhibitor of the exchange allowed the reversal potential of the current to be measured (see Fig. 8.5). It could be shown that the results were consistent with a $3Na^+$–$1Ca^{2+}$ exchange.

8.3 The electrogenic current produced by the Na–Ca exchange

Recent experiments have attempted to measure a Na–Ca exchange current. There are at least two reasons for this. First, such a current may have important consequences for the electrophysiology of the cell (cf. Mullins 1979). Secondly, the current provides a convenient means to measure the activity of the Na–Ca exchange in the intact cell.

8.3.1 Experimental difficulties

In a previous article (Eisner and Lederer 1985) we have pointed out many of the problems involved in trying to find a current which can be related *directly* to Na–Ca exchange:

1. The most obvious problem is the lack of a specific inhibitor of the exchange. If such an inhibitor existed one could obtain a current–voltage relation *immediately* after adding the inhibitor (before any changes in ionic concentrations developed) and compare it with the control. The difference current could then be ascribed to Na–Ca exchange. Various inhibitors have been reported for Na–Ca exchange. However, it has been reported that these also inhibit other membrane currents (Bielefeld *et al.* 1986). A partial resolution of this problem is provided by blocking other membrane currents with inhibitors which do not block Na–Ca exchange. Under such conditions all the electrophysiological effects of non-specific inhibitors of Na–Ca exchange will be on Na–Ca exchange. This technique has been used by Kimura, Miyamae, and Noma (1987).

2. Given the lack of a specific inhibitor many studies have attempted to change the rate of Na–Ca exchange and look for changes in membrane current. The disadvantage of this method is that changes of Na–Ca exchange activity will change $[Ca^{2+}]_i$ and $[Na^+]_i$, and thereby affect membrane currents. There is therefore a problem in establishing which of the changes of membrane current result *directly* from any electrogenic current produced by Na–Ca exchange and which are secondary to secondary changes of ionic concentration.

Examples of these problems are provided by considering some of the earlier attempts to look for current changes produced by manipulating the activity of Na–Ca exchange. A frequently used method is to remove external Na. In cardiac muscle this produces a transient increase of contraction and $[Ca^{2+}]_i$ due to Ca entry on Na–Ca exchange (Lüttgau and Niedergerke 1958). The entry of Ca is accompanied by efflux of Na on the exchange and therefore by a fall of $[Na^+]_i$. This fall of $[Na^+]_i$ will, in turn, decrease Ca influx on the exchange. One therefore expects that Na removal should produce a transient increase of outward current due to Na–Ca exchange. Consistent with this, Coraboeuf *et al.* (1981) found that Na removal produced a transient hyperpolarization. Work on cardiac sarcolemmal vesicles has found that changes of Na–Ca exchange activity produce changes of membrane potential. These were measured using lipophillic cations (Caroni, Reinlib, and Carafoli 1980; Reeves and Sutko 1980). However, a subsequent study (Lederer *et al.* 1984) found that Na removal produced large changes of membrane conductance thus making it difficult to ascribe the changes of membrane potential to Na–Ca exchange. Another study has found that the current produced by Na removal is blocked by 20 mM Cs (Mead and Clusin 1984), an observation which is also not consistent with it resulting from Na–Ca exchange.

Various sophisticated approaches have been adopted to circumvent the problems described above. For convenience we have grouped them into three classes. One can:

1. Find a preparation in which there are no other ionic currents or electrogenic transporters which interfere. This method is limited in its applicability but has been very successful in studies on rod photoreceptors (e.g. Yau and Nakatani 1984).

2. Inhibit interfering currents and manipulate ionic conditions by perfusing the contents of the cell in such a way that the changes of current must be due to Na–Ca exchange current (e.g. Kimura, Noma, and Irisawa 1986). This method is ideally suited to studying the basic properties of the exchange, but the need to block other currents and manipulate the contents of the cell makes it difficult to study the exchange under physiological conditions.

3. Show that the properties of any putative Na–Ca exchange current cannot be accounted for by known ionic currents (e.g. Clusin, Fischmeister, and DeHaan 1983; Fedida *et al.* 1987a,b; Hume and Uehara 1986a,b; Mechmann and Pott 1986). This method can be used under physiological conditions.

8.3.2 Photoreceptors

The outer segment of the vertebrate rod appears to be an ideal tissue in which to study Na–Ca exchange. (This is discussed at length in Chapter 12.) For the present purpose it is enough to mention that the outer segment contains a channel which is permeable to Na and Ca ions, but is closed by light, and a Na–Ca exchange. Thus the rod can be loaded with Ca in a Na-free, Ca-containing solution, in the dark. The method is illustrated in Fig. 8.3. When the cell is illuminated the channels close. Subsequent addition of external Na

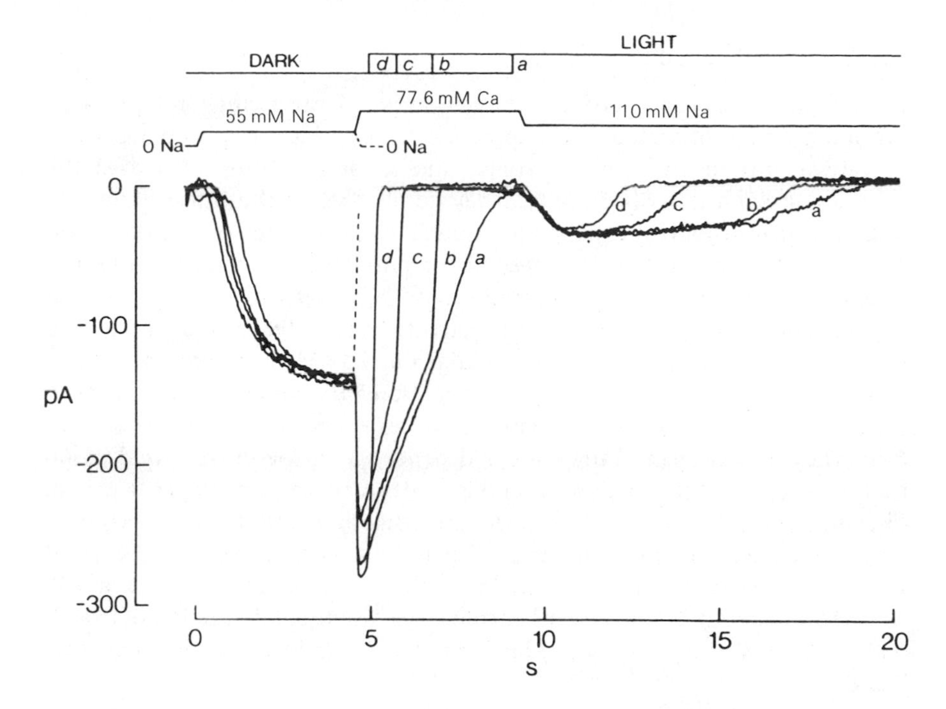

Fig. 8.3. Electrogenic Na–Ca exchange current in rod photoreceptor. The record shows the current recorded from a rod. The rod was initially in the dark. After exposure to a solution containing 55 mM Na, the rod was exposed to 77.6 mM Ca to load with Ca. The light was then switched on at a variable time indicated above to terminate the loading. Finally, external Na was replaced resulting in the development of an inward Na–Ca exchange current. Taken from Yau and Nakatani (1984).

produces an inward current which then decays. Consistent with it resulting from Na–Ca exchange, this current was inhibited by La. The charge moved increased with the degree of Ca loading in the dark and a comparison of the current flowing in the dark with that produced after adding back Na was consistent with a 3Na:1Ca stoichiometry (Yau and Nakatani 1984).

This work on the rod has been extended by Hodgkin, McNaughton, and Nunn (1987) and Hodgkin and Nunn (1987) who used the integral under the inward Na–Ca exchange current as a measure of the quantity of Ca pumped out of the cell. From this they could calculate the amount of Ca pumped out at any time and thence the amount of Ca remaining in the rod. They found that the pump current was a saturating function of estimated $[Ca^{2+}]_i$. Increasing $[Ca^{2+}]_o$ or decreasing $[Na^+]_o$ decreased the magnitude of the exchange current and prolonged its decay.

8.3.3 Dialysed cardiac cells

In the majority of preparations, the large number of interfering ionic channels precludes the use of such a simple approach. This problem has been addressed in cardiac muscle by Kimura, Noma, and Irisawa (1986) who used the second approach described above. They could control the intracellular Na and Ca concentrations in guinea-pig ventricular myocytes (using the intracellular dialysis technique developed by Soejima and Noma 1984) and also inhibited essentially all ionic currents with a variety of pharmacological interventions. Extracellular calcium activated an outward current *only* if intracellular Na (30 mM) and intracellular Ca (73 nM) were present. This is consistent with the current resulting from Na efflux coupled to Ca entry on Na–Ca exchange as this has a requirement for intracellular Ca (see Section 8.5). They also measured the voltage dependence of the current and found that it increased with depolarization (Fig. 8.4). A subsequent paper (Kimura, Miyamae, and Noma 1987) studied the inward current which required extracellular Na and intracellular Ca. The outward Na–Ca exchange current was a saturating function of $[Ca^{2+}]_o$ with a $K_{0.5}$ of 1.2 mM and a Hill coefficient of about 1. The inward current was a sigmoidal function of $[Na^+]_o$ (Hill coefficient 3; $K_{0.5}$ 70 mM). The current was inhibited by Ni^{2+} ions (see Fig. 8.5).

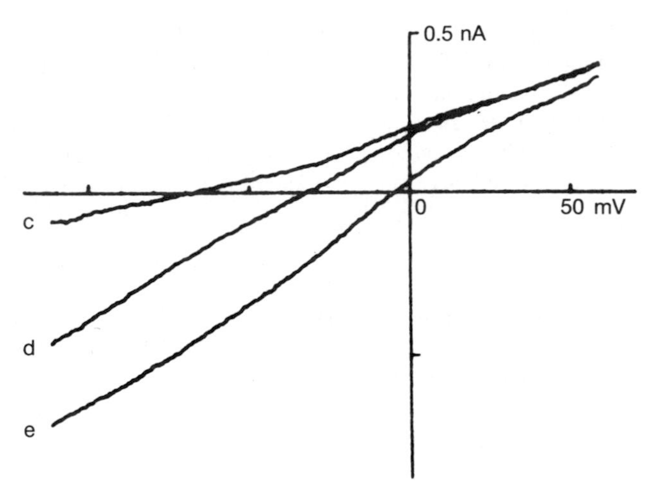

Fig. 8.4. Electrogenic Na–Ca exchange current in perfused cardiac cells. The cell was Na-free and contained 430 nM $[Ca^{2+}]$ (pipette concentration). $[Ca^{2+}]_o$ was 1 mM. Record c shows a current–voltage relationship in the absence of external Na (replaced by Li). Records d and e were obtained after adding back 140 mM Na^+. Record d was before the current had activated completely. The Na–Ca exchange current therefore represents the difference between e and c. Taken from Kimura, Miyamae, and Noma (1987).

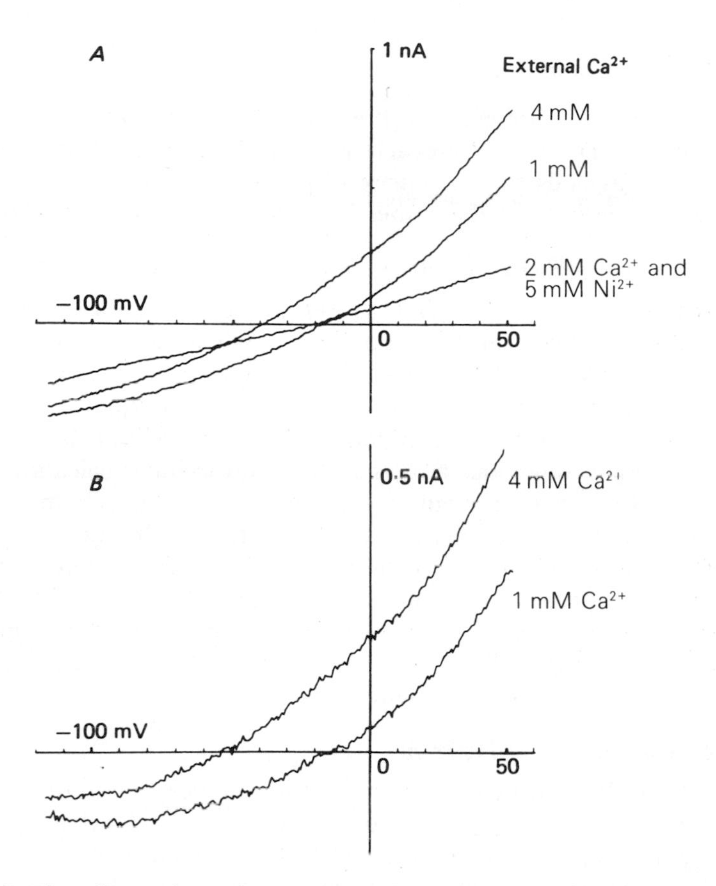

Fig. 8.5. The effects of membrane potential on the Ni-sensitive Na–Ca exchange current. (A) I–V relationships obtained with 140 mM Na_o, 10 mM $[Na^+]_i$, 172 mM $[Ca^{2+}]_i$ with the $[Ca^{2+}]_o$ and $[Ni^{2+}]_o$ indicated. (B) Ni-sensitive difference currents. Note that elevating $[Ca^{2+}]_o$ shifts the reversal potential for the Na-sensitive current to more negative potentials (from −15 to −53 mV). These values are very similar to those expected for a $3Na^+$: $1 Ca^{2+}$ exchange (−20 and −57 mV). Taken from Kimura, Miyamae, and Noma (1987).

As mentioned above, this work on dialysed cardiac cells has the great advantage that it provides very strong evidence that the current changes observed are due to Na–Ca exchange. However, both the presence of many of the substances used to block other ionic currents (Ba, Cs, D-600, TEA, EGTA, aspartate) and the absence of potassium ions may affect the Na–Ca exchange. This, and the fact that the intracellular ionic concentrations were controlled by dialysis, makes it difficult to relate the results directly to the case of the intact cell.

8.3.4 Intact cardiac cells

The final group of experiments which we will consider use the third strategy outlined above. The general approach used is based on the fact that the Na–Ca exchange responds to an increase of $[Ca^{2+}]_i$ by increasing the rate of extrusion of Ca from the cell and thereby increasing net inward current.

8.3.4.1 *Caffeine-induced current*

One way to increase $[Ca^{2+}]_i$ is to apply caffeine to release Ca ions from the sarcoplasmic reticulum. This technique has the advantage that, because the rise of $[Ca^{2+}]_i$ is brief, it is unlikely that the concentrations of other ions change. Clusin (1983) and Clusin, Fischmeister, and DeHaan (1983) examined the effects of caffeine in embryonic chick heart cells. This manœuvre (which is thought to produce a phasic rise in intracellular calcium) gave a phasic increase of inward current over the range of potentials from -110 to $+60$ mV. All known ionic channels apart from the Ca channel would have reversal potentials within this range of potentials and therefore could not produce a current which was always inward. D-600, which blocks the Ca current, had no effect on the caffeine-induced current thus eliminating it as the cause of the current. Therefore it was suggested that the caffeine-induced current resulted from Na–Ca exchange.

8.3.4.2 *Transient inward current*

The results on the caffeine-induced current have a bearing on those on the transient inward current (I_{TI}). This current is responsible for the depolarizations which occur afterwards and which are the cause of many of the arrhythmias produced by digitalis intoxication (Lederer and Tsien 1976). The current is accompanied by a transient contraction and is thought to be a Ca-activated current (Kass *et al.* 1978a). Kass *et al.* (1978a,b) suggested that the current was produced by a Ca-activated non-specific conductance and this conclusion was supported by the direct demonstration of such channels in isolated membrane patches (Colquhoun *et al.* 1981). However, some (but not all; cf. Cannell and Lederer 1986) later studies failed to detect any reversal potential for this current and therefore suggested that it might result from the effects of the change of $[Ca^{2+}]_i$ on Na–Ca exchange (Arlock and Katzung 1985; Karagueuzian and Katzung 1982; Noble 1984). This point was examined in more detail in single cells (Fedida *et al.* 1987a). Depolarization decreased the size of the current but no reversal was seen. This, again, is more consistent with a Na–Ca exchange rather than a Ca-activated conductance.

Related experiments have been performed by Pott and collaborators on guinea-pig atrial myoballs. These display currents produced by spontaneous Ca release from the sarcoplasmic reticulum. Mechmann and Pott (1986) showed that the properties of this current were consistent with it being produced by Na–Ca exchange. Subsequent work (Lipp and Pott 1988) investigated the voltage dependence of this current. Fig. 8.6A shows two spontaneous transient inward currents. Depolarization of the holding potential decreases the magnitude of the current change and slows and renders incomplete the relaxation. Both these effects of depolarization are consistent with depolarization decreasing Ca efflux on a voltage-dependent Na–Ca exchange. Fig. 8.6B shows the voltage dependence of the current both before

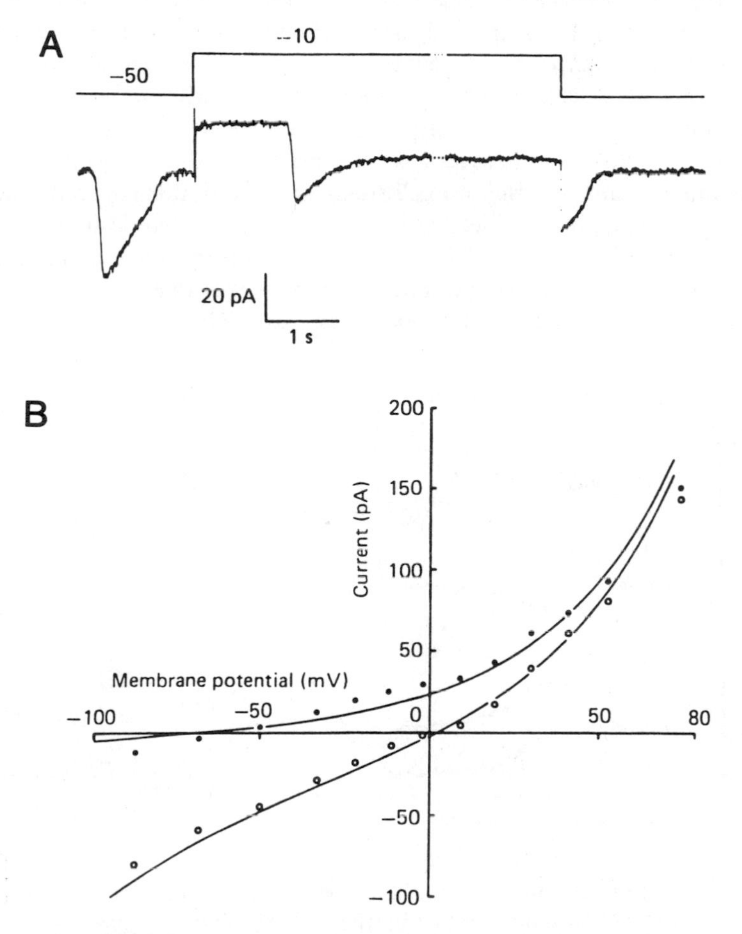

Fig. 8.6. The effects of membrane potential on the transient inward current. (A) Original records. Spontaneous inward currents are shown at -50 and -10 mV. Note (i) the smaller magnitude of the inward current and (ii) the slower and incomplete relaxation of the current at the less negative potential. (B) Voltage-dependence of membrane current before (●) and at the peak of (○) the spontaneous inward currents. The curves were fitted by the authors to a theoretical expression for Na–Ca exchange. Taken from Lipp and Pott (1988).

the spontaneous release and after it. The authors could model these curves in terms of an Na–Ca exchange.

8.3.4.3 *Creep current*

Other studies have examined the effects on membrane current of a rise of $[Ca^{2+}]_i$ produced by depolarization. Hume and Uehara (1986a,b) have studied the so-called 'creep' currents in frog atrial myocytes. These currents were originally described by Eisner and Lederer (1979) in Purkinje fibres in which the Na^+–K^+ pump had been inhibited. Under these conditions depolarization produces an increase of tonic tension which develops over a second or so. There is an inward 'creep' of current with a similar timecourse. On repolarization both current and tension recover (Fig. 8.7A). This correlation between tension and current is consistent with the creep current being Ca activated. Hume and Uehara (1986a,b) showed that these creep currents were not inhibited by a variety of manoeuvres which would be expected to block Na, Ca, or K currents. The creep current was inhibited by La and was sensitive to external Na and Ca. Furthermore, it was markedly increased by elevating $[Na^+]_i$ with the ionophore monensin (Fig. 8.7B). Finally, these

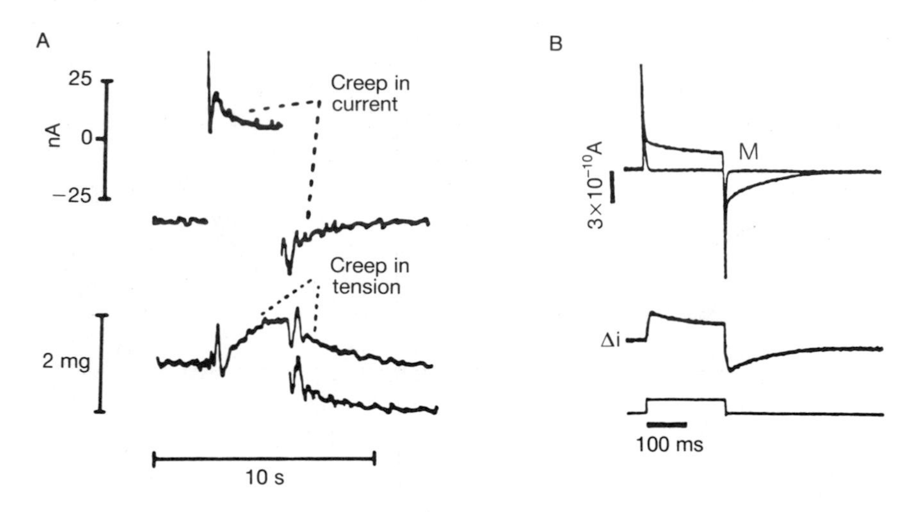

Fig. 8.7. The creep current. (A) Correlation between creep current and contraction. The record was obtained after inhibiting the Na^+–K^+ pump by exposure to a K-free solution. The two upper traces show current and tension in response to a depolarizing pulse. The bottom trace shows the result of inverting the latter part of the current record and delaying it by 100 ms. This has a very similar timecourse to contraction. From Eisner and Lederer (1979). (B) Creep current induced by Na-loading with monensin. The two upper traces show current in response to depolarizing pulse in the absence and presence (M) of monensin. The bottom trace shows the monensin-induced current. Taken from Hume and Uehara (1986a).

authors addressed the question of whether the creep current is produced by Na–Ca exchange or, alternatively, by a Ca-activated conductance. Increasing $[Ca^{2+}]_i$ did not appear to increase the apparent conductance but rather shifted the current–voltage relationship. This is more easily accommodated on the basis of an Na–Ca exchange current rather than a Ca-activated conductance.

Barcenas-Ruiz, Beuckelmann, and Wier (1987) have studied the creep current while simultaneously measuring $[Ca^{2+}]_i$ with the fluorescent indicator fura-2. They confirmed that the creep current was, indeed, accompanied by the expected changes of $[Ca^{2+}]_i$. Their experiments were performed with the Ca current and sarcoplasmic reticulum Ca release inhibited (with verapamil and ryanodine, respectively). The current was a linear function of $[Ca^{2+}]_i$ over the range 0.2 to 1 μM. Interestingly in these experiments the creep current was observed in the absence of deliberate Na-loading.

Elegant as they are, the experiments on the creep current involve somewhat unphysiological conditions and provide no direct information about the contribution of the Na–Ca exchange to membrane current under normal conditions. This point has been addressed by two groups using frog atrium (Hume 1987) and guinea-pig ventricle (Fedida *et al.* 1987a). Both groups found a slowly decaying inward current tail on repolarization. In the guinea-pig the slow inward tail was affected in a similar way to contraction by either changing stimulus rate or by the application of isoproterenol. In the frog atrial cells the tail current was most obvious after the Ca current was enhanced either by isoproterenol or the Ca channel agonist Bay K 8644, and was abolished by removal of extracellular Na (Hume 1987). In addition to the Na-sensitive tail current there was an Na-sensitive current during the depolarizing pulse. The Na-sensitive current consisted of two phases. First, an initial outward current possibly due to the fact that depolarization increases Ca entry–Na efflux and therefore outward current on the exchange. Second, a transient inward current which could result from a transient increase of $[Ca^{2+}]_i$ due to Ca entry through Ca channels. This transient increase of $[Ca^{2+}]_i$ would produce a transient increase of inward current (or decrease of outward current) on the exchange.

8.4 Voltage dependence of the Na–Ca exchange

8.4.1 Theoretical predictions

In Section 8.2 we reviewed the evidence showing that the Na–Ca exchange transports $3Na^+$ ions per Ca^{2+}. The overall reaction therefore involves charge transfer. It is therefore tempting to assume that depolarization will favour Na efflux and Ca entry while hyperpolarization will favour Ca efflux and Na entry. However, as pointed out previously (Eisner and Lederer 1985), the conclusion is over-simplified. The effects of voltage depend not only on the overall reaction but also on the voltage sensitivity of the individual steps. This is illustrated by the scheme of Fig. 8.8 which is a consecutive model in which Na binds at one surface and is transported before Ca binds at the other surface. We assume that the ions are carried across the membrane by a carrier which, when unloaded, has a charge. Two of the reactions involve

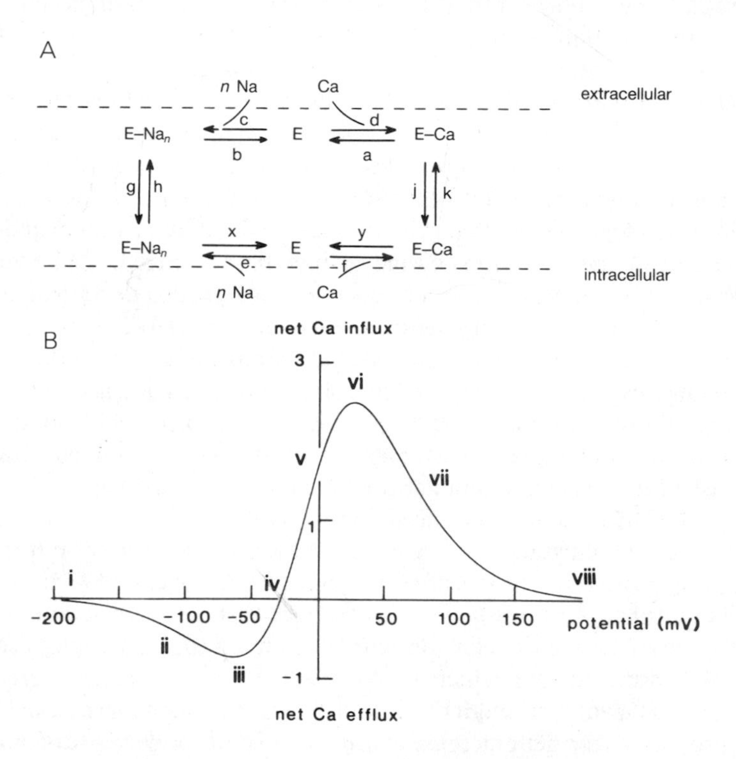

Fig. 8.8. Theoretical model for the effects of membrane potential on the Na–Ca exchange. (A) Model. The scheme is for a sequential reaction scheme in which the carrier can bind either 1 Ca or *n* Na ions. It can only cross the membrane in the bound form. (B) Predictions of the model for the case where $n=3$. We have arbitrarily assumed that the unloaded carrier has one negative charge. The Na^+-loaded form will therefore have two net positive charges and the Ca^{2+}-loaded form one net positive charge. Note in particular that there is a voltage (iv) where the reaction is in equilibrium (E_{NaCa}). Further depolarization from this point at first produced a gradually increasing net Ca influx. This is because the effect of depolarization to increase the Na entry step (here 2 net charges transported) is greater than that to decrease the Ca entry step (here 1 net charge transported). However, with further depolarization the slowing of the Ca entry step dominates and slows the overall reaction to zero. Taken from Eisner and Lederer (1985).

charge translocation. If the unloaded carrier has two negative charges then the Ca translocation step will not involve charge transfer and therefore may not be voltage dependent. Similarly, if the unloaded carrier has three charges then the Na translocation step will not be voltage dependent. If the unloaded carrier has any other charge, however, both translocation steps will be voltage dependent. Figure 8.8B shows a model in which the unloaded carrier has one negative charge. Both translocation reactions involve transfer of positive charge. Therefore, although depolarization will accelerate the Na efflux step, it will inhibit the Ca entry step and, with extreme depolarization, the flux will go to zero as the carrier is 'trapped' at the outside of the membrane. Similarly, extreme hyperpolarization will stimulate the Na entry step but will inhibit the Ca efflux step thus trapping the carrier inside the membrane and reducing the flux to zero.

Thus, in general, if more than one reaction in the cycle is voltage dependent, one would expect the overall voltage dependence of the exchange to be non-monotonic. It will be evident below that this has not been seen. The simplest inference from this is that the Na–Ca exchange reaction involves only one voltage-dependent reaction.

8.4.2 The effects of membrane potential on isotopic fluxes

A convenient means to change the membrane potential is by varying external K concentration. This is, however, not always appropriate for studies on the Na–Ca exchange since there is evidence of modulatory roles of external cations and in particular of external K (Baker *et al.* 1969). This is emphasized by the fact that the effects of external K on Ca_o-dependent Na efflux are greater than those produced by an equivalent depolarization produced by passing current (Allen and Baker 1986a,b; Sjodin and Abercrombie 1978). With this caution in mind we will consider the effects of membrane potential on isotopic fluxes.

Na_o-dependent Ca efflux has been found to be inhibited by depolarization (Bartschat and Lindenmayer 1980; Bers, Philipson, and Nishimoto 1980; Blaustein, Russell, and deWeer 1974; Caroni, Reinlib, and Carafoli 1980; DiPolo *et al.* 1985; Kadoma *et al.* 1982; Mullins and Brinley 1975; Philipson and Nishimoto 1980; Reeves and Hale 1984). The exact sensitivity to membrane potential does, however, vary between experiments. For example, Mullins and Brinley (1975) found that the efflux was increased *e*-fold by a 25 mV hyperpolarization whereas Allen and Baker (1986b) found that a 50 mV change was required. Mullins and Brinley also found that the voltage-dependence was decreased by elevating $[Ca^{2+}]_i$ and this effect might explain some of the discrepancy. This point was also examined more recently by DiPolo *et al.* (1985). They found that changes of $[Ca^{2+}]_i$ (from 0.1 to 230 μM) had little effect on the voltage-sensitivity of the Na_o-dependent Ca efflux.

Similarly, there was little effect of $[Ca^{2+}]_o$ or ATP on the voltage-sensitivity. They did, however, find striking effects of $[Na^+]_i$. In the absence of $[Na^+]_i$ the Ca efflux was scarcely affected by membrane potential (7.4 per cent decrease for 25 mV depolarization). In contrast when $[Na^+]_i$ was elevated to between 20 and 55 mM the same change of membrane potential increased the flux by 32.3 per cent. The apparent K_m for internal Na for this effect was 5 mM. Internal Na inhibits Na_o-dependent Ca efflux, although, for reasons which are not entirely apparent, this effect was largely transient (their Figs. 8.4 and 8.5). It would therefore be tempting to suggest that membrane potential is overcoming the inhibitory effects of internal Na. Such a result might be produced by either a real voltage-dependence of the Na-binding step or the voltage-dependent reaction being more rate-limiting in the presence of elevated $[Na^+]_i$.

8.4.3 The effects of membrane potential on Na–Ca exchange currents

The Na–Ca exchange current appears to be affected by membrane potential (Hume and Uehara 1986a,b; Kimura, Miyamae, and Noma 1987; Kimura, Noma, and Irisawa 1986; Lipp and Pott 1988). Theoretical predictions (Mullins 1981; Noble 1986) suggest that inward current through the exchange (Na influx–Ca efflux) should be increased by hyperpolarization in an approximately exponential manner. Similarly, the outward current (Na efflux–Ca influx) should be increased by depolarization exponentially. These predictions are fulfilled to some extent. Thus Fig. 8.4 (Kimura, Miyamae, and Noma 1987) shows that the inward current increases steeply with hyperpolarization. In Fig. 8.5 the outward current increases approximately exponentially with depolarization. However, in this same figure the inward current seems to reach a limiting value at negative potentials. This can be explained if (as is likely) steps in the reaction other than the voltage-dependent translocations can limit the rate of the reaction.

8.4.4 Voltage dependence of $[Ca^{2+}]_i$

In many preparations depolarization increases intracellular calcium. Many factors are, of course, responsible for this. In muscle one must consider Ca entry through Ca channels and Ca release from the sarcoplasmic reticulum. In addition there is evidence that part of the voltage dependence may, at least under some circumstances, result from Na–Ca exchange. Here we will consider briefly evidence from two intensively studied tissues: squid axon and cardiac muscle.

Early work showed an increase of ^{45}Ca influx in squid axons during stimulation (Hodgkin and Keynes 1957). Direct measurements of $[Ca^{2+}]_i$

(using the photoprotein aequorin) demonstrated that this was partly through Na channels (TTX-sensitive). The TTX-resistant component was more significant for pulses longer than a few milliseconds and was therefore termed the 'late' Ca entry or channel (Baker, Hodgkin, and Ridgway 1971). Subsequent work showed that prolonged depolarization (either with KCl or by passing current) produced a TTX-insensitive increase of $[Ca^{2+}]_i$ which peaked and then declined to a steady level over a few minutes. This was inhibited by Mn^{2+} and was suggested to represent Ca entry through an inactivating channel (Baker, Meves, and Ridgway 1973a,b). More recent work, however, has questioned this conclusion. Mullins and Requena (1981) showed that the response to KCl depolarization was abolished if the axon had a low $[Na^+]_i$. This suggested that the effects of depolarization were mediated via Na–Ca exchange. Furthermore, if the axon was depolarized in the absence of external Ca there was no rise of $[Ca^{2+}]_i$. However, the re-addition of external Ca produced a transient increase of $[Ca^{2+}]_i$ suggesting that any inactivation depended on the Ca entry rather than the depolarization. Mullins, Requena, and Whittembury (1985) extended this result using voltage-clamp depolarizing pulses. They further suggested that the difference with the work of Baker, Meves, and Ridgway (1973a,b) could be explained by the unphysiologically high $[Ca^{2+}]_o$ used in the earlier work. Apparently at normal $[Ca^{2+}]_o$ most of the Ca entry produced by depolarization is via Na–Ca exchange.

In cardiac muscle, depolarization produces an initial twitch which is accompanied by an increase of Ca due to Ca entry via the Ca current and Ca release from the sarcoplasmic reticulum. However, in amphibian cardiac muscle there is also a maintained increase of tension. This tonic tension has been suggested to result from Na–Ca exchange (Chapman and Tunstall 1981; Horackova and Vassort 1979). Consistent with this interpretation, it is not affected by manoeuvres which interfere with the Ca current or sarcoplasmic reticulum function. Tonic tension is not seen under normal conditions in mammalian cardiac muscle. If, however, $[Na^+]_i$ is elevated then tonic tension is increased by depolarization (Eisner and Lederer 1979). As shown in Fig. 8.1, the quantitative dependence of tonic tension on voltage and $[Na^+]_i$ is consistent with an Na–Ca exchange. Furthermore, tonic tension and the accompanying increase of $[Ca^{2+}]_i$ are abolished by exposure to Na-free solution (Cannell *et al.* 1986).

8.4.5 Physiological importance of voltage-dependent Na–Ca exchange

The work reviewed above has shown that Na–Ca exchange is sensitive to changes of membrane potential. It is tempting to conclude from this that changes of membrane potential during the action potential will have significant effects on $[Ca^{2+}]_i$ via their effects on Na–Ca exchanges. While this conclusion may turn out to be correct we would like to inject a note of caution.

Before accepting that the effects of membrane potential on Na–Ca exchange will have large effects on $[Ca^{2+}]_i$, one must know to what extent Na–Ca exchange contributes to Ca efflux and influx under physiological conditions. As mentioned in section 8.2 it is possible that under resting conditions ($[Ca^{2+}]_i = 100$ nM) much of the efflux of Ca is produced by a Ca–ATPase. It is, of course, possible that, when $[Ca^{2+}]_i$ increases during activity, Na–Ca exchange produces a much larger fraction of the Ca efflux. However, a recent study has suggested that, even at a $[Ca^{2+}]_i$ of 6 μM the Ca–ATPase will only produce about 50 per cent of the net Ca efflux (Philipson and Ward 1986). The experiment illustrated in Fig. 8.9 (Cannell *et al.* 1987) was designed to study the effects of voltage-dependent Na–Ca exchange under as

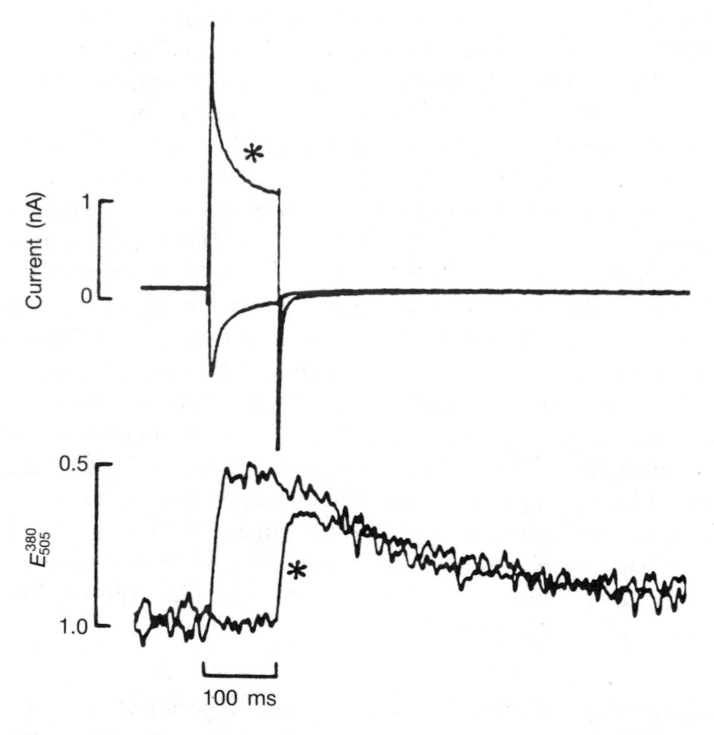

Fig. 8.9. Effects of depolarization on $[Ca^{2+}]_i$ in isolated rat ventricular myocyte. Traces show: top, membrane current, bottom, fura-2 fluorescence, a function of $[Ca^{2+}]_i$. Depolarizing pulses were applied to + 10 mV and + 100 mV(*). Note that the pulse to + 100 mV produces no rise of $[Ca^{2+}]_i$ during the pulse. The increase of $[Ca^{2+}]_i$ on repolarization is presumably due to Ca entry through Ca channels as the membrane potential is returned to levels less positive than E_{Ca}. Taken from Cannell *et al.* (1987).

near as possible to physiological conditions. It shows the effects of depolarization on $[Ca^{2+}]_i$ measured in an isolated rat myocyte. Depolarization to +10 mV produces an increase of $[Ca^{2+}]_i$ due to Ca entry via the Ca current and consequent Ca-activated release of Ca from the sarcoplasmic reticulum. In contrast, a depolarization to +100 mV (no Ca entry through the Ca current) did not increase $[Ca^{2+}]_i$. This suggests that, under physiological conditions, any entry of Ca into the cell via Na–Ca exchange which is produced by solely depolarization is insufficient to materially affect $[Ca^{2+}]_i$.

8.5 The effects of intracellular Ca on the Na–Ca exchange

If the Na–Ca exchange were a simple carrier which exchanged Na for Ca one would predict that the effects of intracellular and extracellular Na and Ca would be accounted for by their actions as substrates and products. Specifically one would expect that an increase of $[Ca^{2+}]_i$ should bias the exchanger in the direction of net Ca efflux and Na influx and away from Ca entry and Na efflux. However, experimental work does not support this hypothesis and has led to the suggestion that intracellular Ca has an activating role even when the exchange produces net Ca influx.

DiPolo (1979) found in dialysed squid axons that the Na_i-dependent Ca influx was abolished by removing all intracellular Ca with the chelator EGTA. Increasing $[Ca^{2+}]_i$ activated Ca influx in an apparently sigmoidal manner. This experimental result has been criticized by Requena and Mullins (1979) who suggested that, in the absence of intracellular EGTA, much of the influxing ^{45}Ca does not enter the dialysis capillary but is, instead, buffered by intracellular buffers. The effect of adding EGTA can then be accounted for as overcoming the cells Ca buffers.

Artefactual explanations for the Ca_i effect are made less tenable by the observation that Ca_o-dependent Na efflux is also increased by intracellular Ca (DiPolo and Beaugé 1986). This result suggests that Ca_i is stimulating 'reverse' Na^+–Ca^{2+} exchange (Na efflux–Ca entry). This result was subsequently confirmed by Rasgado-Flores and Blaustein (1987) on barnacle muscle. In their experiments elevating $[Ca^{2+}]_i$ from 0.01 to 1 μM elevated Na efflux. Indeed, in the absence of intracellular Ca, changes of $[Ca^{2+}]_o$ had no effect on Na efflux. However, not all of this extra Na efflux was abolished by removing external Ca and the authors suggested, therefore, that part of the effect of Ca_i would be to increase a Na conductance.

DiPolo, Rojas, and Beaugé (1982) investigated the concentration dependence on $[Ca^{2+}]_i$. They found that a concentration of $[Ca^{2+}]_i$ of 0.6 μM was required to half-activate Ca influx. Interestingly, in the absence of ATP $[Ca^{2+}]_i$ had much less effect on Ca influx. In experiments measuring the Ca_o-

dependent Na efflux, the addition of ATP decreased the $K_{0.5}$ for intracellular Ca from 15 to 1.8 μM.

The effects of intracellular Ca on Na efflux have been studied in detail by DiPolo and Beaugé (1987). They measured isotopic Na efflux into Na-containing but Ca-free solution (Na–Na exchange) and into a solution with Ca (Na–Ca). They presented evidence, illustrated in Fig. 8.10, that this Na–Na exchange was a mode of operation of the Na–Ca exchange. In the absence of

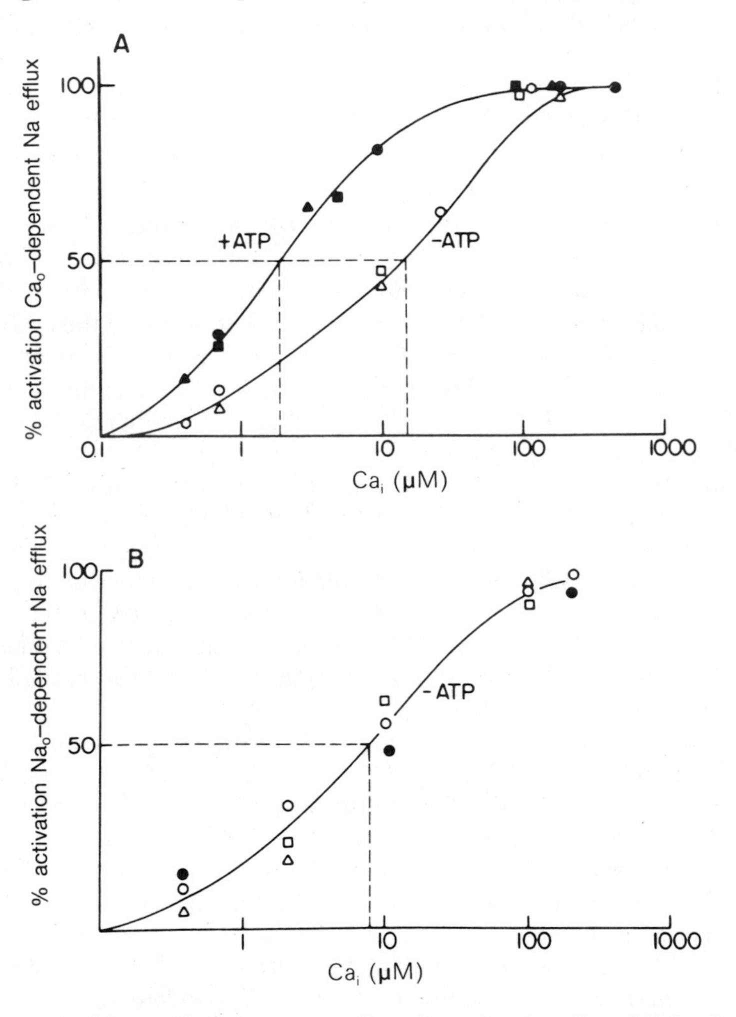

Fig. 8.10. The effects of $[Ca^{2+})_i$ on Ca_o and Na_o-dependent Na efflux. (A) Ca_o-dependent Na efflux (reverse Na–Ca exchange). The open symbols were obtained in the absence and the closed in the presence of ATP. (B) Na_o dependent Na efflux (Na–Ca exchange). The data were obtained on dialysed squid axons. Taken from DiPolo and Beaugé (1987).

ATP Na–Na exchange was also activated by intracellular Ca ($K_{0.5}$ 9 μM) compared with 15 μM for the Na–Ca exchange).

Similar effects of $[Ca^{2+}]_i$ have recently been found in sarcolemmal vesicles (Reeves and Poronnik 1987). Ca uptake was measured in vesicles containing Na. Ca uptake was increased by an increase of intravesicular Ca concentration. The concentrations of Ca required to stimulate were of the order of 0.1 to 0.5 mM—concentrations considerably greater than those needed in intact cells (see above). The authors also found that an increase of intravesicular Ca decreased the apparent K_m for extravesicular Ca. They suggested that these effects of Ca might explain the variability of affinity for Ca found in the literature.

Another phenomenon which may be related to this role of intracellular Ca concerns the effects of intracellular chelating agents. Allen and Baker (1985) found that the addition of Ca chelating agents decreased Na efflux and Ca influx through Na–Ca exchange. This result would, of course, be consistent with those reviewed above were it not for the fact that the injection of Ca–EGTA mixtures designed to maintain $[Ca^{2+}]_i$ near physiological levels also abolished the efflux. This observation might suggest that the effect of the chelator was on another heavy metal ion were it not for the fact that chelators which chelate heavy metals, but not Ca, were ineffective.

Further evidence for a regulatory effect of intracellular Ca is provided by the membrane current measurements of Kimura, Noma, and Irisawa (1986). These workers measured the current produced by adding intracellular Na and extracellular Ca and found that it was abolished by removing intracellular Ca (no added Ca and 42 mM EGTA). However, elevating $[Ca^{2+}]_i$ to 73 nM with an EGTA buffer produced an outward Na–Ca exchange current (Na efflux in exchange for Ca influx). While this result is in qualitative agreement with those reviewed above in the squid axon and barnacle muscle the quantitative differences in the amount of $[Ca^{2+}]_i$ required should be noted (73 nM *vs.* 600 nM).

The mechanism of the effects of changes of $[Ca^{2+}]_i$ is still controversial. DiPolo and Beaugé (1986) suggested that the carrier can only produce Na efflux if it has either intracellular Ca or ATP bound. These authors also pointed out that the intracellular site at which Ca activates the reverse mode (Na efflux for Ca influx) has a much higher affinity than that for which intracellular Ca activates the forward mode (Ca efflux for Na influx) (0.6 *vs.* 10 μM). They therefore suggested that the site at which Ca binds may not be the transport site. Surface charge is known to affect the rate of the exchange and this may be altered by Ca. Caroni and Carafoli (1983) suggested that the exchange activity can be stimulated by a Ca–calmodulin-dependent protein kinase. This may therefore be the mechanism of the intracellular Ca activation. This last mechanism differs from the previous two in requiring enzymes other than just the exchanger.

There are many important consequences if increasing $[Ca^{2+}]_i$ increases Ca influx and Na efflux on Na–Ca exchange. In particular, an increase of $[Ca^{2+}]_i$ will increase Ca influx and potentially act as a positive feedback system. Whether or not the Na–Ca exchange does produce positive feedback depends on the balance of two opposing factors. As reviewed above, an increase of $[Ca^{2+}]_i$ increases Ca influx on Na–Ca exchange. An increase of $[Ca^{2+}]_i$ increases Ca efflux through the exchange. If the exchange is to operate as a useful Ca extrusion mechanism it must respond to an increase of $[Ca^{2+}]_i$ by increasing Ca net efflux. Measurements of net Ca fluxes will be required to establish this point.

Acknowledgements

The work from the authors' laboratories was supported by grants from the British Heart Foundation, Wellcome Trust, and Medical Research Council (D.A.E.), and the N.I.H. and American Heart Association (W.J.L.).

References

Allen, T. J. and Baker, P. F. (1985). Intracellular Ca indicator Quin-2 inhibits Ca^{2+} inflow via Na_i–Ca_o exchange in squid axon. *Nature* **315**, 755–6.

Allen, T. J. A. and Baker, P. F. (1986a). Comparison of the effects of potassium and membrane potential on the calcium-dependent sodium efflux in squid axons. *J. Physiol.* **378**, 53–76.

Allen, T. J. A. and Baker, P. F. (1986b). Influence of membrane potential on calcium efflux from giant axons of *Loligo*. *J. Physiol.* **378**, 77–96.

Arlock, P. and Katzung, B. G. (1985). Effects of sodium substitutes on transient inward current and tension in guinea-pig and ferret papillary muscle. *J. Physiol.* **360**, 105–20.

Axelsen, P. H. and Bridge, J. H. B. (1985). Electrochemical ion gradients and the Na–Ca exchange stoichiometry—Measurements of these gradients are thermodynamically consistent with a stoichiometric coefficient greater than or equal to 3. *J. Gen. Physiol.* **85**, 471–5.

Baker, P. F., Hodgkin, A. L., and Ridgway, E. G. (1971). Depolarization and Ca entry in squid axons. *J. Physiol.* **218**, 709–55.

Baker, P. F., Meves, H., and Ridgway, E. G. (1973a). Effects of manganese and other agents on the calcium uptake that follows depolarization of squid axons. *J. Physiol.* **231**, 511–26.

Baker, P. F., Meves, H., and Ridgway, E. G. (1973b). Calcium entry in response to maintained depolarisation of squid axons. *J. Physiol.* **231**, 527–48.

Baker, P. F., Blaustein, M. P., Hodgkin, A. L., and Steinhardt, R. A. (1969). The influence of calcium on sodium efflux in squid axons. *J. Physiol.* **200**, 431–58.

Barcenas-Ruiz, L., Beuckelmann, D. J., and Wier, W. G. (1987). Sodium–calcium exchange in heart: membrane currents and changes in $[Ca^{2+}]_i$. *Science* **238**, 1720–2.

Bartschat, D. K. and Lindenmayer, G. E. (1980). Calcium movements promoted by vesicles in a highly enriched sarcolemma preparation from canine ventricle. *J. Biol. Chem.* **255**, 9626–34.

Bers, D. M. and Ellis, D. (1982). Intracellular calcium and sodium activity in sheep heart Purkinje fibres. Effects of changes of external sodium and intracellular pH. *Pflüg. Arch.* **393**, 171–8.

Bers, D. M., Philipson, K. D., and Nishimoto, A. Y. (1980). Sodium–calcium exchange and sidedness of isolated sarcolemmal vesicles. *Biochem. Biophys. Acta* **601**, 358–71.

Bielefeld, D. R., Hadley, R. W., Vassilev, P. M., and Hume, J. R. (1986). Membrane electrical properties of vesicular Na–Ca exchange inhibitors in single atrial myocytes. *Circ. Res.* **59**, 381–9.

Blaustein, M. P. and Hodgkin, A. L. (1969). The effect of cyanide on the efflux of calcium from squid axons. *J. Physiol.* **200**, 497–528.

Blaustein, M. P. and Russell, J. M. (1975). Sodium–calcium exchange and calcium–calcium exchange in internally dialyzed squid giant axons. *J. Membr. Biol.* **22**, 285–312.

Blaustein, M. P., Russell, J. M., and deWeer, P. (1974). Calcium efflux from internally dialyzed squid axons: The influence of external and internal cations. *J. Supramol. Struct.* **2**, 558–61.

Bridge, J. H. B. and Bassingthwaighte, J. B. (1983). Uphill sodium transport driven by an inward calcium gradient in heart muscle. *Science*, **219**, 178–80.

Cannell, M. B., Berlin, J. R., and Lederer, W. J. (1987). Effect of membrane potential changes on the calcium transient in single rat cardiac muscle cells. *Science* **238**, 1419–23.

Cannell, M. B. and Lederer, W. J. (1986). The arrhythmogenic inward current I_{TI} in the absence of electrogenic sodium–calcium exchange in sheep cardiac Purkinje fibres. *J. Physiol.* **374**, 201–19.

Cannell, M. B., Eisner, D. A., Lederer, W. J., and Valdeolmillos, M. (1986). Effects of membrane potential on intracellular calcium concentration in sheep Purkinje fibres in sodium-free solutions. *J. Physiol.* **381**, 193–203.

Carafoli, E. (1985). The homeostasis of calcium in heart cells. *J. Mol. Cell. Cardiol.* **17**, 203–12.

Caroni, P. and Carafoli, E. (1980). An ATP-dependent Ca^{2+}-pumping system in dog heart sarcolemma. *Nature* **283**, 765–7.

Caroni, P. and Carafoli, E. (1983). The regulation of the Na^+–Ca^{2+} exchanger of heart sarcolemma. *Eur. J. Biochem.* **132**, 451–60.

Caroni, P., Reinlib, L., and Carafoli, E. (1980). Charge movements during the Na–Ca exchange in heart sarcolemmal vesicles. *Proc. Natl. Acad. Sci. USA* **77**, 6354–8.

Chapman, R. A. (1986). Sodium/calcium exchange and intracellular calcium buffering in ferret myocardium: an ion-sensitive micro-electrode study. *J. Physiol.* **373**, 163–79.

Chapman, R. A. and Tunstall, J. (1980). The interaction of sodium and calcium ions at the cell membrane and the control of contractile strength in frog atrial muscle. *J. Physiol.* **305**, 109–23.

Chapman, R. A. and Tunstall, J. (1981). The tension–depolarization relationship of frog atrial trabeculae as determined by potassium contractures. *J. Physiol.* **310**, 97–115.

Clusin, W. T. (1983). Caffeine induces a transient inward current in cultured cardiac cells. *Nature* **301**, 248–50.

Clusin, W. T., Fischmeister, R., and DeHaan, R. L. (1983). Caffeine-induced current in embryonic heart cells: time course and voltage-dependence. *Am. J. Physiol.* **245**, H528–32.

Colquhoun, D., Neher, E., Reuter, H., and Stevens, C. F. (1981). Inward current channels activated by intracellular Ca in cultured cardiac cells. *Nature* **294**, 752–4.

Coraboeuf, E., Gautier, P., and Guiraudou, P. (1981). Potential and tension changes induced by sodium removal in dog Purkinje fibers: role of an electrogenic sodium–calcium exchange. *J. Physiol.* **311**, 605–22.

Deitmer, J. W. and Ellis, D. (1978). Changes in the intracellular sodium activity of sheep heart Purkinje fibres produced by calcium and other divalent cations. *J. Physiol.* **277**, 437–53.

DiPolo, R. (1979). Calcium influx in internally dialyzed squid giant axons. *J. Gen. Physiol.* **73**, 91–113.

DiPolo, R. and Beaugé, L. (1979). Physiological role of ATP-driven Ca pump in squid axons. *Nature*, 271–3.

DiPolo, R. and Beaugé, L. (1980). Mechanisms of calcium transport in the giant axon of the squid and their physiological role. *Cell Calcium* **1**, 147–69.

DiPolo, R. and Beaugé, L. (1986). Reverse Na–Ca exchange requires internal Ca and/or ATP in squid axons. *Biochem. Biophys. Acta* **854**, 298–306.

DiPolo, R. and Beaugé, L. (1987). Characterization of the reverse Na/Ca exchange in squid axons and its modulation by Ca_i and ATP. *J. Gen. Physiol.* **90**, 505–25.

DiPolo, R., Rojas, H., and Beaugé, L. (1982). Ca entry at rest and during prolonged depolarization in dialyzed squid axons. *Cell Calcium* **3**, 19–41.

DiPolo, R., Bezanilla, F., Caputo, C., and Rojas, H. (1985). Voltage dependence of the Na/Ca exchange in voltage-clamped dialyzed squid axons. *J. Gen. Physiol.* **86**, 457–78.

Eisner, D. A. and Lederer, W. J. (1979). Inotropic and arrhythmogenic effects of potassium depleted solutions on mammalian cardiac muscle. *J. Physiol.* **294**, 255–77.

Eisner, D. A. and Lederer, W. J. (1985). Na–Ca exchange: stoichiometry and electrogenicity. *Am. J. Physiol.* **248**, C189–202.

Eisner, D. A., Lederer, W. J., and Vaughan-Jones, R. D. (1983). The control of tonic tension by membrane potential and intracellular Na activity in the sheep cardiac Purkinje fibre. *J. Physiol.* **335**, 723–43.

Eisner, D. A., Lederer, W. J., and Vaughan-Jones, R. D. (1984). The quantitative relationship between twitch tension and intracellular sodium activity in sheep cardiac Purkinje fibres. *J. Physiol.* **355**, 251–66.

Fedida, D., Noble, D., Rankin, A. C., and Spindler, A. J. (1987a). The arrhythmogenic transient inward current i_{TI} and related contraction in isolated guinea-pig ventricular myocytes. *J. Physiol.* **392**, 523–42.

Fedida, D., Noble, D., Shimoni, Y., and Spindler, A. J. (1987b). Inward current related to contraction in guinea-pig ventricular myocytes. *J. Physiol.* **385**, 565–89.

Garrahan, P. and Glynn, I. M. (1967). The stoichiometry of the sodium pump. *J. Physiol.* **192**, 217–35.

Hill, J. A. Jr., Coronado, R., and Strauss, H. C. (1988). Reconstitution and characterization of a calcium-activated channel from heart. *Circ. Res.* **62**, 411–15.

Hodgkin, A. L. and Keynes, R. D. (1957). Movements of labelled calcium in squid axons. *J. Physiol.* **138**, 253–81.

Hodgkin, A. L. and Nunn, B. J. (1987). The effects of ions on sodium–calcium exchange in salamander rods. *J. Physiol.* **391**, 371–98.

Hodgkin, A. L., McNaughton, P. A., and Nunn, B. J. (1987). Measurement of sodium–calcium exchange in salamander rods. *J. Physiol.* **391**, 347–70.

Horackova, M. and Vassort, G. (1979). Sodium–calcium exchange in regulation of cardiac contractility. Evidence for an electrogenic, voltage-dependent mechanism. *J. Gen. Physiol.* **73**, 403–24.

Hume, J. R. (1987). Component of whole cell Ca current due to electrogenic Na–Ca exchange in cardiac myocytes. *Am. J. Physiol.* **252**, H666–70.

Hume, J. R. and Uehara, A. (1986a). 'Creep currents' in single frog atrial cells may be generated by electrogenic Na/Ca exchange. *J. Gen. Physiol.* **87**, 857–84.

Hume, J. R. and Uehara, A. (1986b). Properties of 'creep currents' in single frog atrial cells. *J. Gen. Physiol.* **87**, 833–55.

Kadoma, M., Froelich, J., Reeves, J., and Sutko, J. (1982). Kinetics of sodium ion induced calcium release in calcium ion loaded cardiac sarcolemmal vesicles: determination of initial velocities by stoppedflow spectrophotometry. *Biochemistry* **21**, 1914–18.

Karagueuzian, H. S. and Katzung, B. G. (1982). Voltage-clamp studies of transient inward current and mechanical oscillations induced by ouabain in ferret papillary muscle. *J. Physiol.* **327**, 255–71.

Kass, R. S., Tsien, R. W., and Weingart, R. (1978a). Ionic basis of transient inward current induced by strophanthidin in cardiac Purkinje fibres. *J. Physiol.* **281**, 209–26.

Kass, R. S., Lederer, W. J., Tsien, R. W., and Weingart, E. (1978b). Role of calcium ions in transient inward currents and aftercontractions induced by strophanthidin in cardiac Purkinje fibres. *J. Physiol.* **281**, 187–208.

Kimura, J., Miyamae, S., and Noma, A. (1987). Identification of sodium–calcium exchange current in single ventricular cells of guinea-pig. *J. Physiol.* **384**, 199–222.

Kimura, J., Noma, A., and Irisawa, H. (1986). Na–Ca exchange current in mammalian heart cells. *Nature* **319**, 596–7.

Lederer, W. J. and Nelson, M. T. (1983). Effects of extracellular sodium on calcium efflux and membrane current in single muscle cells from the barnacle. *J. Physiol.* **341**, 325–39.

Lederer, W. J. and Tsien, R. W. (1976). Transient inward current underlying arrhythmogenic effects of cardiotonic steroids in Purkinje fibres. *J. Physiol.* **263**, 73–100.

Lederer, W. J., Sheu, S. S., Vaughan-Jones, R. D., and Eisner, D. A. (1984). The effects of Na–Ca exchange on membrane currents in sheep cardiac Purkinje fibers. In: *Electrogenic transport: fundamental principles and physiological implications (*eds M. P. Blaustein and M. Lieberman). Raven Press, New York.

Ledvora, R. F. and Hegyvary, C. (1983). Dependence of Na–Ca exchange and Ca–Ca exchange on monovalent cations. *Biochem. Biophys. Acta* **729**, 123–36.

Lipp, P. and Pott, L. (1988). Transient inward current in guinea-pig atrial myocytes reflects a change of sodium–calcium exchange current. *J. Physiol.* **397**, 601–30.

Lüttgau, H. C. and Niedergerke, R. (1958). The antagonism between Ca and Na ion on the frog heart. *J. Physiol.* **143**, 486–505.

Mead, R. H. and Clusin, W. T. (1984). Origin of the background sodium current and effects of sodium removal in cultured embryonic cardiac cells. *Circ. Res.* **55**, 67–77.

Mechmann, S. and Pott, L. (1986). Identification of Na–Ca exchange current in single cardiac myocytes. *Nature* **319**, 597–9.

Mullins, L. J. (1979). The generation of electric currents in cardiac fibers by Na/Ca exchange. *Am. J. Physiol.* **236**, C103–10.

Mullins, L. J. (1981). *Ion transport in heart.* Raven Press, New York.

Mullins, L. J. and Brinley, F. J. (1975). Sensitivity of calcium efflux from squid axons to changes in membrane potential. *J. Gen. Physiol.* **65**, 135–52.

Mullins, L. J. and Requena, J. (1981). The 'late' Ca channel in squid axons. *J. Gen. Physiol.* **78**, 683–700.

Mullins, L. J., Requena, J., and Whittembury, J. (1985). Ca^{2+} entry in squid axons during voltage-clamp pulses is mainly Na^+/Ca^{2+} exchange. *Proc. Natl. Acad. Sci. USA* **82**, 1847–51.

Noble, D. (1984). The surprising heart: a review of recent progress in cardiac electrophysiology. *J. Physiol.* **353**, 1–50.

Noble, D. (1986). Sodium–calcium exchange and its role in generating electric current. In: *Cardiac muscle: the regulation of excitation and contraction* (ed. R. D. Nathan), pp. 171–200. Academic Press, London.

Philipson, K. D. and Nishimoto, A. Y. (1980). Na^+–Ca^{2+} exchange is affected by membrane potential in cardiac sarcolemmal vesicles. *J. Biol. Chem.* **255**, 6880–2.

Philipson, K. D. and Ward, R. (1986). Ca^{2+} transport capacity of sarcolemmal Na^+–Ca^{2+} exchange. Extrapolation of vesicle data to *in vivo* conditions. *J. Mol. Cell. Cardiol.* **18**, 943–51.

Pitts, B. J. R. (1979). Stoichiometry of sodium–calcium exchange in cardiac sarcolemmal vesicles. *J. Biol. Chem.* **254**, 6232–5.

Post, R. L. and Jolly, P. C. (1957). The linkage of sodium, potassium and ammonium active transport across the human erythrocyte membrane. *Biochem. Biophys. Acta* **25**, 118–28.

Rasgado-Flores, H. and Blaustein, M. P. (1987). Na/Ca exchange in barnacle muscle cells has a stoichiometry of 3 Na^+/1 Ca^{2+}. *Am. J. Physiol.* **252**, C499–504.

Reeves, J. P. and Hale, C. C. (1984). The stoichiometry of the cardiac sodium–calcium exchange system. *J. Biol. Chem.* **259**, 7733–9.

Reeves, J. P. and Poronnik, P. (1987). Modulation of Na^+–Ca^{2+} exchange in sarcolemmal vesicles by intravesicular Ca^{2+}. *Am. J. Physiol.* **252**, C17–23.

Reeves, J. P. and Sutko, J. L. (1979). Sodium–calcium ion exchange in cardiac sarcolemmal vesicles. *Proc. Natl. Acad. Sci. USA* **76**, 590–4.

Reeves, J. P. and Sutko, J. L. (1980). Sodium–calcium exchange activity generates a current in cardiac membrane vesicles. *Science* **208**, 1461–4.

Requena, J. and Mullins, L. J. (1979). Calcium movements in nerve fibres. *Quart. Rev. Biophys.* **12**, 371–460.

Reuter, H. and Seitz, N. (1968). The dependence of calcium efflux from cardiac muscle on temperature and external ion composition. *J. Physiol.* **195**, 451–70.

Schatzmann, H. J. (1966). ATP-dependent Ca^{2+} extrusion form human red cells. *Experimentia* **22**, 364–6.

Sheu, S.-S. and Fozzard, H. A. (1982). Transmembrane Na^{+} and Ca^{2+} electrochemical gradients in cardiac muscle and their relationship to force development. *J. Gen. Physiol.* **80**, 325–51.

Sjodin, R. A. and Abercrombie, R. F. (1978). The influence of external cations and membrane potential on Ca-activated Na efflux in *Myxicola* giant axons. *J. Gen. Physiol.* **71**, 453–66.

Slaughter, R. S., Sutko, J. L., and Reeves, J. P. (1983). Equilibrium calcium–calcium exchange in cardiac sarcolemmal vesicles. *J. Biol. Chem.* **258**, 3183–90.

Soejima, M. and Noma, A. (1984). Mode of regulation of the ACh-sensitive K-channel by the muscarinic receptor in rabbit atrial cells. *Pflügers Arch.* **400**, 424–31.

Talor, Z. and Arruda, J. A. L. (1985). Partial purification and reconstitution of renal basolateral Na^{+}–Ca^{2+} exchange into liposomes. *J. Biol. Chem.* **260**, 15473–6.

Wakabayashi, S. and Goshima, K. (1981). Kinetic studies on sodium-dependent calcium uptake by myocardial cells and neuroblastoma cells in culture. *Biochem. Biophys. Acta* **642**, 158–72.

Yau, K.-W. and Nakatani, K. (1984). Electrogenic Na–Ca exchange in retinal rod outer segment. *Nature* **311**, 661–3.

Yellen, G. (1982). Single Ca-activated non-selective cation channels in neuroblastoma. *Nature* **296**, 357–9.

9 Sodium–calcium exchange in mammalian smooth muscles

Mordecai P. Blaustein

9.1 Introduction. Pathways for calcium movement in smooth muscles

The intracellular free-calcium concentration, $[Ca^{2+}]_i$, in vertebrate smooth muscles, as in other types of muscles, is determined by the balance of: plasma membrane (sarcolemma) Ca^{2+} transport systems, intracellular sequestration and release systems (primarily in the sarcoplasmic reticulum, SR), and Ca^{2+} binding to cytoplasmic proteins (including the Ca^{2+}-regulated contractile proteins). The relative volume of the SR varies greatly in different smooth muscles (Devine, Somlyo, and Somlyo 1972; Somlyo 1980; Somlyo and Franzini-Armstrong 1985), and this difference appears to be reflected in the relative role of the SR in contractile activation in different smooth muscles (Ashida *et al.* 1988).

Ca^{2+} can enter the sarcoplasmic space from the extracellular fluid via agonist-operated and/or voltage-gated Ca-selective channels (e.g. Bean *et al.* 1986; Benham, Hess, and Tsien 1987; Nelson and Worley 1988; Yatani *et al.* 1987). Ca^{2+} can also enter from the SR as a result of activation of SR Ca^{2+} channels by various agonists; this release may be mediated by inositol triphosphate (Hashimoto *et al.* 1986; Somlyo *et al.* 1985), or perhaps by Ca-induced Ca^{2+} release (Itoh *et al.* 1983).

Ca^{2+} is removed from the sarcoplasmic space by resequestration in the SR via a calmodulin-insensitive, ATP-driven Ca^{2+} pump (coupling ratio: 2 Ca^{2+}/ATP) and by extrusion across the sarcolemma via a calmodulin-sensitive, ATP-driven Ca^{2+} pump (coupling ratio: 1 Ca^{2+} per ATP) (Carafoli 1987; Eggermont *et al.* 1988). Ca^{2+} sequestration into and release from mitochondria will be ignored because evidence suggests that little Ca^{2+} is stored in mitochondria under normal physiological conditions (Hansford 1985).

The principle pathways involved in the movement of Ca^{2+} into and out of the sarcoplasmic space are diagrammed in Fig. 9.1. Note that, in parallel with

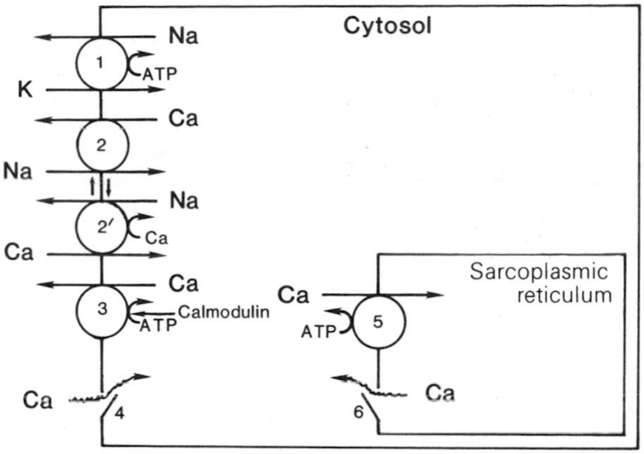

Fig. 9.1. Diagram of a smooth muscle cell showing the plasma membrane and sarcoplasmic reticulum (SR) mechanisms that play a role in the regulation of $[Ca^{2+}]_i$ in the cytosol. 1. Ouabain-sensitive, ATP-driven Na^+pump; 2. Plasma membrane Na–Ca exchange system operating in the Ca^{2+} efflux mode (2) and Ca^{2+} influx mode (2′); ATP-driven, calmodulin-regulated plasma membrane Ca^{2+}pump; 4. Ca-selective channels (both voltage-gated and receptor-operated); 5. ATP-driven, calmodulin-insensitive SR Ca^{2+}pump; and 6. Ca^{2+} channels that mediated Ca^{2+} release from the SR. As illustrated, the Ca^{2+} entry mode of Na–Ca exchange is activated by intracellular free Ca^{2+} in the dynamic physiological concentration range (about 10^{-7} to 10^{-6} M). Although not shown here, the exchanger is also modulated by (internal) ATP, which affects the cation affinities (Blaustein 1977c; DiPolo and Beaugé 1984). From Blaustein (1985), with permission.

the Ca^{2+} channels and ATP-driven Ca^{2+} pump in the sarcolemma, Ca^{2+} can be transported both inwardly and outwardly across this membrane via a coupled Na–Ca countertransport system (Blaustein 1974 and 1977a). Most investigators now agree that the Na–Ca exchanger is present in the sarcolemma of most (if not all) smooth muscle cells. Nevertheless, there is considerable controversy about its physiological significance in these tissues—especially in vascular smooth muscle (e.g. Blaustein 1977a,b; Brading and Lategan 1985; Casteels *et al.* 1985; Mulvany 1985; Somlyo *et al.* 1986; Van Breemen, Aaronson, and Loutzenheiser 1979). We will discuss some of the properties of the exchanger that are relevant to smooth muscle, and some of the problems associated with studies of this transport system in smooth muscles. The possible physiological roles and pathophysiological significance of this exchanger, operating in parallel with the sarcolemmal Ca^{2+} channels and ATP-driven Ca^{2+} pump, are discussed at the end of the chapter, and an effort is made to resolve some of the discordant views.

9.2 Properties of the Na–Ca exchanger

Net movements of Ca^{2+} mediated by the Na^{+}–Ca^{2+} exchanger [$J_{Ca(NaCa)}$] are governed by the difference between the membrane potential, V_M, and the reversal potential of the exchanger, E_{NaCa}:

$$J_{Ca(Na\text{-}Ca)} = k\ (V_M - E_{NaCa}). \tag{1}$$

The variable, k is a complex kinetic parameter that depends upon the number of carriers, the fractional saturation of the carrier binding sites by the activating ions and transported ions, V_M, and ATP, which is not used as a fuel, but does modulate the carrier kinetics. In most preparations where it has been studied, the Na–Ca exchange coupling ratio is about 3 Na^{+}:1 Ca^{2+} (Rasgado-Flores and Blaustein 1987; Sheu and Blaustein 1986). Thus the reversal potential is given by:

$$E_{NaCa} = 3\ E_{Na} - 2\ E_{Ca} \tag{2}$$

where: $$E_{Na} = RT/F\ ln\ ([Na^{+}]_o/[Na^{+}]_i) \tag{2a}$$

and: $$E_{Ca} = RT/2F\ ln\ ([Ca^{2+}]_o/[Ca^{2+}]_i). \tag{2b}$$

The terms in square brackets refer to the respective free ion concentrations in the extracellular (o) and intracellular (i) fluids; R, T, and F have their usual meanings.

Exchanger-mediated Ca^{2+} entry as well as exit is activated by $[Ca^{2+}]_i$ in the dynamic physiological range, about 10^{-7} to 10^{-6} M (Rasgado-Flores and Blaustein 1987; Rasgado, Santiago and Blaustein 1989). The exchanger transports Ca^{2+} very slowly when $[Ca^{2+}]_i$ is below contraction threshold ($<10^{-7}$ M), and much faster when $[Ca^{2+}]_i$ is above the contraction threshold (Sheu and Blaustein 1986; Rasgado, Santiago, and Blaustein 1989). The direction of Ca^{2+} transport is governed by the difference between V_M and E_{NaCa} (1): Net movement is outward when V_M is more negative than E_{NaCa} and inward when V_M is more positive than E_{NaCa}. This means that the exchanger can be turned on and off, and can move Ca^{2+} either inward or outward, as $[Ca^{2+}]_i$, $[Na^{+}]_i$, and V_M vary during the cell activity cycle (Sheu and Blaustein 1986; Rasgado-Flores, Santiago, and Blaustein 1988).

9.3 Pitfalls in identifying Na–Ca exchange in smooth muscles

9.3.1 Use of tension increase to estimate increases in $[Ca^{2+}]_i$

A rise in $[Ca^{2+}]_i$ is an immediate trigger for contraction in vertebrate smooth muscles (Adelstein and Klee 1980; Johansson and Somlyo 1980). Thus, alterations in steady 'resting' tension and in the contractile response to activation may serve as indirect measures of changes in $[Ca^{2+}]_i$ and the size of the SR Ca^{2+} store in smooth muscles (see Ashida *et al.* 1988), as in cardiac muscle (e.g. Chapman 1983). Indeed, the most common method of studying Na–Ca exchange in smooth muscle has been the measurement of resting and (in some cases) agonist-evoked tension in response to alterations in the trans-sarcolemmal Na^+ electrochemical gradient, $\Delta\bar{\mu}_{Na}$. Even before Na–Ca exchange was first recognized (in invertebrate nerve: Baker and Blaustein 1968, Baker *et al.* 1969, Blaustein and Hodgkin 1969; and in mammalian cardiac muscle: Reuter and Seitz 1968), changes in the Na^+ concentration gradient across the plasma membrane were found to affect resting and/or evoked tension in several types of smooth muscles. For example, lowering $[Na^+]_o$, or inhibiting the Na^+ pump by external K^+ removal or treatment with cardiotonic steroids (to raise $[Na^+]_i$), either induced contractions and/or enhanced the responsiveness to agonists in vascular smooth muscle (e.g. Leonard 1957; Mason and Braunwald 1964) and tracheal smooth muscle (Dixon and Brodie 1903). Alternatively, reduction of $[Na^+]_o$ induced reversible contractures in vascular smooth muscle (e.g. Hinke and Wilson 1962) and intestinal smooth muscle (Judah and Willoughby 1964).

These findings fit with the idea that a reduction in the trans-sarcolemmal Na^+ gradient tends to promote a rise in $[Ca^{2+}]_i$ and/or increased storage of Ca^{2+} in the SR. Unfortunately, this interpretation is limited by two factors: inhibition of the Na^+ pump and reduction in $[K^+]_o$ may depolarize the smooth muscle cells and promote Ca^{2+} entry via Ca^{2+} channels, and nerve terminals also have a Na–Ca exchanger (e.g. Blaustein 1987; Sanchez-Armass and Blaustein 1987); thus, reduction of the Na^+ gradient should stimulate endogenous transmitter release (e.g. Bonaccorsi *et al.* 1977; Katsuragi *et al.* 1988; Lorenz *et al.* 1980; Magyar *et al.* 1987) which could also activate contraction. However, at least in some cases, these alternative explanations could be ruled out because the smooth muscle contractions induced by raising $[Na^+]_i$ and/or lowering $[Na^+]_o$ were observed in the presence of Ca^{2+} channel blockers (e.g. Ashida and Blaustein 1987a; Johansson and Hellstrand 1987; Ozaki and Urakawa 1979; Sekine, Yamakawa, and Ogata 1984) and/or in the presence of α-receptor antagonists (e.g. Ashida and Blaustein 1987a; Ozaki, Karaki, and Urakawa 1978).

The aforementioned observations indicate that an increase in tension can serve as a convenient measure of a rise in $[Ca^{2+}]_i$ but some potential pitfalls may cause changes in $[Ca^{2+}]_i$ to be underestimated or even missed. Tension is a non-linear function of $[Ca^{2+}]_i$, and exhibits a threshold. Therefore, a small rise in $[Ca^{2+}]_i$, that may be induced by a small reduction in $\Delta\bar{\mu}_{Na}$, may not be detected if $[Ca^{2+}]_i$ remains below the contraction threshold; some of the entering Ca^{2+} may be sequestered in the SR so that $[Ca^{2+}]_i$ is prevented from reaching threshold, or an increase in tension is attenuated (Ashida and

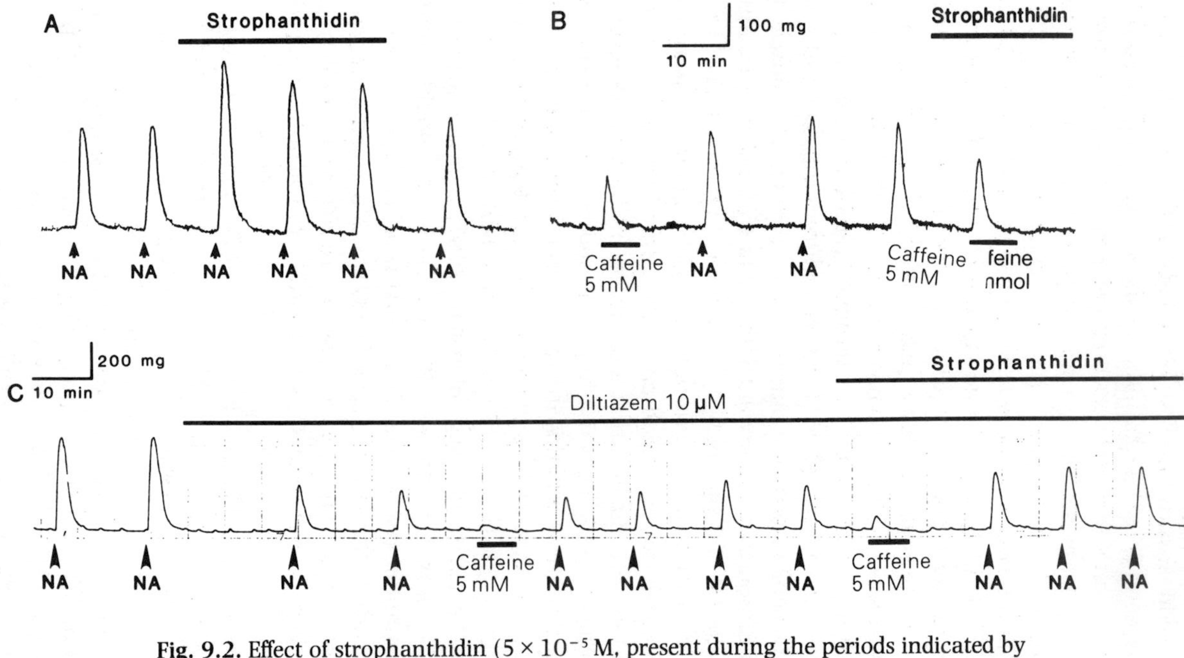

Fig. 9.2. Effect of strophanthidin (5×10^{-5} M, present during the periods indicated by the bars) on the contractile response of rings of rat aorta to brief pulses of noradrenaline (NA, 6×10^{-8} M and 5 mM caffeine (B and C). The calcium channel blocker, diltiazem (10 μM) was present during the period indicated by the bar in Ca. Data from Ashida and Blaustein (1987a), with permission.

Blaustein 1987a). This could account for the apparent inability of some smooth muscles to develop contractions when they are exposed to Na-depleted media (e.g. Toda 1978; Itoh *et al.* 1982). This problem may be circumvented by inhibiting Ca^{2+} sequestration in the SR with ryanodine (Ashida *et al.* 1988) or caffeine (Ashida and Blaustein 1987a), although caffeine may also elevate cyclic AMP, which will then reduce the sensitivity of the contractile machinery to Ca^{2+} (Ruegg and Paul 1982; Schoeffter *et al.* 1987). For example, Figs 9.2 and 9.3 show that caffeine-induced contractions in rat aorta were enhanced when a low dose of strophanthidin was applied (Figs 9.2B and C) or $[Na^+]_o$ was reduced to 30 mM (Fig. 9.3), even though these treatments did not raise the baseline tension. Also, the noradrenaline-induced contractions, which depend, in part, upon Ca^{2+} release from the SR (Ashida *et al.* 1988), were enhanced after cardiac glycoside treatment (Figs 9.2A and B; and see Sekine, Yamakawa, and Ogata 1984)—presumably because of the increased store of Ca^{2+} in the SR.

Most efforts to induce contractions by reduction of $[Na^+]_o$ are carried out on 'resting' smooth muscles. Under these conditions of low $[Ca^{2+}]_i$ and $[Na^+]_i$, the exchanger is expected to be relatively dormant (see Section 9.2 above). This, too, may contribute to the need to reduce $[Na^+]_o$ markedly in order to raise $[Ca^{2+}]_i$ (via Na–Ca exchange) to the contraction threshold—and this would cause the role of the Na–Ca exchanger to be underestimated (e.g. see Aalkjaer and Mulvany 1985; Mulvany, Aalkjaer, and Peterson 1984). Such 'resting' conditions are non-physiological for many tonic smooth muscles such as vascular smooth muscles, which are chronically exposed to the agonists present in blood plasma such noradrenaline, vasopressin, and angiotensin II. *In vivo*, these smooth muscles must be maintained in a state of partial contractile activation in which $[Ca^{2+}]_i$ is continuously held above the contraction threshold. The Na–Ca exchanger should be partially activated under these circumstances (see Section 9.2 above). These conditions can be mimicked experimentally by exposing the tissues continuously to low concentrations of agonists. When this is done, the sensitivity of the tissues to small

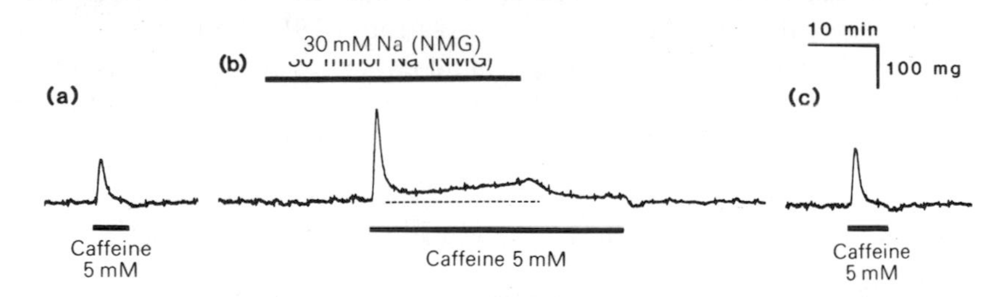

Fig. 9.3. Contractile response of a ring of rat aorta to 5 mM caffeine during incubation in media containing the normal concentration (140 mM) of Na^+ (a and c), or only 30 mM Na^+ (b). From Ashida and Blaustein (1987a), with permission.

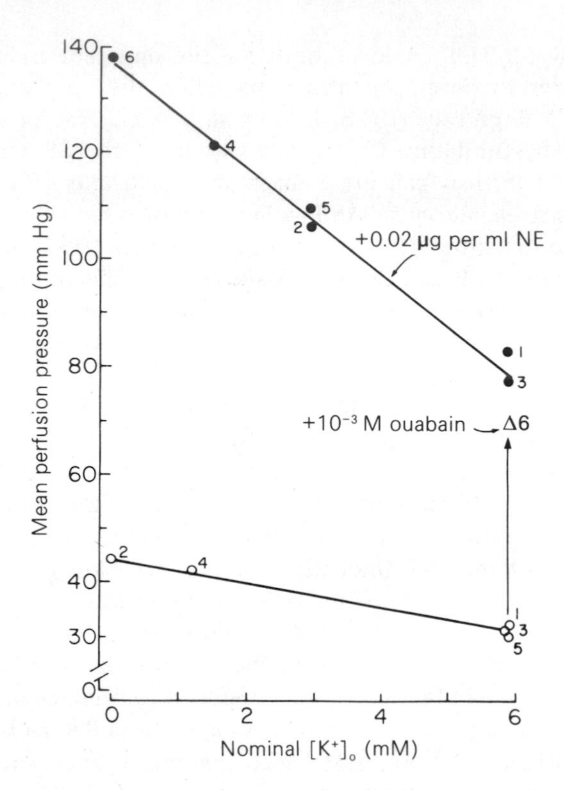

Fig. 9.4. Relationship between the nominal $[K^+]_o$ (abscissa) and the mean arterial perfusion pressure (ordinate) for two arterially-perfused hindquarter preparations from reserpine-treated rats. The open circles are from a catecholamine-free preparation; the filled circles are from a different preparation perfused with fluid containing 0.02 μg ml^{-1}-noradrenaline (NE). The numbers adjacent to the circles indicate the sequence of perfusion fluid changes. From Lang and Blaustein (1980), with permission.

changes in $\Delta\bar{\mu}_{Na}$ appears to be markedly increased. This is illustrated by data from the perfused rat hindlimb (Lang and Blaustein 1980): a low concentration of noradrenaline in the perfusion fluid enhances the vascular tension (measured as perfusion pressure) developed in response to a reduction of $[K^+]_o$, which should inhibit the Na^+ pump and raise $[Na^+]_i$ (see Fig. 9.4).

9.3.2 Na^+ gradient-dependent changes in pH_i; the Na–H exchanger

Many types of smooth muscle possess a Na^+–H^+ exchanger (e.g. Kahn, Shelat, and Allen 1986; Kahn *et al.* 1988a; Kahn, Allen, and Shelat 1988; Wray 1988). A reduction in $\Delta\bar{\mu}_{Na}$ may therefore be expected to lower pH_i, especially when the incubation medium does not contain a bicarbonate buffer so that Cl^-–HCO_3 exchange is unable to participate in proton extrusion. A fall in pH_i should attenuate contraction and promote relaxation (Wray 1988), and thus counteract the effect of a rise in $[Ca^{2+}]_i$. A fall in pH_i may also inhibit both Ca^{2+} entry (Baker and McNaughton 1977) and Ca^{2+} exit (DiPolo and Beaugé 1982) mediated by the Na–Ca exchanger. Nevertheless, as discussed below, the dominant factor in mediating Na^+ gradient-dependent increases in tension appears to be the Na–Ca exchange system.

Another complication of efforts to reduce $\Delta\bar{\mu}_{Na}$ by lowering $[Na^+]_o$ is that this manipulation may also lead to a rapid fall in $[Na^+]_i$ (Aickin 1987), and thus attenuate the change in $\Delta\bar{\mu}_{Na}$. In particular, substitution of Li^+ for Na^+ may produce an especially rapid fall in $[Na^+]_i$ because of Na–Li exchange which may be mediated by the Na–H exchanger (cf. Kahn *et al.* 1988). This might account for differences in the effectiveness of various Na^+ replacements to promote Na–Ca exchange-mediated Ca^{2+} entry and external Ca-dependent contractions in some smooth muscles; Li^+ appears to be a particularly poor Na^+ substitute (e.g. Ashida and Blaustein 1987a; Brading, Burnett, and Sneddon 1980, Ozaki, Karaki, and Urakawa (1978); Ozaki and Urakawa 1981).

9.3.3 Relaxation rate as a measure of a reduction in $[Ca^{2+}]_i$

Another way to detect a change in $[Ca^{2+}]_i$ is to induce a contraction and then use the rate of relaxation to estimate the rate of removal of Ca^{2+} from the sarcoplasmic space. The Na^+ dependence of the relaxation rate may then serve as a measure of the relative contribution of Na–Ca exchange to the reduction in $[Ca^{2+}]_i$. However, Na^+ entry–Ca^{2+} exit mode exchange is a sigmoid function of $[Na^+]_o$ and is half maximally activated by $[Na^+]_o = 20$–30 mM (e.g. Ashida and Blaustein 1987a); nearly all of the external Na^+ must be removed to inhibit Ca^{2+} extrusion via the exchanger. Furthermore, other mechanisms, including Ca^{2+} sequestration in the SR, may also help to remove Ca^{2+} from the cytosol, and these mechanisms must be taken into account when trying to estimate the relative role of the Na–Ca exchanger in Ca^{2+} extrusion (Ashida and Blaustein 1987a). This is exemplified by the rat aorta data in Fig. 9.5 which indicate that relaxation from a low-Na^+ (15 mM) contraction induced in the presence of caffeine, can be promoted either by Ca^{2+} sequestration in the SR when caffeine is removed, or by Ca^{2+} extrusion via Na–Ca exchange, when the normal $[Na^+]_o$ is restored in the presence of caffeine (which inhibits SR Ca^{2+} sequestration).

Several investigators have shown that relaxation of contracted smooth muscle is slowed in Na-depleted media and/or when $[Na^+]_i$ is increased (e.g. rabbit aorta: Bohr, Seidel, and Sobieski 1969; guinea pig taenia coli: Katase and Tomita 1972; rat mesenteric resistance vessels: Petersen and Mulvany 1984; rat aorta: Ashida and Blaustein 1987a, b). However, these relaxation

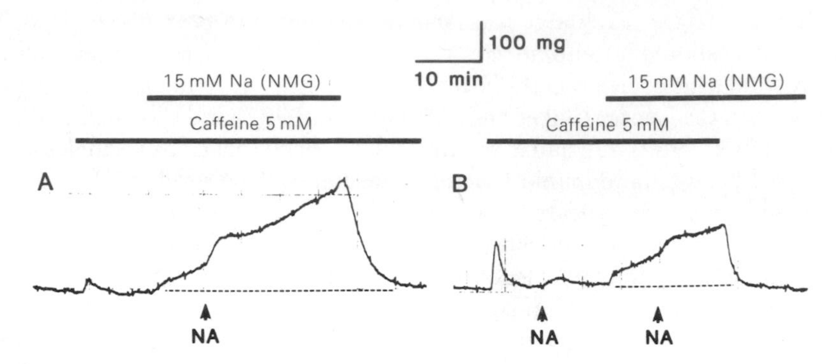

Fig. 9.5. The effect of Na^+ replacement and removal of caffeine on tension after induction of contractions with low $[Na^+]_o$ in a ring of rat aorta. In 'A' and 'B' the ring was superfused with standard (139.2 mM Na^+) media containing 5 mM caffeine, and low-Na^+ contractions were then induced by lowering $[Na^+]_o$ to 15 mM (Na^+ replaced, mole-for-mole by N-methylglucamine). In A the ring relaxed when the normal $[Na^+]_o$ was restored; in B the ring relaxed when caffeine was removed (to enable the SR to sequester Ca^{2+}) despite continued superfusion with media containing only 15 mM Na^+. In A the relaxation was apparently produced by the extrusion of Ca^{2+} by external Na-dependent Na–Ca exchange because Ca^{2+} sequestration in the SR was still inhibited by the continued presence of caffeine. In B relaxation was apparently produced by sequestering the Ca^{2+} in the SR since Ca^{2+} extrusion via the Na–Ca exchanger was still partially inhibited because of the low $[Na^{2+}]_o$. The tissue was stimulated with brief pulses of noradrenaline (NA) at the times indicated by the arrowhead. Data from an unpublished experiment of T. Ashida and M. P. Blaustein.

1984; rat aorta: Ashida and Blaustein 1987a, b). However, these relaxation experiments are complicated by the fact that reduction of $[Na^+]_o$ not only inhibits Na^+ entry–Ca^{2+} exit exchange; it also promotes Na^+ exit–Ca^{2+} entry exchange. Therefore, it is important to use Ca-free media in order to link the slowing of relaxation in low-Na^+ media directly to inhibition of Na_o-dependent Ca^{2+} extrusion (Ashida and Blaustein 1987a; Ma and Bose 1977). When Ca^{2+} entry is prevented and SR Ca^{2+} sequestration is minimized, the Na_o-dependent component of relaxation may serve as a measure of the relative role the Na–Ca exchanger in extruding Ca^{2+} following cell activation (Fig. 9.6).

9.3.4 Direct measurement of Ca^{2+} movements

Tension serves only as an indirect measure of $[Ca^{2+}]_i$, and its usefulness is hindered by the many drawbacks mentioned above. It is therefore essential to verify conclusions about Na–Ca exchange drawn from tension measurements with direct determinations of Ca^{2+} (and Na^+) movements. In 1973, Reuter, Blaustein, and Haeusler examined the influence of Na^+ gradient changes on

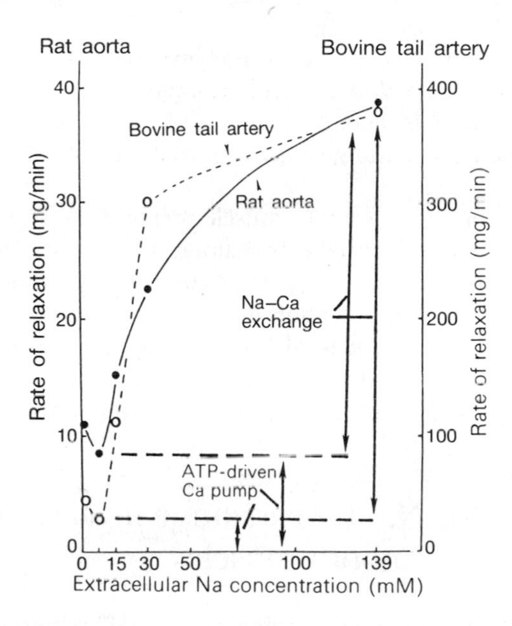

Fig. 9.6. Relationship between $[Na^+]_o$ and the rate of relaxation of rings of rat aorta (closed circles, left-hand ordinate scale) and bovine tail artery (open circles, right-hand ordinate scale). The normal $[Na^+]_o$ was 139.2 mM; in solutions with reduced $[Na^+]$ (indicated on the abscissa), the Na^+ was replaced mole-for-mole by N-methyl glucamine. All solutions contained 10 μM phentolamine (an α-receptor antagonist) and 10 μM verapamil (a Ca^{2+} channel blocker); in the rate aortic ring experiment the superfusion fluid also contained 5 mM caffeine. The Na-dependent (Na–Ca exchange-mediated) and Na-independent (ATP-driven Ca^{2+} pump-mediated) fractions of the relaxation are indicated. From Ashida and Blaustein (1987a) with permission.

Ca^{2+} fluxes, net Ca^{2+} movements, tension, and membrane potential in isolated rabbit blood vessels. They showed that a reduction in $[Na^+]_o$ and/or inhibition of the Na^+ pump promoted Ca^{2+} influx and net Ca^{2+} gain, and caused the tissue to contract. Removal of external Na^+ (in the absence of external Ca^{2+}, to minimize tracer Ca–Ca exchange and net Ca^{2+} uptake) reduced Ca^{2+} efflux reversibly, but did not depolarize the vascular smooth muscle cells. Some of the experiments were carried out in chemically-sympathectomized preparations to rule out the influence of endogenous catecholamine release. Although Ca-dependent Na^+ fluxes and $[Na^+]_i$ were not measured in this study (but see Aaronson and Jones 1988; Kaplan, Kennedy, and Somlyo 1987), the sum of the observations appeared to provide strong evidence that the Ca^{2+} distribution in rabbit vascular smooth muscle is influenced by the Na^+ gradient across the sarcolemma.

Subsequently, Burton and Godfraind (1974) suggested that Na–Ca exchange might account for the increased ^{45}Ca entry and net gain of Ca^{2+} induced by lowering $[K^+]_o$ and thereby raising $[Na^+]_i$ in guinea-pig ileum longitudinal smooth muscle. However, Raeymaekers, Wuytack, and Casteels (1974) obtained evidence that Na–Ca exchange occurs at extracellular sites in guinea-pig taenia coli, and Aaronson and van Breemen (1981, 1982) suggested that such an exchange might occur at intracellular binding sites in this tissue.

Unfortunately, the intact smooth muscle preparations are complex multicellular tissues with a substantial extracellular matrix that can bind ions and serve as a diffusion barrier. This greatly complicates the interpretation of tracer flux and net ion concentration measurements in these tissues. Nevertheless, it is difficult to understand how an exchange in the extracellular matrix could account for the contractile effects of changes in the Na^+ gradient observed in many of the studies mentioned above.

9.4 Evidence for Na–Ca exchange in various mammalian smooth muscles

The preceding discussion covers some of the difficulties encountered in evaluating the presence and physiological role of Na–Ca exchange in smooth muscles. With these considerations in mind we will review a number of reports, including many recent ones, concerning Na–Ca exchange in a variety of smooth muscles. Some of the more recent studie‑ ‑ıvolve new tissue preparations and methodologies in which many of the aforementioned difficulties are circumvented.

9.4.1 Vascular smooth muscle

As indicated above (and see Brading and Lategan 1985), tension serves only as an indirect measure of $[Ca^{2+}]_i$ in vascular smooth muscle (VSM). However, in a number of recent studies, $[Ca^{2+}]_i$ and tracer fluxes of both Ca^{2+} and Na^+ have been measured directly in freshly dissociated VSM cells and in cultured VSM cells. Isolated VSM sarcolemmal vesicles have also been employed for tracer flux studies. For example, in cultured rat aortic VSM cells, there is a Ca_o-dependent ^{22}Na efflux and a Na_i-dependent ^{45}Ca influx, both of which are inhibited by the amiloride analogue, 2′,4′-dimethylbenzamil (Smith *et al.* 1987); the latter agent has been reported to inhibit Na–Ca exchange ten times more effectively than it inhibits Na–H exchange (Smith *et al.* 1987). The enhanced efflux of Ca^{2+} evoked by angiotensin II in these cells also is largely external Na^+-dependent (Nabel *et al.* 1988; Smith and Smith 1987).

Amiloride and its analogues are not selective inhibitors of Na–Ca exchange (e.g. Bielefeld *et al.* 1986). Nevertheless, it seems noteworthy that amiloride inhibits both the contraction induced by reducing $\Delta\bar{\mu}_{Na}$ *and* relaxation induced by increasing this gradient in strips of guinea-pig aorta—even in the presence of a Ca^{2+} channel blocker (Bova, Cargnelli, and Luciani 1988). Reduction of the Na^+ gradient should also lower pH_i which should antagonize the rise in tension, so that the inhibitory effect of amiloride on tension development could also be explained by its inhibition of Na–H exchange. However, amiloride's inhibition of relaxation, when the normal Na^+ gradient is restored, cannot be explained by its effect on pH_i because a fall in pH_i should promote relaxation.

While most studies of the Na–Ca exchanger in smooth muscle have focussed on Na^+ gradient-dependent Ca^{2+} movements, it is essential to show that these movements are coupled to countermovements of Na^+. Recent studies indicate that a component of ouabain-insensitive ^{22}Na efflux in rabbit portal vein is Ca_o-dependent (Kaplan, Kennedy, and Somylo 1987). Also, a component of the ^{22}Na influx in rabbit aorta is Ca_i-dependent (Aaronson and Jones 1988). Although these experiments were not designed to study Na–Ca exchange, this counterion dependence is circumstantial evidence that these ^{22}Na fluxes may be manifestations of Na–Ca exchange. Moreover, these findings complement earlier observations on the Na_i-dependence of ^{45}Ca influx and Na_o-dependence of ^{45}Ca efflux in rabbit VSM (Reuter, Blaustein, and Haeusler 1973).

Several investigators have employed tracer flux studies to examine the properties of the Na–Ca exchanger in sarcolemmal vesicles from VSM (e.g. Daniel 1985; Kahn, Allen, and Shelat 1988; Matlib 1988; Matlib and Reeves 1987; Matlib, Schwartz, and Yamori 1985; Morel and Godfraind 1984). Most of these studies have been limited to the investigation of Na^+ gradient-dependent ^{45}Ca fluxes; however, Kahn, Allen, and Shelat (1988) also demonstrated Ca gradient-dependent ^{22}Na movements both into and out of the vesicles. They and Matlib (1988) also showed that the uptake of ^{45}Ca into the vesicles was not only dependent upon an outwardly-directed Na^+ gradient, but was promoted by depolarization of the vesicles with an inward K^+ gradient and valinomycin. This parallels observations on the effects of depolarization in intact arterial preparations (Ashida and Blaustein 1987a), and fits with the idea that the exchange stoichiometry is 3 Na^+:1 Ca^{2+} (Sheu and Blaustein 1986; Rasgado-Flores and Blaustein 1987).

One problem with the VSM sarcolemmal vesicle studies is that the maximum velocity of the Na^+ gradient-dependent Ca^{2+} flux in most (but not all) studies is low, and the apparent half-maximal concentration for activation by Ca^{2+} is high (Morel and Godfraind 1984). However, the validity of this type of experiment depends upon precise knowledge of the yield of non-leaky vesicles (cf. Daniel 1985). Recent attempts have yielded vesicles with high exchanger activity (e.g. Kaczorowski *et al.* 1987; Matlib 1988) that are more compatible with the physiological observations on intact artery preparations (Ashida and Blaustein 1987a). The latter report clearly indicates that the exchanger operates when $[Ca^{2+}]_i$ and $[Ca^{2+}]_o$ are in the normal physiological range. Indeed, this is suggested by recent measurements of $[Ca^{2+}]_i$ with the Ca^{2+}-sensitive fluorochrome, fura-2, in dissociated single VSM cells (Blaustein *et al.* 1986) and in cultured aortic cells (Smith and Smith 1987). Digital fluorescent images of fura-2-loaded bovine tail artery cells directly demonstrate that $[Ca^{2+}]_i$ increases occur in association with contraction when $[Na^+]_i$ is increased and $[Na^+]_o$ is reduced (Fig. 9.7).

This extensive evidence that a Na–Ca exchanger is present and can transport Ca^{2+} at a substantial rate in many VSMs suggests that this transport system plays an important physiological role. Indeed, it seems reasonable to conclude that the exchanger, operating in parallel with the sarcolemmal Ca^{2+} channels and ATP-driven Ca^{2+} pump, helps to modulate $[Ca^{2+}]_i$. In VSMs that are tonically activated, the exchanger must, therefore,

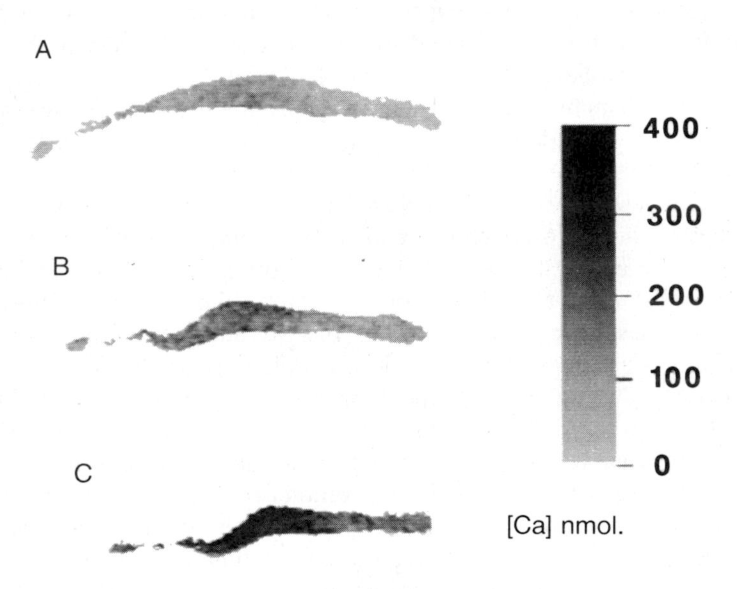

Fig. 9.7. 'Free calcium images' of a single, isolated, fura-2 loaded bovine tail artery cell. The normal resting free Ca^{2+} (A), and the effects of K-free media (B) and K-free, low-Na^+ media (C) on $[Ca^{2+}]_i$ are illustrated. All solutions contained 10 μM verapamil; Na^+ (139.2 mM in the control solution) was replaced mole-for-mole by N-methyl glucamine in the low-Na^+ solution ($[Na^+]_o = 1.2$ mM). The calculated mean $[Ca^{2+}]_i$ in the resting cell (A) was 98 nM. When the cell was superfused with K-free solution, the $[Ca^{2+}]_i$ increased to about 160 nM in the central segment of the cell. Subsequent reduction of $[Na^+]_o$ to 1.2 mM caused $[Ca^{2+}]_i$ to increase to about 310 nM in the central segment. The images were constructed by digital analysis of video-taped fura-2 fluorescent images. Data from W. F. Goldman, W. G. Wier, and M. P. Blaustein (see Blaustein *et al.* 1986).

participate in the modulation of vascular tone and peripheral vascular resistance (Lang and Blaustein 1980; Reuter, Blaustein, and Haeusler 1973). There is much evidence (e.g. Blaustein 1977a,b; Blaustein and Hamlyn 1983, 1984, 1985; deWardener and MacGregor 1980; Haddy and Overbeck 1976; Hamlyn *et al.* 1982; Songu-Mize, Bealer, and Caldwell 1983) that elevated levels of a Na^+ pump inhibitor (endogenous digitalis) are present in the plasma of many hypertensive human subjects and animals. Accordingly, this Na^+ pump inhibitor may play a role in the pathogenesis of those forms of hypertension (including essential hypertension) that are associated with salt and water retention and volume expansion (Blaustein 1977a,b; Haddy and Overbeck 1976; Hamlyn and Blaustein 1986) by contributing to the increased vascular reactivity and peripheral vascular resistance that is the hallmark of this disease (or group of diseases).

9.4.2 Tracheal smooth muscle

It has long been known that respiratory tract smooth muscle contracts in response to inhibition of the Na^+ pump (Chideckel *et al.* 1987; Marco, Park and Aviado 1968; Souhrada and Souhrada 1982). Reduction of external Na^+ also promotes contraction in tracheal smooth muscle from a variety of mammalian species, including man, and these findings have been ascribed to an Na–Ca exchange mechanism (see Bullock, Fettes, and Kirkpatrick 1981; Chideckel *et al.* 1987; Kawanishi, Baba, and Tomita 1984). This view is supported by recent evidence that sarcolemmal vesicles from tracheal smooth muscle exhibit a large capacity Na^+ gradient-dependent Ca^{2+} influx and efflux (Slaughter, Welton, and Morgan 1987). The presence of an Na–Ca exchange mechanism in airway smooth muscle leads to the suggestion that it may play a role in the control of bronchomotor tone (Chideckel *et al.* 1987). Whether or not this contributes to the production of bronchospasm in diseases such as asthma remains to be investigated.

9.4.3 Uterine smooth muscle

Reduction of Na^+ in the bathing medium has been shown to induce contractions in both the non-pregnant and pregnant rat myometrium (Masahashi and Tomita 1983; Matsuzawa *et al.* 1987). In circular uterine smooth muscle from non-pregnant animals, reduction of $[Na^+]_o$ produces a small stimulation of contraction, but this effect is greatly enhanced during pregnancy (Matsuzawa *et al.* 1987). Also, relaxation from K-induced contractures of rat uterine smooth muscle is markedly slowed in Na-deficient media (Ma and Bose 1977). These effects have all been attributed, at least in part, to Na–Ca exchange. This view is supported by Daniel and his colleagues (Daniel 1985; Grover, Kwan, and Daniel 1981; Grover *et al.* 1983), who have shown that sarcolemmal vesicles from rat myometrium exhibit Na^+ gradient-dependent Ca^{2+} influx and efflux.

The role played by Na–Ca exchange in the control of myometrial tone and in the development of uterine contractions during delivery of the foetus is unexplored.

9.4.4 Intestinal smooth muscle

Judah and Willoughby (1964) first showed that guinea-pig ileum develops contractures in Na^+-depleted solutions; subsequent relaxation is Na^+-dependent and is inhibited by strophanthin-G. However, Taniyama (1974) found that low $[Na^+]_o$ inhibits agonist-activated contractions, although the mechanism is uncertain. Low-Na^+ or Na-free media also induce contractions in rat ileal smooth muscle strips (Huddart 1981; Huddart and Saad 1978). Furthermore, Ca^{2+} efflux is inhibited by removal of external Na^+ in the absence of Ca^{2+} (to inhibit tracer Ca^{2+} exchange) (Huddart 1981). Morel and Godfraind (1984) subsequently identified Na–Ca exchange activity (Na^+ gradient-dependent Ca^{2+} fluxes) in sarcolemmal vesicles prepared from guinea-pig ileum, but the specific activity was quite low.

Replacement of external Na^+ by Tris-induced contractions but antagonized agonist-induced contractions in guinea-pig taenia coli (Brading, Burnett, and Sneddon 1980). Relaxation following K^+ contractions was inhibited by removal of external Na^+ (Katase and Tomita 1972; Ma and Bose 1977), and the Na^+-dependent relaxation was slowed when the tissue was treated with ouabain or K^+-free media to inhibit the Na^+ pump (Katase and Tomita 1972). Indeed, the latter treatments may also induce contractions in taenia (Ozaki, Shida, and Urakawa 1978). These findings may be correlated with the observation that Ca^{2+} influx is enhanced in low $[Na^+]_o$ solutions, and re-admission of Na^+ stimulates Ca^{2+} efflux in this tissue (Aaronson and van Breemen 1981, 1982). In sum, these data indicate that the availability of Ca^{2+} for contraction is influenced by the Na^+ gradient across the taenia sarcolemma. More direct evidence comes from the fura-2 studies of Pritchard and Ashley (1986, 1987) on suspensions of dissociated taenia coli cells. In these studies, $[Ca^{2+}]_i$ increased when the cells were exposed to ouabain or Na^+-free media, and the rate of $[Ca^{2+}]_i$ increase was speeded up when ouabain-treated cells were subsequently exposed to Na^+-free media.

These data suggest that Na–Ca exchange may contribute to the control of tone and the Ca^{2+} content of the SR store in intestinal smooth muscle. The precise physiological and pathophysiological significance, in terms of its influence on intestinal motility, is unknown. It seems logical, however, to consider the possibility that Na–Ca exchange may contribute to the maintenance of tension in sphincters, and to the pathogenesis of intestinal spasm and other such gastrointestinal motility disturbances.

9.4.5 Urinary tract smooth muscle

9.4.5.1 *Ureter*

Smooth muscle from guinea-pig ureter is a phasic muscle that normally develops contractions following action potential generation (Aickin, Brading, and Burdyga 1984). Measurements of tension, membrane potential, $[Na^+]_i$, and ^{22}Na efflux in this tissue (Aickin 1987; Aickin, Brading, and Burdyga 1984; Aickin, Brading, and Walmsley 1987) provide strong evidence that it contains a Na–Ca exchange system. The exchanger appears to be relatively dormant in the resting tissue (although its effect on the SR store of Ca^{2+} has not been investigated): exposure to low-Na^+ solutions, or to K^+-free solutions or ouabain, produces little tension in relaxed tissue (see Sections 9.2 and 9.3 above). After Na^+ loading with K^+-free solutions or ouabain, however, the tissue does develop contractures when $[Na^+]_o$ is reduced (Aickin, Brading, and Burdyga 1984); the tension is enhanced when the Na^+-depleted solutions contain elevated K^+ concentrations (Aickin, Brading, and Burdyga 1984), as expected for contractions caused by Ca^{2+} entry via an exchanger with a stoichiometry of 3 Na^+:1 Ca^{2+}. As in the heart (Deitmer and Ellis 1978), the Na–Ca exchanger helps to keep $[Na^+]_i$ from rising above about 20 mM by promoting Ca_o-dependent Na^+ efflux when the tissue is treated with ouabain (Aickin 1987; Aickin, Brading, and Walmsley 1987). These findings raise the possibility that the Na–Ca exchanger is able to extrude Na^+ at the same rate as it enters the cells when $[Na^+]_i$ is about 20 mM. Alternatively, another mechanism of Na^+ entry (such as Na–H exchange) may be activated when the Na^+ pump is inhibited, and this mechanism may subsequently be turned off by the Na–Ca exchange-mediated rise in $[Ca^{2+}]_i$ (e.g. Cala, Mandel, and Murphy 1986; and P. M. Cala, personal communication).

Although Aickin, Brading, and Burdyga, and Aickin, Brading, and Walmsley (1984, 1987, respectively) conclude that Na–Ca exchange plays a major role in regulating Ca^{2+} and Na^+ in ureter smooth muscle when the Na^+ pump is inhibited, the question of its role under normal physiological conditions remains unanswered. Some possibilities are that low rate of exchanger turnover under resting conditions is nevertheless sufficient to modulate the store of Ca^{2+} in the ureter smooth muscle SR, and that the exchanger may help to extrude Ca^{2+} following cell activation (cf. Fig. 3 in Aickin, Brading, and Burdyga 1984), as it does in the heart (Sheu and Blaustein 1986).

9.4.5.2 *Vas deferens*

Data on this tissue are less clear-cut. For example, Katsuragi, Fukushi, and Suzuki (1978) observed that, although ouabain induced contractions in guinea-pig vas deferens, this effect was associated with enhanced noradrenaline release from the tissue; moreover, the contractions were markedly attenuated (but not abolished) by α-blockers. However, ouabain also potentiated the Ca-dependent contractions activated by 40 mM K^+ (Katsuragi and Ozawa 1978). In rat vas deferens, Na-free solutions induce spontaneous phasic contractions and enhance the amplitude and duration of the initial contractions induced by noradrenaline (Wakui and Inomata 1984). These findings raise the possibility that vas deferens may have an Na–Ca exchanger, but additional data are required to resolve the uncertainty.

9.5 Summary and general conclusions

The data described above indicate that a Na–Ca exchange system is present in most (if not all) types of mammalian smooth muscle that have been examined to date. However, some questions about its role in the physiology and pathophysiology of these tissues remain unanswered.

Recent studies on the regulation of the Na–Ca exchanger in various types of cells including cardiac muscle (Kimura, Noma, and Irisawa 1986; Kimura, Miyamae, and Noma 1987; Sheu and Blaustein 1986), squid axons (Blaustein 1977c; DiPolo 1976; DiPolo and Beaugé 1984, 1986, 1987), and barnacle muscle (Rasgado-Flores and Blaustein 1987; Rasgado-Flores, Santiago and Blaustein 1989) indicate that the exchanger is much more complex than originally expected. It is regulated by $[Ca^{2+}]_i$ and ATP. It may be relatively dormant at resting $[Ca^{2+}]_i$ and may be activated when $[Ca^{2+}]_i$ rises as a result of cell activation (Sheu and Blaustein 1986). Moreover, exchanger-mediated increases in $[Ca^{2+}]_i$ may be attenuated by rapid sequestration of the entering Ca^{2+} in intracellular organelles (Ashida and Blaustein 1987a). Confusion about the presence and role of Na–Ca exchange in smooth muscles (e.g. Brading and Lattegan 1985; Casteels *et al.* 1985; Mulvany 1985; Van Breemen, Aaronson, and Loutzenheiser 1979) has resulted from 1) Lack of data about these phenomena, many of which have been discovered only recently. 2) From the difficulty in separating the effects of membrane depolarization (which may open Ca^{2+} and or Na^+ channels) from the effects of changing $\Delta\bar{\mu}_{Na}$. 3) The difficulty in avoiding changes in $[Na^+]_i$ and pH_i when $[Na^+]_o$ is reduced because of the rapid efflux of Na^+ (e.g. Aickin 1987; Kaplan, Kennedy, and Somlyo 1987) and the inhibition of Na–H exchange, respectively. Each of these problems has been taken into account or circumvented in at least some of the studies reviewed above. The sum of the data provide convincing evidence that substantial Na–Ca exchange activity is present in the sarcolemma of all of the types of smooth muscle mentioned above. In many instances, little exchanger-mediated Ca^{2+} movement can be identified in these tissues under resting conditions (when $[Ca^{2+}]_i$ is below the contraction threshold); but even with a slow turnover rate, the exchanger may regulate (indirectly) the amount of Ca^{2+} stored in the SR and available for phasic contractions. Moreover, the exchanger may directly modulate $[Ca^{2+}]_i$ and tone when $[Ca^{2+}]_i$ is maintained above the contraction threshold. Consequently, Na–Ca exchange appears to be very important not only because of its role in normal smooth muscle function, but also because of its likely contribution to altered smooth muscle function in various pathological conditions, some of which were mentioned above.

Acknowledgements

I thank Dr. J. Schaeffer for critical comments on the manuscript.

Supported by NIH grant AM32276 and a grant from the Muscular Dystrophy Association of America.

References

Aalkjaer, C. and Mulvany, M. J. (1985). Effect of ouabain on tone, membrane potential and sodium efflux compared with [^{3}H] ouabain binding in rat resistance vessels. *J. Physiol.* **362**, 215–31.

Aaronson, P. I. and Jones, A. W. (1988). Ca dependence of Na influx during treatment of rabbit aorta with NE and K solutions. *Am. J. Physiol.* **254**, C75–83.

Aaronson, P. and van Breemen, C. (1981). Effects of sodium gradient manipulation upon cellular calcium, ^{45}Ca fluxes and cellular sodium in the guinea-pig taenia coli. *J. Physiol* **319**, 443–61.

Aaronson, P. and van Breemen, C. (1982). Effect of Na re-admission on cellular ^{45}Ca fluxes in Na-depleted guinea pig taenia coli. *J. Membr. Biol.* **65**, 89–98.

Adelstein, R. S. and Klee, C. B. (1980). Smooth muscle myosin light chain kinase. In: *Calcium and cell function.* Vol. I. *Calmodulin* (ed. W. Y. Cheung) pp. 167–82. Academic Press, New York.

Aickin, C. C. (1987). Investigation of factors affecting the intracellular sodium activity in the smooth muscle of guinea-pig ureter. *J. Physiol.* **385**, 483–505.

Aickin, C. C., Brading, A. F., and Burdyga, Th. V. (1984). Evidence of sodium–calcium exchange in the guinea-pig ureter. *J. Physiol.* **347**, 411–30.

Aickin, C. C., Brading, A. F., and Walmsley, D. (1987). An investigation of sodium–calcium exchange in the smooth muscle of guinea-pig ureter. *J. Physiol.* **391**, 325–46.

Ashida, T. and Blaustein, M. P. (1987a). Regulation of cell calcium and contractility in mammalian arterial smooth muscle: the role of sodium–calcium exchange. *J. Physiol.* **392**, 617–35.

Ashida, T. and Blaustein, M. P. (1987b). Control of contractility and the role of Na/Ca exchange in arterial smooth muscle. *J. Cardiovasc. Pharmacol.* **10** (Suppl. 10), S65–7.

Ashida, T., Schaeffer, J., Goldman, W. F., Wade, J. B., and Blaustein, M. P. (1988). Role of sarcoplasmic reticulum in arterial contraction: comparison of ryanodine's effect in a conduit and a muscular artery. *Circ. Res.* **62**, 854–63.

Baker, P. F. and Blaustein, M. P. (1968). Sodium-dependent uptake of calcium by crab nerve. *Biochim. Biophys. Acta* **150**, 167–70.

Baker, P. F., Blaustein, M. P., Hodgkin, A. L., and Steinhardt, R. A. (1969). The influence of calcium ions on sodium efflux in squid axons. *J. Physiol.* **200**, 431–58.

Baker, P. F. and McNaughton, P. A. (1977). Selective inhibition of the Ca-dependent Na efflux from intact squid axons by a fall in intracellular pH. *J. Physiol.* **269**, 78P–79P.

Bean, B. P., Sturek, M., Puga, A., and Hermsmeyer, K. (1986). Calcium channels in muscle cells isolated from rat mesenteric arteries: modulation by dihydropyridine drugs. *Circ. Res.* **59**, 229–35.

Benham, C. D., Hess, P., and Tsien, R. W. (1987). Two types of calcium channels in single smooth muscle cells from rabbit ear artery studies with whole-cell and single-channel recordings. *Circ. Res.* **61** (Suppl. I): I-10–I-16.

Bielefeld, D. R., Hadley, R. W., Vassilev, P. M., and Hume, J. R. (1986). Membrane electrical properties of vesicular Na/Ca-exchange inhibitors in single atrial myocytes. *Circ. Res.* **59**, 381–9.

Blaustein, M. P. (1974). The interrelationship between sodium and calcium fluxes across cell membranes. *Rev. Physiol. Biochem. Pharmacol.* **70**, 32–82.

Blaustein, M. P. (1977a). Sodium ions, calcium ions, blood pressure regulation and hypertension: a reassessment and a hypothesis. *Am. J. Physiol.* **323** (Cell Physiol. 1), C165–73.

Blaustein, M. P. (1977b). The role of Na–Ca exchange in the regulation of tone in vascular smooth muscle. In: *Excitation-contraction coupling in smooth muscle* (eds R. Casteels, T. Godfraind, and J. C. Reugg) pp. 101–8. Elsevier North-Holland Biomedical Press B.V., Amsterdam.

Blaustein, M. P. (1977c). Effects of internal and external cations and ATP on sodium–calcium exchange and calcium–calcium exchange in squid axons. *Biophys. J.* **20**, 79–111.

Blaustein, M. P. (1985). How salt causes hypertension: the natriuretic hormone–Na–Ca exchange–hypertension hypothesis. *Klin. Wochenschr.* **63** (Suppl. III), 82–85.

Blaustein, M. P. (1987). Calcium and synaptic function. *Handbook Exp. Pharmacol.* **83**, 275–304.

Blaustein, M. P., Ashida, T., Goldman, W., Wier, W., and Hamlyn, J. (1986). Sodium/calcium exchange in vascular smooth muscle: a link between sodium metabolism and hypertension. *Ann. NY Acad. Sci.* **488**, 199–216.

Blaustein, M. P. and Hamlyn, J. M. (1983). Role of natriuretic factor in essential hypertension. *Ann. Intern. Med.* **98** (No. 5, Pt. 2), 785–92.

Blaustein, M. P. and Hamlyn, J. M. (1984). Sodium transport inhibition, cell calcium, and hypertension. The natriuretic hormone —Na^+/Ca^{2+} exchange- hypertension hypothesis. *Am. J. Med.* **77**(4A), 45–59.

Blaustein, M. P. and Hamlyn, J. M. (1985). Role of endogenous inhibitor of Na pumps in the pathophysiology of essential hypertension. In: *The sodium pump* (eds I. M. Glynn and C. Ellory) pp 629–39. The Company of Biologists Ltd, Cambridge (UK).

Blaustein, M. P. and Hodgkin, A. L. (1969). The effect of cyanide on the efflux of calcium from squid axons. *J. Physiol* **200**, 497–527.

Bohr, D. F., Seidel, C., and Sobieski, J. (1969). Possible role of sodium–calcium pumps in tension development of vascular smooth muscle. *Microvasc. Res.* **1**, 335–43.

Bonaccorsi, A., Hermsmeyer, K., Smith, C. B., and Bohr, D. F. (1977). Norepinephrine release in isolated arteries induced by K-free solution. *Am. J. Physiol.* **232**, H140–5.

Bova, S., Cargnelli, G., and Luciani, S. (1988). Na/Ca exchange and tension development in vascular smooth muscle: Effect of amiloride. *Br. J. Pharmacol.* **93**, 601–8.

Brading, A. F., Burnett, M., and Sneddon, P. (1980). The effect of sodium removal on the contractile responses of the guinea-pig taenia coli to carbachol. *J. Physiol.* **306**, 411–29.

Brading, A. F. and Lategan, T. W. (1985). Na–Ca exchange in vascular smooth muscle. *J. Hyperten.* **3**, 109–16.

Bullock, C. G., Fettes, J. J. F., and Kirkpatrick, C. T. (1981). Tracheal smooth muscle—second thoughts on sodium–calcium exchange. *J. Physiol.* **318**, 46P.

Burton, J. and Godfraind, T. (1974). Sodium–calcium sites in smooth muscle and their accessibility to lanthanum. *J. Physiol.* **241**, 287–98.

Cala, P. M., Mandel, L. J., and Murphy, E. (1986). Volume regulation by *Amphiuma* red blood cells: cytosolic free Ca and alkali metal-H exchange. *Am. J. Physiol.* **250**, C423–9.

Carafoli, E. (1987). Intracellular calcium homeostasis. *Ann. Rev. Biochem.* **56**, 395–433.

Casteels, R., Raeymaeker, Droogmans, G., and Wuytack, F. (1985). Na^+–K^+ ATPase, Na–Ca exchange, and excitation-contraction coupling in smooth muscle. *J. Cardiovasc. Pharmacol.* **7**, S103–10.

Chapman, R. A. (1983). Control of cardiac contractility at the cellular level. *Am. J. Physiol.* **245**, H535–52.

Chideckel, E. W., Frost, J. L., Mike, P., and Fedan, J. S. (1987). The effect of ouabain on tension in isolated respiratory tract smooth muscle of humans and other species. *Br. J. Pharmacol.* **92**, 609–14.

Daniel, E. E. (1985). The use of subcellular membrane fractions in analysis of control of smooth muscle function. *Experientia*, **41**, 905–13.

Deitmer, J. W. and Ellis, D. (1978). Changes in the intracellular sodium activity of sheep heart Purkinje fibres produced by calcium and other divalent cations. *J. Physiol.* **277**, 437–53.

Devine, C. E., Somlyo, A. V., and Somlyo, A. P. (1972). Sarcoplasmic reticulum and excitation-contraction coupling in mammalian smooth muscle. *J. Cell Biol.* **52**, 690–718.

deWardener, H. E. and MacGregor, G. A. (1980). Dahl's hypothesis that a saluretic substance may be responsible for a sustained rise in arterial pressure: its possible role in essential hypertension. *Kidney Intl.* **18**, 1–9.

DiPolo, R. (1976). The influence of nucleotides upon calcium fluxes. *Fed. Proc.* **35**, 2579–82.

DiPolo, R. and Beaugé, L. (1982). The effect of pH on Ca^{2+} extrusion mechanisms in dialyzed squid axons. *Biochim. Biophys. Acta* **688**, 237–45.

DiPolo, R. and Beaugé, L. (1984). Interaction of physiological ligands with the Ca pump and Na/Ca exchange in squid axons. *J. Gen. Physiol.* **84**, 895–914.

DiPolo, R. and Beaugé, L. (1986). Reverse Na–Ca exchange requires internal Ca and/or ATP in squid axons. *Biochim. Biophys. Acta* **854**, 298–306.

DiPolo, R. and Beaugé, L. (1987). Characterization of the reverse Na/Ca exchange in squid axons and its modulation by Ca_i and ATP. Ca-dependent Na_i/Ca_o and Na_i/Na_o exchange modes. *J. Gen. Physiol.* **90**, 505–26.

Dixon, W. E. and Brodie, T. G. (1903). Contributions to the physiology of the lungs. Part 1. The bronchial muscles, their innervation and the action of drugs upon them. *J. Physiol.* **29**, 97–173.

Eggermont, J. A., Vrolix, M., Raeymaekers, L., Wuytack, F., and Casteels, R. (1988). Ca^{2+}-transport ATPases of vascular smooth muscle. *Cir. Res.* **62**, 266–78.

Grover, A. K., Kwan, C. Y., and Daniel, E. E. (1981). Na-Ca exchange in rat myometrium membrane vesicles highly enriched in plasma membrane. *Am. J. Physiol.* **340**, C175–82.

Grover, A. K., Kwan, C. Y., Rangachari, P. K., and Daniel, E. E. (1983). Na–Ca exchange in smooth muscle plasma membrane enriched fraction. *Am. J. Physiol.* **244**, C158–65.

Haddy, F. J. and Overbeck, H. J. (1976). The role of humoral agents in volume expanded hypertension. *Life Sci.* **19**, 935–48.

Hamlyn, J. M., Ringel, R., Schaeffer, J., Levinson, P. D., Hamilton, B. P., Kowarski, A. A., and Blaustein, M. P. (1982). A circulating inhibitor of (Na + K)-ATPase associated with essential hypertension. *Nature* **300**, 650–2.

Hamlyn, J. M. and Blaustein, M. P. (1986). Sodium chloride, extracellular fluid volume, and blood pressure regulation. *Am. J. Physiol.* **251**, F563–75.

Hansford, R. G. (1985). Relation between mitochondrial calcium transport and control of energy metabolism. *Rev. Physiol. Biochem. Pharmacol.* **102**, 1–72.

Hashimoto, T., Hirata, M., Itoh, T., Kanamura, Y., and Kuriyama, H. (1986). Inositol 1,4,5-triphosphate activates pharmacomechanical coupling in smooth muscle of the rabbit mesenteric artery. *J. Physiol* 370, 605–18.

Hinke, J. A. M. and Wilson, M. L. (1962). Effects of electrolytes on contractility of artery segments *in vitro*. *Am. J. Physiol.* **203**, 1161–6.

Huddart, H. (1981). Calcium regulation in ileal smooth muscle I. Sodium–calcium counterexchange. *Gen. Pharmacol.* **12**, 155–60.

Huddart, H. and Saad, K. H. M. (1978). The effect of sodium and magnesium and their interaction with quinine and lanthanum on spontaneous activity and related calcium movements of rat ileal smooth muscle. *J. Comp. Physiol.* **126**, 233–40.

Itoh, T., Kajiwara, M., Katimura, K., and Kuriyama, H. (1982). Roles of stored calcium on mechanical response evoked in smooth muscle cells of the porcine coronary artery. *J. Physiol.* **322**, 107–25.

Itoh, T., Kuriyama, H., and Suzuki, H. (1983). Differences and similarities in the noradrenaline- and caffeine-induced mechanical responses in the rabbit mesenteric artery. *J. Physiol.* **337**, 609–29.

Johansson, B. and Hellstrand, P. (1987). Contracture induced by reversed Na^+/Ca^{2+} exchange in rat portal vein: Effects of calcium antagonists. *J. Cardiovasc. Pharmacol.* **10**, S75–S81.

Johansson, B. and Somlyo, A. (1980). Electrophysiology and excitation-contraction coupling. In: *Handbook of physiology: vascular smooth muscle* (eds D. Bohr, A. Somlyo, and H. Sparks) pp 301–24. American Physiological Society Bethesda, MD. USA.

Judah, J. D. and Willoughby, D. A. (1964). Inhibitors of sodium dependent relaxation of guinea-pig ileum. *J. Cell Comp. Physiol.* **64**, 363–70.

Kaczorowski, G. J., Slaughter, R. S., Garcia, M. L., and King, V. F. (1987). Sodium–calcium exchange in mammalian aortic smooth muscle sarcolemma. *First Int. Meeting on Sodium-Calcium Exchange* (Abstr. No. 4).

Kahn, A. M., Allen, J. C., Cragoe, E. J., Jr., Zimmer, R., and Shelat, H. (1988). Sodium–lithium exchange in sarcolemmal vesicles from canine superior mesenteric artery. *Circ. Res.* **62**, 478–85.

Kahn, A. M., Allen, J. C., and Shelat, H. (1988). Na-Ca exchange in sarcolemmal vesicles from bovine superior mesenteric artery. *Am. J. Physiol.* **254**, C441–9.

Kahn, A. M., Shelat, H., and Allen, J. C. (1986). Na^+–H^+ exchange is present in sarcolemmal vesicles from dog superior mesenteric artery. *Am. J. Physiol.* **250**, H313–19.

Kaplan, J. H., Kennedy, B. G., and Somlyo, A. P. (1987). Calcium-stimulated sodium efflux from rabbit vascular smooth muscle. *J. Physiol.* **388**, 245–60.

Katase, T. and Tomita, T. (1972). Influence of sodium and calcium on the recovery process from potassium contracture in the guinea-pig taenia coli. *J. Physiol.* **224**, 489–500.

Katsuragi, T. and Ozawa, H. (1978). Ouabain-induced potentiation of Ca^{2+} contraction in the depolarized vas deferens of guinea-pig. *Arch. Intl. Pharmacodyn. Ther.* **231**, 243–48.

Katsuragi, T., Fukushi, Y., and Suzuki, T. (1978). Neuronal norepinephrine as mediator for ouabain-induced smooth muscle contraction. *Eur. J. Pharmacol.* **47**, 407–13.

Kawanishi, M., Baba, K., and Tomita, T. (1984). Effects of Na removal and readmission on the mechanical response in the guinea-pig tracheal smooth muscle. *Jap. J. Physiol.* **34**, 127–39.

Kimura, J., Miyamae, S., and Noma, A. (1987). Identification of sodium–calcium exchange current in single ventricular cells of guinea-pig. *J. Physiol.* **384**, 199–222.

Kimura, J., Noma, A., and Irisawa, H. (1986). Na–Ca exchange current in mammalian heart cells. *Nature*, **319**, 596–7.

Lang, S. and Blaustein, M. P. (1980). The role of the sodium pump in the control of vascular tone. *Circ. Res.* **46**, 463–70.

Leonard, E. (1957). Alteration of contractile responses of artery strips by a potassium-free solution, cardiac glycosides and changes in stimulus frequency. *Am. J. Physiol.* **189**, 185–90.

Lorenz, R. R., Powis, D. A., Vanhoutte, P. M., and Shepherd, J. T. (1980). The effects of acetylstrophanthidin and ouabain on the sympathetic adrenergic neuroeffector junction in canine vascular smooth muscle. *Circ. Res.* **47**, 845–54.

Ma, T. S. and Bose, D. (1977). Sodium in smooth muscle relaxation. *Am. J. Physiol.* **232**, C59–66.

Magyar, K., Nguyen, T. T., Torok, T. L., and, Toth, P. T. (1987). [^{3}H]noradrenaline release from rabbit pulmonary artery: sodium-pump-dependent sodium–calcium exchange. *J. Physiol.* **393**, 29–42.

Marco, V., Park, C. D., and Aviado, D. M. (1968). Bronchopulmonary effects of digitalis in the anaesthetized dog. *Dis. Chest*, **54**, 437–44.

Masahashi, T. and Tomita, T. (1983). The contracture produced by sodium removal in the non-pregnant rat myometrium. *J. Physiol.* **334**, 351–63.

Mason, D. T. and Braunwald, E. (1964). Studies on digitalis. X. Effects of ouabain on forearm vascular resistance and venous tone in normal subjects and in patients with heart failure. *J. Clin. Invest.* **43**, 532–43.

Matlib, M. M. (1988). Functional characteristics of a Na^+–Ca^{2+} exchange system in sarcolemmal membrane vesicles of vascular smooth muscle. *Biophys. J.* **53** (No. 2, Pt. 2), 346a.

Matlib, M. A. and Reeves, J. P. (1987). Solubilization and reconstitution of the sarcolemmal Na^+–Ca^{2+} exchange system of vascular smooth muscle. *Biochim. Biophys Acta* **904**, 145–8.

Matlib, M. A., Schwartz, A., and Yamori, Y. (1985). A Na^+–Ca^{2+} exchange process in isolated sarcolemmal membranes of mesenteric arteries from WKY and SHR rats. *Am. J. Physiol.* **249**, C166–72.

Matsuzawa, K., Masahashi, T., Kihira, M., and Tomita, T. (1987). Contracture caused by sodium removal in the pregnant rat myometrium. *Jap. J. Physiol.* **37**, 19–31.

Morel, N. and Godfraind, T. (1984). Sodium/calcium exchange in smooth-muscle microsomal fraction. *Biochem. J.* **218**, 421–7.

Mulvany, M. J. (1985). Changes in sodium pump activity and vascular contraction. *J. Hyperten.* **3**, 429–36.

Mulvany, M. J., Aalkjaer, C., and Petersen, T. T. (1984). Intracellular sodium, membrane potential, and contractility of rat mesenteric small arteries. *Circ. Res.* **54**, 740–9.

Nabel, E. G., Berk, B. C., Brock, T. A., and Smith, T. W. (1988). Na^+–Ca^{2+} exchange in cultured vascular smooth muscle cells. *Circ. Res.* **62**, 486–93.

Nelson, M. T. and Worley, J. F., III. (1988). Dihydropyridine inhibition of single calcium channels and contraction in rabbit mesenteric artery depends upon voltage. *J. Physiol.* (in press).

Ozaki, H., Ishida, Y., and Urakawa, N. (1978). Properties of contractions induced by ouabain and K^+-free solution in guinea pig taenia coli. *Eur. J. Pharmacol.* **50**, 9–15.

Ozaki, H., Karaki, H., and Urakawa, N. (1978). Possible role of Na–Ca exchange mechanism in the contractions induced in guinea-pig aorta by potassium-free solutions and ouabain. *Naunyn-Schmiedeberg's Arch. Pharmakol.* **304**, 203–9.

Ozaki, H. and Urakawa, N. (1979). Na–Ca exchange and tension development in guinea-pig aorta. *Naunyn-Schmiedeberg's Arch. Pharmakol.* **309**, 171–8.

Ozaki, H. and Urakawa, N. (1981). Involvement of a Na–Ca exchange mechanism in contraction induced by low-Na solution in isolated guinea-pig aorta. *Pflügers Arch.* **390**, 107–112.

Petersen, T. T. and Mulvany, M. J. (1984). Effect of sodium gradient on the rate of relaxation of rat mesenteric small arteries from potassium contractures. *Blood Vessels* **21**, 279–89.

Pritchard, K. and Ashley, C. C. (1986). Na^+/Ca^{2+} exchange in isolated smooth muscle cells demonstrated by the fluorescent calcium indicator fura-2. *FEBS Lett.* **195**, 23–7.

Pritchard, K. and Ashley, C. C. (1987). Evidence for Na^+/Ca^{2+} exchange in isolated smooth muscle cells: a fura-2 study. *Pflügers Arch.* **410**, 401–7.

Raeymaekers, L., Wuytack, F., and Casteels, R. (1974). Na–Ca exchange in taenia coli of the guinea-pig. *Pflügers Arch.* **347**, 329–40.

Rasgado-Flores, H., and Blaustein, M. P. (1987). Na/Ca exchange in barnacle muscle cells has a stoichiometry of 3 Na^+: 1 Ca^{2+}. *Am. J. Physiol.* **252**, C499–C504.

Rasgado-Flores, H., Santiago, E. M., and Blaustein, M. P. (1989). Kinetics and stoichiometry of coupled Na efflux and Ca influx (Na/Ca exchange) in barnacle muscle cells. *J. Gen. Physiol.* (in press).

Reuter, H., Blaustein, M. P., and Haeusler, G. (1973). Na–Ca exchange and tension development in arterial smooth muscle. *Phil. Trans. Roy. Soc. Lond. B* **265**, 87–94.

Reuter, H. and Seitz, H. (1968). The dependence of calcium efflux from cardiac muscle on temperature and external ion composition. *J. Physiol.* **195**, 451–70.

Ruegg, J. C. and Paul, R. J. (1982). Calmodulin and cyclic AMP-dependent protein kinase alter calcium sensitivity in porcine carotid skinned fibers. *Circ. Res.* **50**, 394–99.

Sanchez-Armass, S. and Blaustein, M. P. (1987). Role of Na/Ca exchange in the regulation of intracellular Ca^{2+} nerve terminals. *Am. J. Physiol.* **252**, C595–C603.

Schoeffter, P., Lugnier, C., Demesy-Waeldele, F., and Stocklet, J. C. (1987). Role of cyclic AMP- and cyclic GMP-phosphodiesterases ion the control of cyclic nucleotide levels and smooth muscle tone in rat isolated aorta. A study with inhibitors. *Biochem. Pharmacol.* **36**, 3965–72.

Sekine, K., Yamakawa, K., and Ogata, E. (1984). Na^+,K^+-ATPase activity and responsiveness of vascular smooth muscle to norepinephrine, angiotensin II and calcium ionophore A23187 in guinea pig aortic strips. *Clin. Exp. Hyperten. Theory and Practice* **A6**, 1267–80.

Sheu, S. S. and Blaustein, M. P. (1986). Sodium/calcium exchange and the regulation of cell calcium and contractility in cardiac muscle, with a note about vascular smooth muscle. In: *The heart and cardiovascular system* (eds H. Fozzard, E. Haber, R. Jennings, A. Katz, and H. Morgan) pp 509–35. Raven Press, New York.

Slaughter, R. S., Welton, A. F., and Morgan, D. W. (1987). Sodium–calcium exchange in sarcolemmal vesicles from tracheal smooth muscle. *Biochim. Biophys. Acta* **904**, 92–104.

Smith, J. B., Cragoe, E. J., Jr., and Smith, L. (1987). Na^+/Ca^{2+} antiport in cultured arterial smooth muscle cells. Inhibition by magnesium and other divalent cations. *J. Biol. Chem.* **262**, 11988–94.

Smith, J. B. and Smith, L. (1987). Extracellular Na^+ dependence of changes in free Ca^{2+}, $^{45}Ca^{2+}$ efflux, and total cell Ca^{2+} produced by angiotensin II in cultured arterial muscle cells. *J. Biol. Chem.* **262**, 17455–60.

Smith, L., Higgins, B. L., Cragoe, E. J., Jr., and Smith, J. B. (1987). Comparison of the effectiveness of seven amiloride congeners as inhibitors of Na/H and Na/Ca aintiport in cultured smooth muscle cells. *Fed. Proc.* **46**, 2028.

Somlyo, A. P., Broderick, R., and Somlyo, A. V. (1986). Calcium and sodium in vascular smooth muscle. *Ann. NY Acad. Sci.* **488**, 228–39.

Somlyo, A. V. (1980). Ultrastructure of vascular smooth muscle. In: *The handbook of physiology: the cardiovascular system. Vol. 2. Vascular smooth muscle* (eds D. F. Bohr, A. P. Somlyo, and H. V. Sparks) pp 33–67. American Physiological Society, Bethesda, USA.

Somlyo, A. V., Bond, M., Somlyo, A. P., and Scarpa, A. (1985). Inositol triphosphate-induced calcium release and contraction in vascular smooth muscle. *Proc. Natl. Acad. Sci. USA.* **82**, 5231–5.

Somlyo, A. V. and Franzini-Armstrong, C. (1985). New views of smooth muscle structure using freezing, deep-etching and rotary shadowing. *Experientia*, **41**, 841–56.

Songu-Mize, E., Bealer, S. L., and Caldwell, R. W. (1982). Effect of AV3V lesions on development of DOCA-salt hypertension and vascular Na-pump activity *Hypertension* **4**, 575–80.

Souhrada, M. and Souhrada, J. F. (1982). Potentiation of Na^+-electrogenic pump of airway smooth muscle by sensitization. *Resp. Physiol.* **47**, 69–81.

Taniyama, K. (1974). Modification of the tonus and drug-induced contraction of the isolated rat ileum by removal of Na from bathing media. *Jap. J. Pharmacol.* **24**, 433–8.

Toda, N. (1978). Mechanical responses of isolated dog cerebral arteries to reduction of external K, Na, and Cl. *Am. J. Physiol.* **234**, H404–11.

Van Breemen, C., Aaronson, P., and Loutzenheiser, R. (1979). Sodium–calcium interactions in mammalian smooth muscle. *Pharmacol. Rev.* **30**, 167–208.

Wakui, M. and Inomata, H. (1984). Requirements of external Ca and Na for the electrical and mechanical responses to noradrenaline in the smooth muscle of guinea-pig vas deferens. *Jap. J. Physiol.* **34**, 199–203.

Wray, S. (1988). Smooth muscle intracellular pH: measurement, regulation and function. *Am. J. Physiol.* **254**, C213–25.

Yatani, A., Seidel, C. L., Allen, J., and Brown, A. M. (1987). Whole-cell and single-channel calcium currents of isolated smooth muscle cells from saphenous vein. *Circ. Res.* **60**, 523–33.

10 Sodium-calcium exchange and smooth muscle function in non-vascular smooth muscles

C. C. Aickin and A. F. Brading

10.1 Introduction

As illustrated in the previous chapter, a great deal of effort has gone into finding the possible role of Na–Ca exchange in vascular smooth muscles. The stimulus for this has been the belief that this process may be involved in some forms of hypertension in man (Blaustein 1977). There is, however, considerable evidence that the exchange is present in, and may play a role in determining the behaviour of, smooth muscles other than vascular ones. In this chapter we shall briefly discuss some of the earlier experimental results and the problems in their interpretation, and then describe more recent research which establishes the existence of the mechanism in non-vascular smooth muscles. Finally we will speculate about the physiological significance of the exchange.

10.2 Historical aspects

There was an initial enthusiasm among smooth muscle physiologists for the concept of Na–Ca exchange, stemming from observations of the contractile behaviour of smooth muscles in response to alterations in the transmembrane Na gradient. For example, an increase in resting tone frequently occurs when the extracellular Na concentration is reduced, although the contraction is often transient, tissue specific, and dependent on the Na substitute used. Similarly, tension often develops after elevation of intracellular Na by inhibition of the Na pump with cardiac glycosides or removal of extracellular K (e.g. guinea-pig taenia: Brading 1978, Brading, Burnett, and Sneddon 1980, Katase and Tomita 1972, Tomita and Yamamoto 1971; guinea-pig ileum: James and Roufogalis 1977; small intestine various species: Shimizu *et al.* 1979; mouse and rat uterus: Masahasi and Tomita 1983, Osa 1971, review: Van Breemen, Aaronson, and Loutzenheiser 1978). There is sufficient evidence of this type to make it likely that in some circumstances the exchange does play a role. A particularly striking example is the ability of Na re-admission to relax smooth muscles that have developed tension in Na-free or low-Na media, even in the presence of ouabain. However, the study of contractile responses in different smooth muscles make one aware of the very wide variations in their behaviour, suggesting that there may be considerable tissue differences in the relative importance of the mechanisms involved in Ca homoeostasis.

Interpretation of tension experiments is in any case difficult because there are many ways in which alterations of the Na gradient could directly or indirectly influence free Ca levels in the cells. Apart from the postulated Na–Ca exchange, intracellular Ca is controlled in smooth muscles by a complex interplay of many systems, including Ca entry down its electrochemical gradient through several types of voltage-dependent Ca channel, Ca release from and uptake into Ca stores (again probably involving several mechanisms), and Ca extrusion by ATP-dependent Ca pumps. Any of these might be altered by the experimental manipulations. Further, in tissues with spontaneous electrical activity, tension is linked to spike frequency (probably involving Ca entry and Ca release) and a small change in membrane potential can modify the frequency and thus the tone. Changes in membrane potential can result from alteration of the transmembrane ionic gradients, from changes in membrane permeability, and from alterations in the contributions of electrogenic pumping mechanisms. They can in turn also cause sustained changes in membrane permeability to Ca ions. Unfortunately there is, as yet, no blocker of Na–Ca exchange which is specific for the mechanism, and it is thus difficult to demonstrate unequivocally that alterations in Na–Ca exchange are responsible for any of the contractile responses monitored.

It might be supposed that more direct evidence for the existence of Na–Ca exchange would be obtained from ion flux experiments, which have proved valuable in studying the exchange in such tissues as the squid giant axon. However, it is difficult to interpret results in smooth muscles. The main problems result from the very large surface area: volume ratio, the numerous intra- and extracellular cation binding sites, the multiplicity of the transmembrane pathways involved in Na and Ca movements, and again, no specific blocker for Na–Ca exchange (for review see Brading 1981). Such problems were highlighted in 1974 by the demonstration of Raemaekers, Wuytack, and Casteels that a Na-dependent Ca efflux (which might be indicative of the exchange) could be demonstrated in strips of smooth muscle from the taenia of the guinea-pig caecum in which normal membrane function had been destroyed by boiling! In this case, the effect was due to competition between Na and Ca ions for tissue binding sites. Some evidence does exist in the taenia for fluxes characteristic of the Na–Ca exchange in the presence of extracellular La^{3+}, which is thought to saturate extracellular binding sites (Brading and Widdicombe 1976). These results are, however, equivocal since La has been shown to have inhibitory effects on Na–Ca exchange under some conditions. Recently similar evidence has been obtained under more physiological conditions, using isolated cell suspensions of the taenia (Hirata, Itoh, and Kuriyama 1981).

Over the last 10 years it has become possible to prepare microsomes from the plasma membrane of smooth muscles, and use these to examine ion transport mechanisms (for review see Daniel 1985). Clear evidence for a Na-dependent Ca uptake by microsomes prepared from rat myometrium has been obtained by Grover, Kwan, and Daniel (1981), and further studies have examined the ionic requirements of the uptake. Rb, K, and Cs have been found to be ineffective as Na replacements, while Li substitutes only poorly (Grover *et al* 1983). The capacity of the exchange in isolated membranes is very much lower than that found in similar studies in heart microsomes. In a comparative study of Na- and ATP-dependent Ca uptake in guinea-pig heart and ileal microsomes, Morel and Godfraind (1982) found that Na-induced Ca uptake in smooth muscle had a lower Ca affinity and a lower capacity than ATP-dependent uptake, whereas in heart, the reverse situation was seen. Daniel (1985) has also compared the two systems, and finds the smooth muscle Na-dependent Ca uptake to have an initial velocity of only one hundredth of that seen in cardiac microsomes, although the comparative leakiness to Na of the smooth muscle microsomes might have led to an underestimate. These studies are promising, and certainly confirm the presence of Na–Ca exchange in smooth muscle membranes, but they have not yet advanced as far in smooth muscles as they have in cardiac muscle. They do, however, suggest that the ATP-dependent Ca pump in smooth muscle membranes may play a more dominant role than Na–Ca exchange in controlling intracellular Ca.

More concrete information about Na–Ca exchange should be provided by measurements of intracellular Na and Ca in intact cells, and by changes in these parameters in response to alterations of the extracellular environment. Pritchard and Ashley (1987) have recently measured intracellular free Ca in cells from guinea-pig taenia using fura-2. They have shown that free Ca is elevated when extracellular Na is removed, and that this effect is enhanced when the Na pump is blocked. These results are certainly encouraging, but more detailed information can be obtained with ion-sensitive microelectrodes, since it is possible to follow changes continuously, with simultaneous measurement of membrane potential. In the next section we will describe results obtained in the guinea-pig ureter using double-barrelled Na-sensitive microelectrodes. We used parallel experiments monitoring tension to give some estimates of changes in intracellular Ca^{2+}.

10.3 Studies in guinea-pig ureter

10.3.1 Initial indications of the presence of Na–Ca exchange

Early work that we carried out on the ureter (Aickin, Brading, and Burdyga 1984) led us to suspect that this tissue had a well-developed Na–Ca exchange system. In normal conditions, removal of extracellular Na does not induce contraction. If the Na pump is blocked either by addition of ouabain or by removal of extracellular K, the tissue remains relaxed. It does, however, then begin to respond to application of low-Na solutions with a contracture. On each application of low-Na, the response reaches a peak within about a minute and then fades. The size of the peak response grows with time of blockade of the Na pump, reaching a maximum in about 40 minutes with 10^{-4} M ouabain, and about 90 minutes in K-free solution. The low-Na contracture revealed by Na-pump inhibition is very little affected by concentrations of Ca-antagonist drugs which abolish the contraction induced by high-K solutions in normal tissues. The behaviour of this smooth muscle in response to blocking the Na pump and to alterations in the external Na, K, and Ca concentrations is very similar to that of mammalian cardiac muscle, in which the exchange is believed to play a significant role (e.g. see Coraboeuf, Gautier, and Guiraudou 1981). The behaviour of this smooth muscle resembles mammalian cardiac muscle in many other respects. Action potentials in the intact animal arise in a pacemaker region, and are propagated through the tissue. These have a well-developed plateau, and both Na and Ca ions seem to play a role in generating the inward currents. Each action potential initiates a phasic contractile response whose duration and size is related to that of the plateau (for review see Shuba 1981).

10.3.2 Experiments in tissues with an inhibited Na pump

More recently, we have monitored intracellular Na activity and membrane potential with double-barrelled ion-sensitive microelectrodes, in which one barrel was filled with a non-selective ion-exchange resin to monitor membrane potential, and the other with a Na-sensitive cocktail (Aickin 1987; Aickin, Brading, and Walmsley 1987). In normal tissues intracellular Na activity is low (7.4 ± 2.9 mM, mean ± S.D., $n=28$), but most interestingly it remains relatively low (20.6 ± 5.5 mM after inhibition of the Na pump with 10^{-4} M ouabain (see Fig. 10.1), even after at least 9 hour exposure. The transmembrane Na gradient is maintained remarkably well in the absence of the Na pump so too is the membrane potential. In fact, after about one hour, the potential stabilizes at a value of -64.0 ± 3.0 mV ($n=9$) compared with -50.8 ± 4.6 mV ($n=32$) in normal solution (Aickin 1987). Inhibition of the

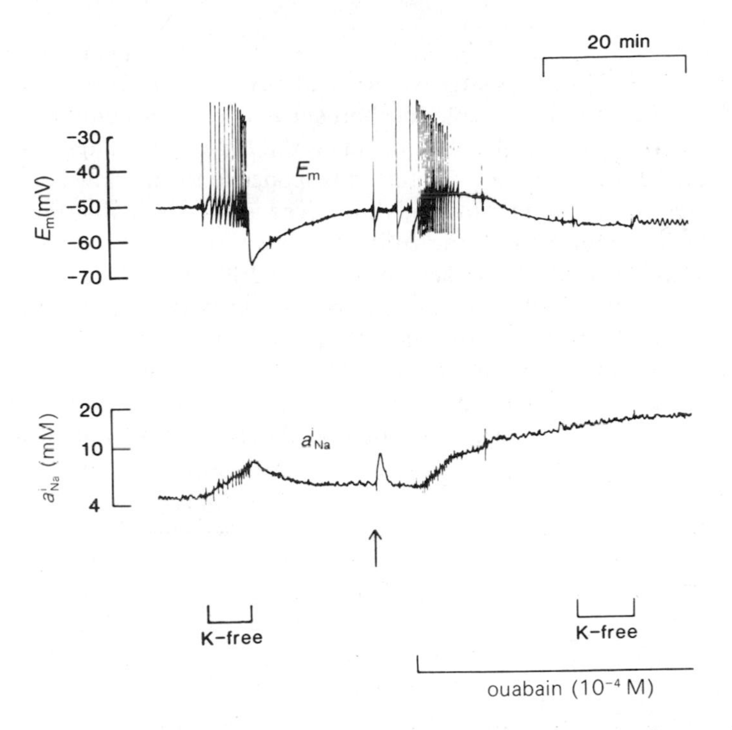

Fig. 10.1. The effects of K^+-free solution and ouabain 10^{-4} M) on membrane potential and intracellular Na activity in the guinea-pig ureter. Both methods of Na-pump inhibition initially generated spike activity (distorted by the low frequency response of the microelectrode). Activation of the pump by re-admission of K, on the left, was associated with a marked hyperpolarization due to the electrogenic nature of the pump. This effect was, however, abolished in the presence of ouabain. At the arrow, a small rise in Na activity occurred, presumably caused by phasic contraction inducing a transient leak round the electrode. Recovery was spontaneous (from Aickin 1987).

Na pump by removal of extracellular K has similar effects (see Fig. 10.1), although the rate of rise of intracellular Na is slower. Stabilization of the intracellular Na so far from equilibrium ($E_{Na} = +39.6$ mV) is strong evidence for the presence of another mechanism for extrusion of Na. Residual Na-pump activity cannot be held responsible, since removal of the external K in the presence of ouabain caused no further rise in intracellular Na, and the hyperpolarization characteristic of the electrogenic effect of the Na pump, observed on reintroduction of K in the absence of ouabain, was abolished, as illustrated in Fig. 10.1.

In an attempt to establish the factors responsible for the membrane potential in the ouabain-poisoned tissues, we measured the total K content of the tissues with flame photometry. Calculation of the intracellular K concentration indicates that E_K will be similar to the measured resting potential. This suggests either that the passive Na permeability of these tissues is very low indeed, or that an electrogenic mechanism is operating continuously and adds to the resting potential. This could be the Na–Ca exchange.

To establish that Na–Ca exchange is responsible for the low-Na contractures when the Na pump is blocked, it is necessary to show that the increase in cellular Ca (indicated by the contractile response) is accompanied by a reduction in intracellular Na activity. Fig. 10.2 illustrates the results of an experiment in which the top and the bottom tracings are the membrane potential and Na activity from one tissue, and the middle record is the contractile activity of another tissue under identical conditions. The tissues had been exposed to 10^{-4} M ouabain for more than an hour before the records were taken. The resting potential is about -60mV, and intracellular

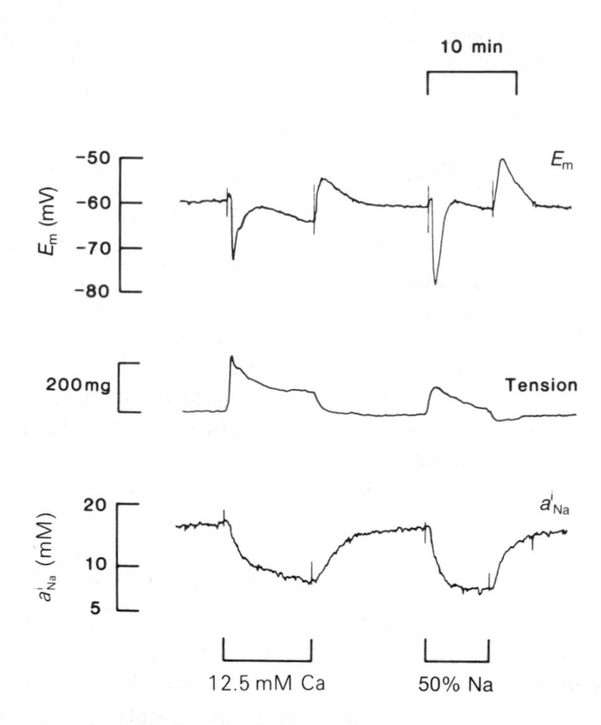

Fig. 10.2. Pen recordings of the simultaneous measurement of membrane potential (top) and Na activity (bottom) recorded in one preparation of guinea-pig ureter, and of tension (middle) recorded under the same conditions in a separate experiment. Ouabain (10^{-4} M) was present throughout, and for at least an hour before the traces were recorded. Except where indicated, the bathing solution was a modified Krebs solution containing 106 mM Na^+, 2.5 mM Ca^{2+}, and 31 mM Tris. Alterations in the Na and Ca concentration were made with isosmotic alteration in the concentration of Tris (from Aickin, Brading and Walmsley 1987.)

Na activity about 18 mM. Several important features can be seen. Lowering extracellular Na to 50 per cent causes a rapid fall in intracellular Na, accompanied by a transient membrane hyperpolarization which follows the differential of the rate of Na loss. Ca is also transiently elevated, as indicated by the tension response. Re-addition of Na to the bathing solution restores intracellular Na and causes a transient depolarization related to the rate of Na re-accumulation. Very similar, but slower responses are seen when extracellular Ca is transiently elevated from 2.5 to 12.5 mM (keeping extracellular Na and osmolarity constant). The loss of intracellular Na in both cases is against an electrical and a chemical gradient, and therefore must be active. Since the Na pump is blocked, this active efflux must be mediated by another mechanism. Na–Ca exchange is an obvious candidate. Interestingly, the transient changes in membrane potential that occur during loss and re-accumulation of Na are both in the direction and of a time course that would be consistent with the electrogenic nature of the Na–Ca exchange (although we cannot at present entirely rule out the possibility that changes in membrane K permeability secondary to the changes in intracellular Ca may also be involved). An electrogenic coupling of 3:1 would mean that Na–Ca exchange could be responsible for the maintenance of low intracellular Na in the presence of ouabain.

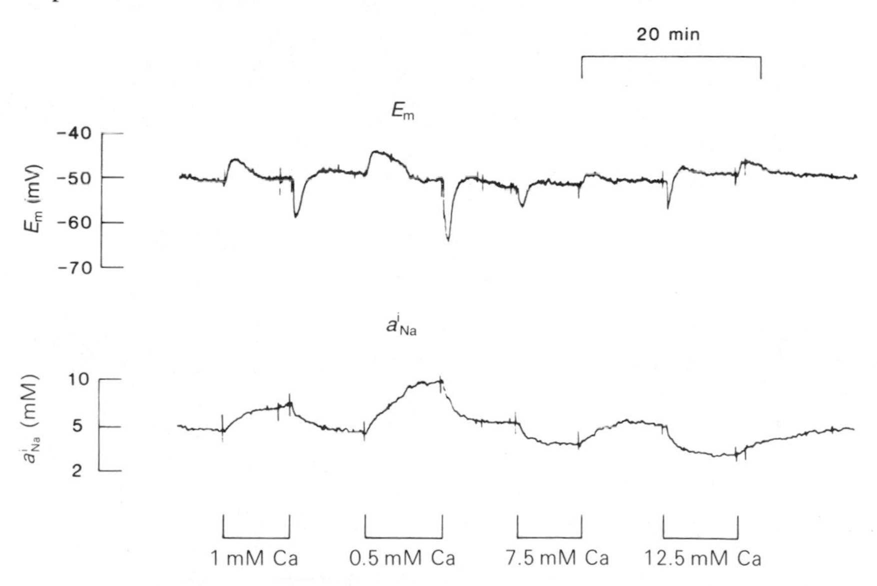

Fig. 10.3. The effect of extracellular Ca^{2+} on intracellular Na activity and membrane potential after prolonged (1.5 hours) inhibition of the Na pump with 10^{-4} M ouabain. Ca^{2+} was 2.5 mM except where indicated, and was altered by isosmotic changes with $Tris^+$. Extracellular Na was 50 per cent of that in normal Krebs, and ouabain was present throughout (from Aickin 1987).

If this suggestion is correct, intracellular Na should be dependent upon extracellular Ca. This is indeed observed as illustrated in Fig. 10.3. Lowering extracellular Ca results in an elevation of intracellular Na, and again, the changes in membrane potential are in the direction predicted for Na movement on an electrogenic Na–Ca exchange. Ion analysis of ouabain-treated tissues confirms that they can only retain a low Na content in the presence of extracellular Ca. This behaviour does not appear to be restricted to the ureter since preliminary experiments have shown that intracellular Na in guinea-pig vas deferens remains low in the presence of 10^{-4} M ouabain and is then lowered by elevation of extracellular Ca.

In an attempt to confirm the role of Na–Ca exchange, we examined the effects of several putative blockers of Na–Ca exchange. Fig. 10.4 shows the effects of 10 mM Mn on the tension, membrane potential, and Na activity

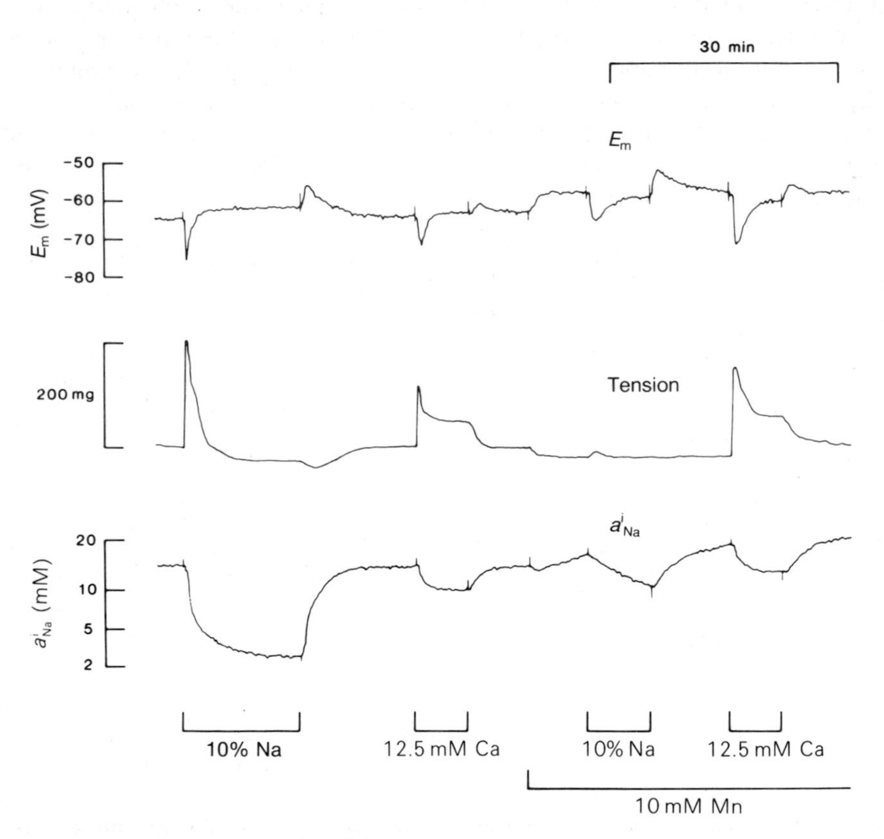

Fig. 10.4. The effect of 10 mM Mn^{2+} on the fall in intracellular Na activity and concomitant contracture induced by reduction of extracellular Na^+ or elevation of extracellular Ca^{2+} after prolonged inhibition of the Na pump. Conditions were the same as in Fig. 10.2 (from Aickin, Brading and Walmsley 1987).

changes induced by lowering extracellular Na and raising extracellular Ca in ouabain-treated tissues. Addition of Mn caused a slow rise in intracellular Na, again suggesting that Na–Ca exchange is involved in the maintenance of the relatively low intracellular Na. It also had a clear inhibitory effect on both the rate of loss of Na and the tension development in response to lowering extracellular Na. To our initial surprise, however, we found little effect on Na loss, and a slight increase in tension development in response to raising extracellular Ca. This seems to be due to a competitive interaction between Ca and Mn demonstrated, for example, by the effect of Mn on the Na-withdrawal contractures in the presence of different concentrations of Ca, as shown in Fig. 10.5. Table 10.1 illustrates the effects of putative blockers of the exchange on the Na-withdrawal contracture. All of these substances were able to block the contracture, but their lack of selectivity is illustrated by comparison of the concentrations necessary to produce a 50 per cent block of the high-K contracture of normal tissues, and of the Na-withdrawal contracture of ouabain-treated tissues. In every case the drugs are far more potent blockers of the depolarization-induced response.

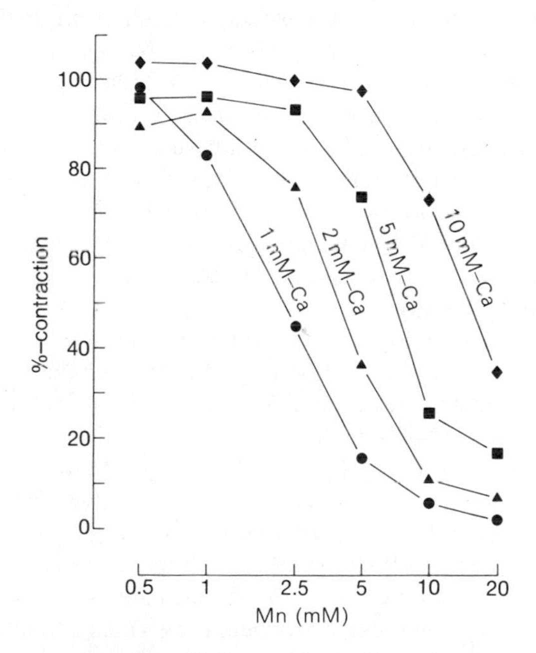

Fig. 10.5. Dose–response curves of the Mn^{2+}-induced inhibition of the Na^{+}-withdrawal contracture at different extracellular Ca^{2+} concentrations. Tissues equilibrated in 10^{-4} M ouabain. The results are expressed as a percentage of the contracture observed at each Ca^{2+} concentration in the absence of Mn^{2+} (from Aickin, Brading, and Walmsley 1987).

Table 10.1 Concentrations (M) of putative inhibitors of Na–Ca exchange required to give a 50 per cent block of the ONa (K) contractures of normal tissues, and the ONa(Tris) contractures of tissues with the Na pump blocked with 10^{-4} M ouabain. Means ± S.D. of three tissues

	Normal tissues K^+ contracture	Tissues in ouabain Na-withdrawal contracture
Quinacrine	$2.5 \pm 1.2 + 10^{-6}$	$2.1 \pm 1.4 \times 10^{-4}$
La^{3+}	$5.4 \pm 4.7 \times 10^{-5}$	$5.4 \pm 2.2 \times 10^{-4}$
Dicholorobenzamil	$5.6 \pm 1.2 \times 10^{-6}$	$5.6 \pm 0.3 \times 10^{-5}$
Dodecylamine	$2.0 \pm 1.0 \times 10^{-5}$	$7.3 \pm 2.8 \times 10^{-5}$
Mn^{2+}	4.0×10^{-4}	3.5×10^{-3}

10.3.3 Studies in normal tissues

We believe that the results described above show conclusively that the ureter does possess a well-developed Na–Ca exchange mechanism. Is this exchange operative under normal conditions, when the Na pump is active? The evidence is at present equivocal. Lowering extracellular Na in this case does not immediately increase tension, so it is unlikely that the intracellular Ca activity rises very far. However, intracellular Na does fall at a rate which is inconsistent with a reduction in the passive leakage of Na into cell. This could be indicative of extrusion on the Na–Ca exchange, but the rate of extrusion is almost unaltered by 10 mM Mn, which drastically slows the loss in ouabain-treated tissues. Increasing extracellular Ca also decreases intracellular Na in unpoisoned tissues. However, increasing extracellular Mg, and even Mn, have similar effects suggesting that this could be due to a non-specific effect on the Na permeability. This contrasts with the behaviour of ouabain-poisoned tissues, in which elevation of Mg has no effect, and addition of Mn causes a slow rise of intracellular Na.

Although we cannot unequivocally resolve the question of whether Na–Ca exchange operates in normal tissues, it does seem intuitively unlikely that the exchange is actually switched off. Whether the exchange mediates net fluxes of Na and Ca ions, and in which direction, remains to be determined. In Na pump-poisoned tissues, the exchange clearly mediates a net Na extrusion at the expense of Ca entry down its electrochemical gradient. In other words it is operating to regulate intracellular Na, not Ca. The tissue must therefore possess a powerful, energy-driven Ca pump to keep the intracellular Ca activity below the threshold for contraction. The necessity for this can be seen by the ability of the tissue to recover from repeated applications of Na-free solution, and the calculation that the drop in Na that occurs on removal of extracellular Na in poisoned tissues would cause an entry of Ca equivalent to 8.8 mmol/l^{-1} cell water, if it were entirely through Na–Ca exchange with a coupling ratio of 3:1 Na:Ca.

In preliminary experiments we have found that ureters of rabbits and humans behave like guinea-pig ureter, except that there is a small rise in tone in the presence of ouabain. What advantage is it to the ureter to possess two well-developed mechanisms for extrusion of both Ca and Na ions? The function of the ureter would be seriously compromised if the tissue was allowed to build up any sustained tension, and the ability to rapidly extrude Ca is clearly an advantage in preventing this. The plateau of the action potential requires an inward Na current, and is responsible for the phasic contractions involved in moving urine through the ureter. Thus it is necessary to ensure a good inwardly directed driving force for Na. Perhaps in a tissue with a well-developed Ca pump, a Na–Ca exchange which could act as a 'second string' for controlling intracellular Na would be an advantage. It is conceivable that the smooth muscle of the ureter may be more subject to changes in extracellular ionic composition than other tissues, since the urine may often be very different in composition from the blood plasma, particularly in K content. A well-developed Na–Ca exchange could then help preserve a low intracellular Na and thus high intracellular K even if the Na pump were less active than normal.

10.4 Studies in other tissues

Examination of the behaviour of other non-vascular smooth muscles demonstrates a wide variation in their responses. There is some indirect evidence to suggest that smooth muscles that are able to maintain tone for prolonged periods may have less well-developed Ca extrusion mechanisms, and a less developed Na–Ca exchange. The more phasic the smooth muscle, the more difficult it is to induce maintained tone, and the better developed these mechanisms appear to be. For example the guinea-pig taenia can maintain a reasonable amount of tension in the presence of muscarinic agonists for several hours. In this tissue, blocking the Na pump with ouabain causes a complete rundown of the Na and K gradients in about 5 hours (Casteels 1966) and the tissues can remain relaxed in these conditions. Since there is no Na gradient and the membranes are depolarized, net Ca entry would occur both through Na–Ca exchange (if present), and through voltage-sensitive Ca channels. The fact that the tissue remains relaxed implies that any Ca entry is easily compensated for by the ATP-driven Ca pump. If extracellular Na is lowered, however, Ca entry exceeds Ca extrusion, and accumulation of Ca and loss of Na occurs (Brading 1978), but an outward Na gradient of at least 10:1 is needed for this to occur. This suggests that these tissues do possess a

Na–Ca exchange, but that it plays only a minor role in Ca or Na homeostasis.

The behaviour of the urinary bladder is intermediate between taenia and ureter. It is rather phasic, but can build up a tetanus, and maintain tone for a few minutes in the presence of agonists. The Na and K gradients do not run down completely when the Na pump is blocked (unpublished observations), but more K is lost than in the ureter. Na–Ca exchange is probably involved in the maintenance of a rather low Na gradient. Interestingly, the normal bladder contracts powerfully when the extracellular Na is removed, but this contraction is associated with rapid spike activity, and is blocked by much lower concentrations of Mn than are needed to block Na–Ca exchange. The contractions are almost certainly caused by depolarization due to reduction in the activity of the electrogenic Na pump.

10.5 Conclusion

There is good evidence that non-vascular smooth muscles do possess a Na–Ca exchange mechanism, but there is considerable variability among tissues in the extent to which it is developed. In abnormal conditions, the exchange can contribute to the control of intracellular Ca concentrations, but it is possible that in tissues in which it is well-developed like the ureter, it may be more important in regulating intracellular Na under conditions when the Na pump is not functioning maximally.

References

Aickin, C. C. (1987). Investigation of factors affecting the intracellular sodium activity in the smooth muscle of guinea-pig ureter. *J. Physiol.* **385**, 483–505.

Aickin, C. C., Brading A. F., and Burdyga, Th. V. (1984). Evidence for sodium–calcium exchange in the guinea-pig ureter. *J. Physiol.* **347**, 411–30.

Aickin, C. C., Brading, A. F., and Walmsley, D. (1987). An investigation of sodium–calcium exchange in the smooth muscle of guinea-pig ureter. *J. Physiol.* **391**, 325–46.

Blaustein, M. P. (1977). Sodium ions, calcium ions. Blood pressure regulation and hypertension: a reassessment and a hypothesis. *Am. J. Physiol.* **232**, C165–73.

Brading, A. F. (1978). Calcium-induced increase in membrane permeability in the guinea-pig taenia coli: evidence for involvement of a sodium–calcium exchange mechanism. *J. Physiol.* **275**, 65–84.

Brading, A. F. (1981). Ionic distribution and mechanisms of transmembrane ion movements in smooth muscle. In: *Smooth muscle: an assessment of current knowledge* (eds. E. Bulbring, A. F. Brading, A. W. Jones, and T. Tomita) pp. 65–92. Edward Arnold, London.

Brading, A. F., Burnett, M., and Sneddon, P. (1980). The effect of Na removal on the contractile responses of the guinea-pig taenia coli to carbachol. *J. Physiol.* **306**, 411–29.

Brading, A. F., and Widdicombe, J. H. (1976). Interaction between sodium and calcium movements in smooth muscle. In: *Smooth muscle pharmacology and physiology* (eds M. Worcel, and G. Vassort) pp. 235–45. INSERM, Paris.

Casteels, R. (1966). The action of ouabain on the smooth muscle cells of the guinea-pig's taenia coli. *J. Physiol.* **184**, 131–42.

Coraboeuf, E., Gautier, P., and Guiraudou, P. (1981). Potential and tension changes induced by sodium removal in dog purkinje fibres: role of an electrogenic sodium–calcium exchange. *J. Physiol.* **311**, 605–22.

Daniel, E. E. (1985). The use of subcellular membrane fractions in analysis of control of smooth muscle function. *Experientia* **41**, 905–13.

Grover, A. K., Kwan, C. Y., and Daniel E. E. (1981). Na–Ca exchange in rat myometrium membrane vesicles highly enriched in plasma membranes. *Am. J. Physiol.* **240**, C175–82.

Grover, A. K., Kwan, C. Y., Rangachari, P. K., and Daniel, E. E. (1983). Na–Ca exchange in smooth muscle plasma membrane-enriched fraction. *Am. J. Physiol.* **244**, C158–65.

Hirata, M., Itoh, T., and Kuriyama, H. (1981). Effects of external cations on calcium efflux from single cells of the guinea-pig taenia coli and porcine coronary artery. *J. Physiol.* **310**, 321–36.

James, M. R., and Roufogalis, B. D. (1977). The effect of ouabain on the guinea-pig ileum longitudinal muscle 2. Intracellular levels of Ca, Na, K and Mg during ouabain response and the dependence of the response on extracellular Ca. *Can. J. Physiol. Pharmacol.* **55**, 1197–203.

Katase, T., and Tomita, T. (1972). Influences of sodium and calcium on the recovery process from potassium contracture in the guinea-pig taenia coli. *J. Physiol.* **224**, 489–500.

Masahashi, T. and Tomita, T. (1983). The contracture produced by sodium removal in the non-pregnant rat myometrium. *J. Physiol.* **334**, 351–63.

Morel, N., and Godfraind, T. (1982). Na–Ca exchange in heart and smooth muscle microsomes. *Arch. Intl. Pharmacodyn.* **258**, 319–21.

Osa, T. (1971). Effect of removing the exernal sodium on the electrical and mechanical activities of the pregnant mouse myometrium. *Jap. J. Physiol.* **21**, 607–25.

Pritchard, K., and Ashley, C. C. (1987). Evidence for Na^+/Ca^+ exchange in isolated smooth muscle cells: a fura-2 study. *Pflügers Arch.* **410**, 401–7.

Raeymaekers, L., Wuytack, F., and Casteels, R. (1974). Na–Ca exchange in taenia coli of the guinea-pig. *Pflüg Arch.* **347**, 329–40.

Shuba, M. F. (1981). Smooth muscle of the ureter: the nature of excitation and the mechanisms of action of catecholamines and histamine. In: *Smooth muscle: an assessment of current knowledge* (eds. E. Bulbring, A. F. Brading, A. W. Jones, and T. Tomita) pp. 377–84. Edward Arnold, London.

Shimzu, K., Koroso, Y., Nakajyo, S., and Urakawa, N. (1979). Species differences in the ouabain sensitivity of the small intestine in contractile response. *Jap. J. Vet. Sci.* **41**, 139–49.

Tomita, T., and Yamamoto, T. (1971). Effects of removing the external potassium on the smooth muscle of guinea-pig taenia coli. *J. Physiol.* **212**, 857–68.

Van Breemen, C., Aaronson, P., and Loutzenheiser, R. (1978). Sodium–calcium interactions in mammalian smooth muscle. *Pharmacol. Rev.* **30**, 167–208.

11 Comparing sodium–calcium exchange as studied with isotopes and measurements of $[Ca]_i$

L. J. Mullins and J. Requena

11.1 Introduction

Information about ionic movements across cell membranes has traditionally been obtained either by bioelectric measurements of ion currents, or by the use of isotopes to measure influx and efflux. Electrical measurements of net flux require that special means be available to identify the ion producing the current while isotope measurements uniquely identify the ion species under study.

Problems arise when the ion species under study is Ca^{2+} because there is an overwhelming tendency for this ion to become bound to intracellular compounds or structures. Additional problems arise because [Ca] in cells is of the order of 100 nM and fluxes across the membrane are often quite small. Because of these difficulties, measurements of $[Ca]_i$ by optical indicators have been made with the idea that changes in Ca_i can be related to changes in Ca fluxes across the membrane.

The purpose of this review is to contrast the results obtained with flux measurements with those obtained using intracellular indicators of free Ca. Each method has its advantages and disadvantages and these can be summarized as follows. Isotopes give fluxes for Ca that do not require worrying about Ca buffering—the measured influxes may have significant errors because collecting the ^{45}Ca entering the cell presents some difficulties. Optical indicators have been greatly improved in recent years and are on the whole, reliable; their use does require, however, a detailed knowledge of buffering and any possible changes in this parameter that experimental treatments may induce. On the whole the two methods of measuring Ca fluxes are at least qualitatively in agreement, although there are some significant areas where measured results differ substantially.

11.2 Methods

11.2.1 Techniques for flux measuring

The measurement of Na and K fluxes in squid axons initially involved soaking the axon in radioactive sea-water for a short time and then rinsing and extruding axoplasm, and counting this for influx measurements. If soaking in the radioisotope were prolonged, then the axon became a chamber with flowing sea-water and the efflux of the loaded ion could be measured. A refinement in efflux measurement was to inject the isotope with a microinjector and thus avoid extracellular isotopic contamination. Using this technique, Hodgkin and Keynes were able to make efflux measurements of ^{45}Ca (1957). A further improvement in flux measurements in squid axon was to internally dialyse the axon (Brinley and Mullins 1967) since this technique allowed the experimenter to control the internal environment from which the ion was moving. Both influx and efflux of monovalent cations were satisfactorily accomplished, but for divalent cations such as Ca_i there was some worry that not all the Ca entering the axon was collected by the internal dialysis capillary. The first measurements of Ca fluxes in dialysed axons were by DiPolo (1973). These showed that Ca efflux could be readily measured and was sensitive to internal substrates such as ATP. Other studies (Brinley *et al.* 1977) had shown that if Ca_i was measured during an episode of stimulation of the axon, then only one-thousandth of the Ca known to enter the axon appeared as an increase in concentration. Presumably the rest was buffered by the known buffering systems of the axon. If inhibitors of electron transport in mitochondria were used, buffering was reduced, but not abolished, and, at best, 5 per cent of the entering Ca appeared as an increase in [Ca]. Thus, in a dialysed axon, very little Ca can be collected in the dialysis capillary unless Ca buffers that will compete with the endogenous buffering of the axon are used. The problem of how good the collection of entering counts of ^{45}Ca is in dialysed axons has not been quantitatively studied, but it is a fact that in axons dialysed with EGTA, very little Ca influx is measured. This finding could have two explanations—first, that EGTA as a buffer is not as strong as the endogenous, non-mitochondrial buffers, and, second, that Ca as free Ca ions are required for the activation of Ca influx. These possibilities will be considered later.

11.2.2 Techniques for free Ca measurement

An early study by Baker, Hodgkin, and Ridgway (1971) used aequorin injected into squid axons and concluded that free Ca was very low (*c.* 300 nM) in axons that had been isolated from squid mantles stored in iced sea-water. Since subsequent studies have shown that axons gain Ca rapidly under such conditions, the actual Ca could be much lower in freshly isolated axons. Such was found to be the case in a further study (DiPolo *et al.* 1976) where a dialysis capillary was inserted into the axon and a drop of aequorin introduced into the capillary. Measurements showed that free Ca in the centre of the axon was ca. 30 nM while when arsenazo III was used as an indicator, free Ca was about 60 nM. In this latter measurement Ca was integrated over the whole of the axon by the nature of the spectrophotometric measurement that was made (arsenazo was injected into the axon since it is not possible to confine small molecules to a dialysis capillary). More recently, a variety of fluorescent dyes have been used to make quantitative measurements of free Ca in cells. All these optical techniques show that free Ca is very low and that it is at best a small fraction of the known Ca content of the cell.

When a small change in Ca influx is made (for example, by increasing the Ca in sea-water from 10 to 50 mM, optical indicators show that the measured $[Ca]_i$ increases with time until it reaches a new steady value. The interpretation placed on this observation is that the new value of $[Ca]_i$ is that where the influx of Ca just matches the rate at which Ca is being withdrawn into intracellular buffers. The initial slope with which [Ca] rises is proportional to influx in this sort of model. Note that the optical measurement of Ca in cells does not distinguish between Ca entry via any particular mechanism, but it does rely on the idea that buffering remains constant during any particular experimental measurement. Ca buffering does seem to decline during a long experiment so that larger increases in $[Ca]_i$ can be expected for the same Ca influx. This process of buffering decline may influence flux measurements as well if not all the Ca entering the fibre is captured by the EGTA used in the dialysis fluid.

11.2.3 Buffering of Ca and the relationship between flux and concentration

Unlike monovalent cations in axoplasm, the evidence is very clear that although the Ca content of fresh axons is about 50 μmol kg^{-1} axoplasm, all but about a thousandth of this Ca is sequestered or bound. For a comprehensive review of Ca buffering in axoplasm see Brinley 1978. Given this complication, it is clear that both flux and Ca concentration measurements need to take into account the realities of Ca buffering. Brinley *et al.* (1977) have shown that in a freshly isolated squid axon, poisoning the mitochondria in a variety of way leads to only very small changes in free or ionized Ca thus implying that mitochondria are not a significant storage site for Ca at physiological levels of $[Ca]_i$. Buffering by mitochondria, in the study cited above, occurs only at concentrations of Ca exceeding 1000 nM. The actual buffer for Ca at low $[Ca]_i$ appears to be, at least in part, the smooth endoplasmic reticulum and selective inhibitors for this sort of buffering have not been discovered. Metabolic inhibitors that block electron transport or collapse the proton gradient in mitochondria have little effect on Ca buffering at Ca_i concentrations of 50 nM; they greatly reduce buffering at concentrations of 1000 nM. Long-term treatment of axons with such inhibitors will reduce ATP levels and this change has a most significant effect in reducing Ca buffering at all concentrations of Ca_i.

11.3 Experimental measurement of Ca fluxes

11.3.1 Ca efflux in injected axons

After the highly successful measurements of Na and K fluxes in squid axons by isotopes, it was natural that these sorts of measurements would be extended to Ca. The simplest sort of measurement was that of injecting radioactive Ca into squid axons and measuring the release of counts. Such measurements (Baker *et al.* 1967; Blaustein and Hodgkin 1969) showed that the efflux of Ca was dependent on external Na and suggested that an earlier proposed mechanism (the exchange of Na entering for Ca exiting) was applicable to squid giant axons.

Additional measurements showed that if cyanide were applied to deplete axoplasmic ATP, then Ca efflux was greatly enhanced, suggesting that inhibitors released Ca from mitochondrial stores and thus led to a vastly increased efflux from a mechanism that required the presence of Na^+ external to the axon.

11.3.2 Ca efflux in dialysed axons

The technique of internal dialysis (Brinley and Mullins 1967) was first applied to Ca efflux studies by DiPolo (1973); results showed that it was feasible to use CaEGTA/EGTA buffers to specify a free [Ca] in axoplasm, to poison the mitochondria so that Ca uptake by such organelles did not take place, and to show that in the absence of ATP there was a Na_o-dependent Ca efflux.

In a series of studies (DiPolo 1974, 1976, 1977) it was further shown that ATP could increase the Na_o-dependent Ca efflux and that as free Ca in the dialysis solution was reduced, less and less of the Ca efflux was sensitive to external Na.

The interpretation placed on these studies was that there were two transport systems in squid axons for the extrusion of Ca, one Na_o-dependent (the Na–Ca exchange) and a second system, an ATP-dependent pump. The actual situation is likely to be much more complicated than this and the nature of these complications will be discussed later.

11.3.3 Ca influx in dialysed axons

The technique of internal dialysis can be used to measure influx if isotopes are placed outside the axon and counts are collected in the internally flowing dialysis solution. While there are more complications with this method, it has worked well for studies using monovalent cations such as Na and K. The Ca ion, after crossing the membrane, must diffuse over a path of some 200 μm where organelles compete for Ca with the endogenous Ca buffer supplied by the experimenter in the dialysis solution. Mitochondria can be poisoned so that they cannot accumulate Ca, but other endogenous Ca buffering systems (see Section 11.2) are not inhibited and are the principal Ca buffer at physiological levels of $[Ca]_i$. Ideally, one might hope to saturate these buffers by prolonged dialysis at a particular CaEGTA/EGTA ratio, but even here, if conditions are changed, the change in flux may not be reflected in a new value for a very long time. There is also reason to believe that some Ca buffering requires ATP and some does not—thus the effect of dialysing ATP into the fibre might be a change in flux that merely reflects changes in Ca buffering.

Measurements that have been made (DiPolo 1979) show that the Ca influx is sensitive to external and internal Na in a manner expected for a Na–Ca exchange system and that there is also a component of Ca influx that is cut off by TTX, reflecting presumably Ca influx via Na channels. Ca influx is also sensitive to ATP, suggesting that the Na–Ca exchange, operating in either direction, requires this substance for maximal flux. There is no indication that ATP causes a net flux via the Na–Ca exchanger, although the data is somewhat sketchy on this point.

Recent measurements (DiPolo and Beaugé 1986) have emphasized that $[Ca]_i$ is required if Ca influx brought about by Na–Ca exchange is to occur and that steady changes in membrane potential can also activate Ca influx with depolarization.

11.3.4 Na efflux dependent on Ca_o

The original observation that part of the Na efflux from squid giant axons depended on external Ca was made by Baker *et al.* (1969). The measurements showed that such a Na efflux was ouabain-insensitive and that, unlike Na efflux via the Na pump, which was about linear with internal Na concentration; the efflux that was Ca_o-sensitive showed an increase in efflux as the square of internal Na concentration. These findings suggested that the Na efflux was the same Na^+/Ca^{2+} exchange mechanism that produced Ca efflux but that the mechanism was now running backward. To maximize the observed Na efflux, Na_o needed to be made zero and Na_i high.

Further investigation of the activating effects of monovalent cations on Ca efflux were made by Blaustein (1977). He showed several important features of the Na–Ca system, particularly that Li applied externally could produce a Ca efflux that was largely cut off when Ca_o was removed from seawater, and that that efflux was about twice the maximum Ca efflux when Na was the external cation (and Ca_i was saturating). A further significant finding was that when K in the internal dialysis fluid was replaced by TMA^+, Ca efflux was reduced more than 10-fold, but only if the external sea-water contained Li. In Na sea-water there was a very much smaller decrease. Changes in membrane potential occasioned by the removal of internal K were compensated by a voltage clamp. The conclusion therefore is that Li promotes a Ca–Ca exchange, but only if the internal cation is K. When a similar experiment was done using Li instead of K as the internal cation, Ca efflux remained high as long as the external sea-water contained Na or Li; in choline sea-water, Ca efflux was very small.

A more detailed study of the Ca_o-dependent Na efflux (Allen and Baker 1986a) showed that Na efflux was sensitive to membrane potential and increased with depolarization. The flux showed no obvious sign of saturating even at membrane potentials of $+70$ mV.

An exceedingly interesting feature of this study was the demonstration that various monovalent cations had large effects on the affinity of Ca_o binding. The concentration of external Ca varied from 1 mM Ca_o to 50 mM depending, respectively, on whether Li or choline was the principal cation in the sea-water. The order of cation activation was: Li, $K > Na$, $Rb >$ guanidinium-$^+$, $>>$ Tris,$^+$ choline.

In a companion study, Allen and Baker (1986b) examined the efflux of Ca from squid axons as influenced by membrane potential; their findings were that the Ca efflux that is dependent on Na_o was enhanced by increases in membrane potential (confirming the findings of Mullins and Brinley 1975), and that the Ca efflux that was independent of external Na was decreased. These important findings may have some application to measurements where it has been tacitly assumed that ATP pump-driven Ca efflux is independent of membrane potential.

11.4 Experimental measurement of free and total Ca in axons

11.4.1 Free Ca injected axons

The pioneering study of Baker *et al.* (1971) showed that an upper limit of 300 nM could be measured by using aequorin injected into axons and comparing the light emission of such axons with aequorin in glass capillaries containing CaEGTA–EGTA buffers. The axons used had total analytical Ca that averaged about 400 μM. Since this is about 8-fold larger than truly fresh axon Ca (see below), the actual Ca might be expected to be about 40 μM in a freshly-dissected axon.

11.4.2 Free Ca in the axon centre

A study using aequorin in a dialysis capillary (DiPolo *et al.* 1976), did in fact show an ionized Ca in the axon centre. The dialysis capillary was localized about 200 μm from the membrane and light emission indicated a Ca_i of about 30 nM in freshly isolated axons kept in sea-water with 4 mM Ca. Experiments where the Ca in sea-water was raised to 10 mM showed that the aequorin glow increased with time. However, at a Ca_o lower than 4 mM there was a decline in glow.

11.4.3 Free Ca averaged over the axon

If instead of aequorin as a Ca detector the dye aresenazo III was used, then (DiPolo *et al.* 1976) one gets a value of about 60 nM suggesting that there may be gradients in Ca in the axon such that the core has a lower and the periphery a higher Ca.

11.4.4 Free Ca in phenol-red injected axons

If axons are injected with phenol red in addition to aequorin, the phenol red acts as a screen to absorb aequorin light from the axon interior (Mullins and Requena 1979). Recent experiments, where the light emission of a phenol-red injected axon (Requena *et al.* 1986) was compared before and after a Ca buffer injection, have led to the conclusion that the concentration of free Ca is 75 nM in the axon periphery. One has, therefore, a consistent set of values: core, 30 nM; periphery, 75 nM, and averaged 60 nM.

11.4.5 Total Ca in axons

Recent measurements of analytical Ca in axons (Requena, Mullins, and Brinley 1979) have shown a value of 60 μmol kg^{-1} axoplasm in contrast with earlier estimates of 400 μmol kg^{-1} axons that had been isolated and stored in sea-water for many hours. Total Ca depends greatly on the experimental conditions imposed on the axon.

Axons can be depolarized with high-K sea-water and shown to undergo large increases in total Ca, or they can be treated with Na-free external solutions and also shown to gain Ca. These changes are completely reversible with a return to normal Ca sea-water leading to a loss of the gained Ca over the course of an hour. Fluxes can approximate 10 pmol cm^{-2} s^{-1} in terms of net Ca gain or an increase in terms of concentration of 3 μmol kg^{-1} h^{-1}; because of buffering the change in free Ca concentration will of course be very much smaller with an increase of the order of 50 nM free Ca. Nevertheless, the axon promptly recovers its original Ca content upon restoration of the conditions that favour Na–Ca exchange.

11.4.6 Effects of metabolic poisons on total and free Ca

As originally noted by Blaustein and Hodgkin (1969) the application of cyanide to a squid axon has little initial effect on Ca efflux, but, after a time consistent with the rundown of ATP in the axon, a large increase in Ca efflux occurs. This finding was interpreted to indicate that mitochondria release their stored Ca when deprived of ATP. Subsequent studies have confirmed this conclusion.

By contrast, Ca influx is little affected by substances such as cyanide until sufficient time has passed to reduce ATP to low levels. Then it should decrease Ca influx since both Ca efflux and influx are enhanced by ATP. It was considerations such as these that led Blaustein and Hodgkin to suggest that the energy for Ca extrusion came from the Na gradient and not from ATP directly. By contrast, DiPolo and Beaugé (1979) have advanced the idea from an ATP-driven Ca efflux pump is the principal way that Ca flux balance is maintained in axons at physiological levels of Ca_i. It has proved very dificult to test such a proposal because lowering ATP has effects on Ca buffering in the axon, Na–Ca exchange, and on the purported ATP-driven Ca pump. The argument needs to be a quantitative one, since many cells have both ATP-driven and Na gradient-driven Ca extrusion systems and one needs to know which mechanism is really controlling Ca under given sets of conditions. The large net fluxes of Ca reported by Requena, Mullins, and Brinley (1979) with only modest changes in Ca_i are only consistent with a Na–Ca exchange extrusion of Ca since the maximum velocity of the ATP-driven pump is small, and studies with inhibitors that would reduce ATP_i to low levels showed that Ca extrusion still occurred when the ATP-driven Ca pump was deprived of its substrate. Admittedly, the inhibitors artificially raised Ca_i to high levels as buffering was poisoned, but one would expect that ultimately the axon would gain Ca if ATP were a requirement when Ca_i were low.

11.4.7 Effects of membrane potential on free Ca

One of the most interesting areas of study during the past decade has been the measurement of free Ca in cells at various membrane potentials. Flux studies have shown that in squid axons hyperpolarization decreases and depolarization increases Ca influx, with the efflux of Ca responding in an opposite sense. Such measurements would predict, therefore, that depolarization would increase free-Ca concentrations in axons and indeed this is a generally accepted notion. In some cells, depolarization opens Ca-selective channels in the membrane so that the increase in free Ca occurs without any relationship to Na–Ca exchange. In squid axons, however, while steady depolarization causes an increase in free Ca (Baker, Meves, and Ridgway 1973), this increase can be totally abolished if $[Na]_i$ is reduced to about half its normal value (Mullins and Requena 1981). Thus it is difficult to suppose that any appreciable fraction of Ca entry with depolarization results from the use of Ca channels. An objection to this conclusion could be that the Ca channels of the squid axon inactivate rapidly so that in steady depolarizations not enough Ca enters over a period of minutes to contribute to the Ca signal. Other experiments (Mullins, Requena, and Whittembury 1985) where depolarization was effected by voltage-clamp pulses in TTX-treated axons show that even under the conditions of pulsing (with the pulse duration and amplitude closely approximating those of the action potential) Ca entry is still dependent on internal Na concentration—a conclusion that again argues against any appreciable concentration of Ca channels in the membrane of the squid giant axon.

11.5 Areas where flux measurements and free Ca disagree

11.5.1 Activation of Ca influx by internal Na

One of the startling findings of the seminal paper by Baker *et al.* (1969) was that the Na efflux from squid axons could be fractionated into a part that was ouabain-sensitive and linearly dependent on $[Na]_i$, and a ouabain-insensitive fraction that was dependent on Ca_o; furthermore, the efflux rose as a power function of $[Na]_i$. The direct measurement of Ca influx itself showed that it, too, was vastly increased by increases in Na_i.

A further study (Baker, Meves, and Ridgway 1973) showed that when squid axons were depolarized by replacing most of the Na in sea-water with K, there was an impressive increase in light emission from aequorin-injected axons. This increase in Ca_i was rapidly reversed when Na was replaced in the sea-water.

Following this observation, Mullins and Requena (1981) were able to show that this impressive increase in $[Ca]_i$ could be abolished by a preliminary stimulation of the axon for about 30 minutes in Li sea-water. The main effect of Li sea-water is to allow generation of the action potential while reducing Na_i, since Li flows inward and Na outward with each impulse. Furthermore, the effect (loss of Ca entry with depolarization) could be readily reversed by re-stimulating the same axon in Na-containing sea-water. Very large responses could be elicited from axons that had had their Na_i increased to high levels. Two interpretations of these findings were possible: that changes in Na_i somehow changed the buffering of Ca by axoplasm, or that the Ca entry with steady depolarization was the result of Na–Ca exchange and, as a result, small changes in Na_i resulted in large changes in Ca entry. Measurements of Ca entry during a short train of impulses in 50 mM Ca sea-water led to an increase in Ca in the axon mainly by Ca flowing through Na channels, and records of such test stimulations in Na-loaded, and Na-depleted axons showed that there was the same increase in Ca_i. Hence, a change in buffering could not have produced the effect.

Further studies (Mullins *et al.* 1983) using Na-sensitive electrodes to measure Na_i continuously, showed that Ca entry with depolarization is a very steep function of Na_i and that the foot of the curve is about 4th power with respect to Na_i.

Measurements of Ca influx as a function of Na_i have also been made by DiPolo, Rojas, and Beaugé (1982) and these also show a component of Ca influx that is dependent on Na_i. The measurements are complicated by the fact that at different internal Ca concentrations the shape of the Ca influx curve is different. Figure 11.1 shows a comparison between the measurements of Ca entry that we have made and those of DiPolo. Note that the

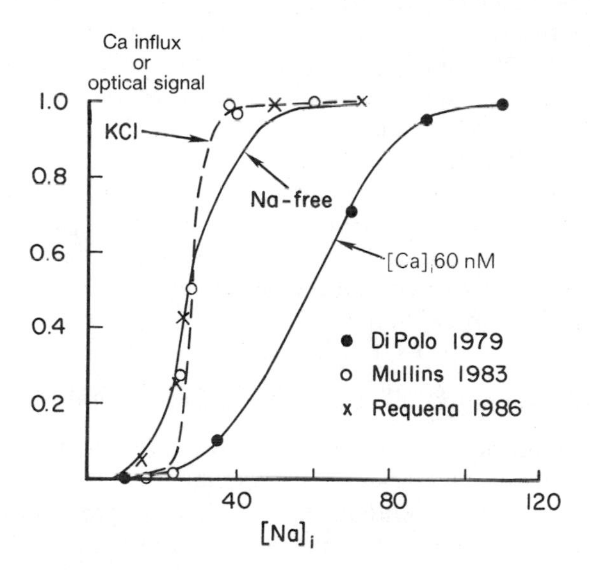

Fig. 11.1. The dependence of Ca influx on Na_i (from DiPolo 1979, Fig. 4) is compared with the optical signal generated from arsenazo-injected squid axons (Mullins *et al.* 1983) when axons are depolarized with high KC1 solutions, and with the optical signal produced by aequorin (Requena *et al.* 1986) when Na-free solutions are applied to squid axons.

isotope measurements indicate a much flatter curve than the optical measurements.

The difference between these two sets of measurements is quite real and, hence, the way that they are made needs to be examined closely. The DiPolo measurements were made at resting flux levels of the order of 0.1 pmol $cm^{-2}s^{-1}$ while those using optical signals as a basis for the measurement were made with fluxes perhaps ten times larger. Even so, there is a disagreement since flux measurement by DiPolo at 800 nM Ca_i show a half saturation at 40 mM Na_i and a curve with a low slope so that Ca influx is still rising at Na_i concentrations of 80–100 mM.

Furthermore, if Ca_i were a real variable, one would expect that the two optical curves would be shifted on the Na_i axis since Ca entry with Na-free conditions is only about a quarter of that for KCl depolarization, hence, the corresponding Ca_i should have been much smaller—instead, the two curves are essentially identical.

An additional finding of DiPolo is that if Ca_i is made zero then there is no Na_i-dependent Ca influx. If we extrapolate from this, we might conclude that Ca_i is necessary to make influx measurements by dialysis work since it supplies cold Ca to exchange with the radioactive Ca that enters and binds to axoplasmic buffers. On this basis it would be possible to explain the leftward shift of the activation curve as a greater and greater fraction of the entering Ca is extracted from the axoplasm. A second point is that raising Ca_i activates Ca efflux Na–Ca exchange and if return of the carrier to the outside is necessary for a rapid Ca influx, then the activating effect of Ca_i is understandable. Under the conditions of the experiments reported, Ca efflux ought to have been several times larger than Ca influx and the activation curve for Ca_i-promoted Ca influx looks very much like the curve for Ca efflux *vs.* Ca_i.

By contrast, optical methods measure net Ca flux (as a gain in influx over efflux) so that the difference in responses between isotopic and optical methods may reflect a difference in the behaviour of Na–Ca exchange when the net flux is a gain in Ca *vs.* a loss.

11.5.2 Membrane potential sensitivity of free Ca and Ca flux

The demonstration of Mullins and Brinley (1975) that hyperpolarization increased Ca efflux was one reason for suspecting that Na–Ca exchange was electrogenic. Indeed, studies with membrane vesicles, cardiac cells, and many other cell types have confirmed that more Na charges move per cycle of the carrier than Ca charges.

A recent study (DiPolo *et al.* 1985) has shown that Ca efflux is only sensitive to membrane potential when the internal Na is finite. Here the half-activation concentration for Na_i is only 5 mM (in contrast with the 40 mM for Ca influx mentioned above) or a value comparable to that found when Ca influx is producing a net gain in Ca by the fibre. There is no obvious reason why Ca efflux (which is actually inhibited by Na_i) should lose its voltage sensitivity in the absence of Na_i, but it may be when the Na concentration gradient is infinite, there is sufficient energy to extrude Ca with a stoichiometry of 2 Na–Ca. We have, therefore, to consider whether variable coupling between Na and Ca might exist. The idea implies that control mechanisms exist in the membrane that can sample the gradient and then decide whether a 2:1 coupling is permissible. We can rule out the idea that the exchanger can always operate 2:1—if it did so, the equilibrium level of free Ca in axons would be unacceptably high.

If we look at net Ca flux as measured with arsenazo (Mullins *et al.* 1983) we find that net flux increases about 4-fold for a 25 mV change in membrane potential while DiPolo *et al.* (1985) found that Ca *efflux* increased *e*-fold for about a 50 mV increase in membrane potential. A similar voltage dependence for Ca efflux has been observed by Allen and Baker (1986b) who also noted that the ATP-driven Ca pump is inhibited by hyperpolarization, a point not considered in previous measurements. The voltage sensitivity of Ca efflux is curiously variable and dependent on many factors. Again, this state of affairs would be understandable if certain conditions that we do not understand could produce both a 3:1 and a 2:1 coupling.

11.5.3 The ability of free Ca (inside) to promote Ca entry

A surprising finding (DiPolo 1979) was that Ca_i appears to have a catalytic effect in increasing Na_i-dependent Ca influx. This effect has been studied in more detail (DiPolo and Beaugé 1986, 1987) using the Ca_o-dependent Na efflux as an indicator of Ca influx. These later studies demonstrated that both ATP and Ca_i have a catalytic effect in increasing the Ca_o-dependent Na efflux (reverse Na–Ca exchange) that was originally described by Baker *et al.* 1969.

The activation curve for Ca_i concentration *vs.* Ca influx is half maximal at 0.3 μM and closely approximates the curve found for Ca efflux *vs.* Ca_i, again suggesting that maximal reverse Na–Ca exchange may require that the Ca efflux system be operating at saturating levels. A second point is that Ca_i may be required to exchange for radioactive Ca complexed with the axoplasmic buffers. Ca_o-dependent Na efflux measurements get around this difficulty but add other difficulties since there is a large component of Na–Na exchange that is not sensitive to Ca_i. In such experiments it is necessary to block the Na–K pump with ouabain—this introduces a new Na efflux (Beaugé and Mullins 1976; Brinley and Mullins 1968) that appears to be an exchange of an external cation for Na_i.

Another important study of the Na efflux that is coupled to Ca influx was carried out by Allen and Baker (1986a). They show that the affinity of the Na–Ca exchange for Ca_o is very low when choline or tris is used as a Na substitute and that ions that activate the system are Li or K. Unfortunately, the DiPolo and Beaugé studies relied on Tris as a Na substitute, hence, the system that they were studying was poorly activated and required about 10 mM Ca_o for half activation of the Na efflux; Mullins *et al.* 1983 showed that half activation of Ca influx when K was the principal external cation occurred at 0.67 mM Ca_o and the results of Allen and Baker agree with this latter estimate.

11.6 General conclusions

Na–Ca exchange has been shown to have many of the properties expected of a stoichiometry of 3 Na–Ca, but there are experimental conditions where it is possible that an electroneutral exchange may take place. Such conditions involve working at low Ca electrochemical potentials.

The membrane potential is a real variable in setting the level of Ca_i as is the Na chemical gradient. Other ions have what might be called activating effects on the Na–Ca exchange—these include K and Li.

Ca influx by Na–Ca exchange appears to be differently activated from Ca efflux since Ca influx requires either ATP_i or Ca_i or both; Ca efflux, while activated by ATP, is not affected by Ca_o except insofar as a Ca–Ca exchange might be promoted. Ions such as Li that promote such an exchange may do so because they become so strongly bound to the Na part of the carrier that they do not dissociate at appreciable rates.

Activation of Na–Ca exchange does require Na_i for Ca influx and Na_o for Ca efflux—the concentrations required are 25 mM for half activation of Ca influx and perhaps 80 mM for half activation of the efflux. Considering the quite different ionic conditions on the two sides of the squid axon membrane, these concentrations may be considered to be much the same. For internal Ca, half-activating Ca_i is of the order of 1 μM while externally it is ca. 700 μM which does appear to indicate a real difference in Ca affinity. ATP affects both Ca efflux and influx but has not been shown to produce a net flux by Na–Ca exchange.

Net flux studies of Ca by optical indicators show that the relationship between Na_i and Ca net influx is very steep with a half-activating concentration of about 25 mM; flux studies show a much shallower dependence on Na_i and a much larger half-activating concentration. Some caution is required in accepting the results of these flux studies.

Pulses of depolarization or hyperpolarization have the same dependence of Ca entry or exit on Na_i as do steady depolarizations.

While isotope studies of the action of Ca_i in promoting Ca entry have the difficulties enumerated above, a recent study (Kimura, Noma, and Irisawa 1986) has obtained measurements of the current generated by Na–Ca (i.e.

measurements of the net Ca flux) and show that such a current is only measurable if Ca_i is finite (*c.* 70 nM) as well as having a Na_i of 30 mM. The measurements were made on cardiac cells where Na–Ca exchange is highly developed. Therefore, it would appear that at physiological levels of Ca_i there is a clear effect of Ca_i on Ca influx. This is in contrast to the flux studies of DiPolo and Beaugé (1986) which claim that at 70 nM Ca_i there is little or no activation of reverse Na–Ca exchange.

References

Allen, T. J. A. and P. F. Baker (1986a). Comparison of the effects of potassium and membrane potential on the calcium-dependent sodium efflux in squid axons. *J. Physiol.* **378**, 53–76.

Allen, T. J. A. and P. F. Baker (1986b). Influence of membrane potential on calcium efflux from giant axons of *Loligo. J. Physiol.* **378**, 77–96.

Baker, P. F., Blaustein, M. P., Hodgkin, A. L., and Steinhardt, R. A. (1967) The effect of sodium concentration in calcium movements in giant axons of *Loligo forbesi. J. Physiol.* **192**, 43–44P.

Baker, P. F., Blaustein, M. P., Hodgkin, A. L., and Steinhardt, R. A. (1969). The influence of calcium on sodium efflux in squid axons. *J. Physiol.* **200**, 431–58.

Baker, P. F., Hodgkin, A. L., and Ridgway, E. B. (1971). Depolarization and calcium entry in squid axons. *J. Physiol.* **218**, 709–55.

Baker, P. F., Meves, H., and Ridgway, E. G. (1973). Calcium entry in response to maintained depolarization of squid axons. *J. Physiol.* **231**, 527–48.

Beaugé, L. and Mullins, L. J. (1976). Strophanthidin-induced sodium efflux. *Proc. Trans. Roy. Soc. Lond. B* **194**, 279–84.

Blaustein, M. P. (1977). Effects of internal and external cations and of the ATP on sodium–calcium and calcium–calcium exchange in squid axons. *Biophys. J.* **20**, 79–111.

Blaustein, M. P. and Hodgkin, A. L. (1969). The effect of cyanide on the efflux of calcium from squid axons. *J. Physiol.* **200**, 497–527.

Brinley, F. J. Jr. (1978). Calcium buffering in squid axons. *Ann. Rev. Biophys. Bioeng.* **7**, 363–92.

Brinley, F. J. Jr. and Mullins, L. J. (1967). Sodium extrusion by internally dialyzed squid axons. *J. Gen. Physiol.* **50**, 2303–31.

Brinley, F. J. Jr., and Mullins, L. J. (1968). Sodium fluxes in internally dialyzed squid axons. *J. Gen. Physiol.* **52**, 181–211.

Brinley, F. J. Jr., Tiffert, J. T., Scarpa, A., and Mullins, L. J. (1977). Intracellular calcium buffering capacity in isolated squid axons. *J. Gen. Physiol.* **70**, 355–84.

DiPolo, R. (1973). Calcium efflux from internally dialyzed squid giant axons. *J. Gen. Physiol.* **62**, 575–89.

DiPolo, R. (1974). Effect of ATP on the calcium efflux in dialyzed squid giant axons. *J. Gen. Physiol.* **64**, 503–17.

DiPolo, R. (1976). The influence of nucleotides on calcium fluxes. *Fed. Proc.* **35**, 2579–82.

DiPolo, R., Requena, J., Brinley, F. J. Jr., Mullins, L. J., Scarpa, A., and Tiffert, T. (1976). Ionized calcium concentrations in squid axons. *J. Gen. Physiol.* **67**, 433–67.

DiPolo, R. (1977). Characterization of the ATP-dependent calcium efflux in dialyzed squid giant axons. *J. Gen. Physiol.* **69**, 795–813.

DiPolo, R. (1979). Calcium influx in internally dialyzed squid giant axons. *J. Gen. Physiol.* **73**, 91–113.

DiPolo, R., and Beaugé, L. (1979). Physiological role of ATP-driven calcium pump in squid axon. *Nature* **278**, 271–3.

DiPolo, R., and Beaugé, L. (1986). Reverse Na–Ca exchange requires internal Ca and/or ATP in squid axons. *Biochim. Biophys. Acta.* **854**, 298–306.

DiPolo, R., and Beaugé, L. (1987). Characterization of the reverse Na/Ca exchange in squid axons and its modulation by Ca_i and ATP. *J. Gen. Physiol.* **90**, 505–25.

DiPolo, R., Bezanilla, R., Caputo, C., and Rojas, H. (1985). Voltage dependence of the Na/Ca exchange in voltage-clamps, dialyzed squid axons. *J. Gen. Physiol.* **86**, 457–78.

DiPolo, R., Rojas, H., and Beaugé, L. (1982). Ca entry at rest and during prolonged depolarization in dialyzed squid axons. *Cell Calcium* **3**, 19–41.

Hodgkin, A. L., and Keynes, R. D. (1957). Movements of labelled calcium in squid axons. *J. Physiol.* **138**, 253–81.

Kimura, J., Noma, A., and Irisawa, H. (1986). Na–Ca exchange current in mammalian heart cells. *Nature* **319**, 596–7.

Mullins, L. J., and Brinley, F. J. Jr., (1975). Sensitivity of calcium efflux from squid axons to changes in membrane potential. *J. Gen. Physiol.* **65**, 135–52.

Mullins, L. J., and Requena, J. (1979). Ca measurement in the periphery of an axon. *J. Gen. Physiol.* **74**, 393–413.

Mullins, L. J., and Requena, J. (1981). The 'late' Ca channel in squid axon. *J. Gen. Physiol.* **78**, 683–700.

Mullins, L. J., Tiffert, T., Vassort, G., and Whittembury, J. (1983). Effects of internal sodium and hydrogen ions and of external calcium ions and membrane potential on calcium entry in squid axons. *J. Physiol.* **338**, 295–319.

Mullins, L. J., Requena, J., and Wittembury, J. (1985). Ca^{2+} entry in squid axons during voltage-clamp pulses is mainly Na^{+}/Ca^{2+} exchange. *Proc. Natl. Acad. Sci. USA* **82**, 1847–51.

Requena, J., Mullins, L. J., and Brinley, F. J. Jr. (1979). Calcium content and net fluxes in squid giant axons. *J. Gen. Physiol.* **73**, 327–42.

Requena, J., Mullins, L. J., Whittembury, J., and Brinley, F. J. Jr. (1986). Dependence of ionized and total Ca in squid axons on Na_o-free or high-K_o conditions. *J. Gen. Physiol.* **87**, 143–59.

12 The sodium–calcium exchange in photoreceptors

Leon Lagnado and Peter A. McNaughton

12.1 Introduction

The advantage of the squid giant axon to those who began working on the Na–Ca exchange were obvious: it was large, and therefore was readily injected or perfused with radioactive tracers, with aequorin, or with intracellular ions or metabolites. Now that the whole-cell clamp has made it possible to voltage-clamp and perfuse small cells it has become clear that the rod outer segment may in many respects be a more favourable preparation for studying the Na–Ca exchange. The outer segment is small, and therefore diffusion of small ions and molecules across the cell is rapid: the solution perfusing the cell interior can be changed by perfusing the lumen of the whole-cell pipette; and the solution bathing the exterior can be changed in 50 ms or less, unhindered by geometrical complications such as the Schwann cell layer. The rod outer segment has some unexpected advantages: the light-sensitive channels are permeable to Ca^{2+}, and therefore can be used to load the cell with calcium before being closed by a bright flash, and the ionic current mechanisms in the membrane are remarkably simple, consisting so far as we can tell only of light-sensitive channels and the Na–Ca exchange mechanism itself. The main disadvantage of the rod outer segment is the limitation on the measurement of concentrations of internal ions imposed by its small size.

Early experiments using this preparation have characterized some basic properties of the exchange: the binding characteristics of Na^+ at the external membrane surface and of Ca^{2+} at the internal surface have been determined, and the forward rate of the exchange in normal Ringer solution has been found to be reduced e-fold by a 70 mV depolarization. Surprisingly, this sensitivity to membrane potential appears not to be the result of the translocation of a charged species across the membrane field, but rather to be due to a voltage-sensitive binding of Na ions at the external binding site, probably because the binding site is located at the foot of a channel traversing part of the membrane field. Even more unexpected is the discovery that K^+

ions are cotransported with Ca^{2+} in an electrogenic exchange with a stoichiometry of $4Na^{+}$–$1Ca^{2+}$, $1K^{+}$, so that the transmembrane K^{+} gradient provides an important contribution to the driving force necessary to transport Ca^{2+} out of the cell against its electrochemical gradient. The cotransport of K^{+} ions with Ca^{2+} has been suspected since the days of the first experiments on the exchange in the squid giant axon (see Baker *et al.* 1969), yet has only now been demonstrated conclusively. The exchange should now be renamed the Na–Ca, K exchange, but for consistency with other articles in this book we shall continue to use the usual nomenclature 'Na–Ca exchange'.

12.2 Characteristics of the light-sensitive current in photoreceptors

The picture of the circulating light-sensitive current in a vertebrate rod appearing in many elementary textbooks is shown in Fig. 12.1A (see Tomita 1970; Hagins and Yoshikami 1975). This picture is not incorrect in its essential features, but some points need elaboration. First, the idea that the light-sensitive current crossing the outer segment membrane is carried purely by Na^{+} ions is now known to be incorrect: a small but significant contribution to the light-sensitive current (probably about 10–15 per cent) comes from Ca^{2+} ions passing through the light-sensitive channel (Hodgkin, McNaughton, and Nunn 1985; Nakatani and Yau 1988a; Yau and Nakatani 1984a). The light-sensitive channel is in fact not highly selective for Na^{+} ions, and the light-sensitive current is carried principally by Na^{+} for the simple reason that Na^{+} is the major external cation.

The second contribution to the outer segment membrane current comes from the operation of the electrogenic Na–Ca exchange responsible for extruding the Ca^{2+} which flows in through the light-sensitive channel. The exchange stoichiometry is $4Na^{+}$–$1Ca^{2+}$, $1K^{+}$ (see Section 12.7), so a current j of calcium ions through the light-sensitive channel contributes an additional inflow of current $j/2$ in the process of its extrusion from the outer segment. The electrogenic Na–Ca exchange is not directly light-sensitive, although its activity declines after a bright flash of light because the level of intracellular ionized Ca declines with a time constant of about 0.5 seconds when the influx of Ca through light-sensitive channels is shut off (Yau and Nakatani 1985; McNaughton, Cervetto, and Nunn 1986, see Section 12.10).

A revised picture of the flow of current in darkness in a rod is shown in Fig. 12.1B. The light-sensitive current and the Na–Ca exchange are the only mechanisms contributing in any significant degree to outer segment membrane current. The inner segment membrane contains a variety of conductance mechanisms which shape the response to light of the cell membrane potential (see review by Owen 1987), and it contains, in addition, the Na–K

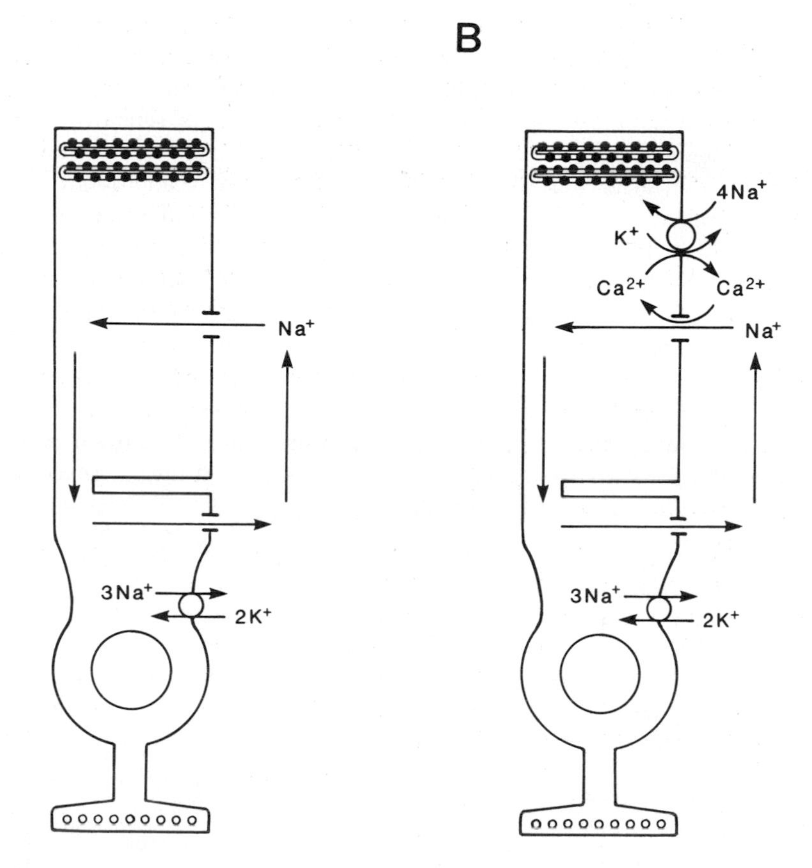

Fig. 12.1. Part A shows the picture of the circulating dark current in a single rod as presented in many textbooks. The light-sensitive current, carried by Na ions, flows in through the outer segment membrane and passes along the cell before leaving through the inner segment membrane. The principal current carrier at the inner segment is K^+, but in view of the complex time and voltage-dependent current-carrying systems in the inner segment (see review by Owen 1987) it might be misleading to show the current as being carried by K^+ alone on a schematic diagram of this type. The ionic gradients of Na and K are maintained by the operation of an ATP driven Na–K pump which seems to be located only in the inner segment membrane. Two discs are shown schematically at the top: in reality, of course, the entire rod outer segment is filled with discs.

Part B shows some necessary modifications to the scheme shown in A. The light-sensitive channel is imperfectly selective, and about 10–15 per cent of the light sensitive current is carried by Ca^{2+} ions. There is probably also an efflux of K^+ ions at the resting potential, although this has not yet been conclusively demonstrated. The calcium ions flowing in through the light sensitive channel are extruded by a $4Na^+$–$1Ca^{2+}$, $1K^+$ exchange which does not appear to depend on energy sources other than that supplied by the transmembrane Na gradient.

pump responsible for maintaining the normal gradients of Na and K. The conductance mechanisms in the inner segment have little effect on the circulating dark current flowing between inner and outer segments, since over the normal range of intracellular voltages the light-sensitive current is almost independent of membrane potential (Baylor and Nunn 1986).

In most invertebrate photoreceptors, light activates an inward current, thereby depolarizing the cell and generating action potentials. Light also releases Ca from stores within the cell (Brown and Blinks 1974). The ionic selectivity of the light-sensitive channel is apparently similar to that of the vertebrate (Brown and Mote 1974), and activation of the light-sensitive current therefore causes a Ca influx which, together with the Ca released internally, is extruded by a Na–Ca exchange mechanism. In this review we concentrate more on the Na–Ca exchange in photoreceptors from vertebrates than from invertebrates, mainly because much more is known about the former, but the increasing body of evidence demonstrating the importance of Na–Ca exchange in invertebrate photoreceptors is also briefly reviewed.

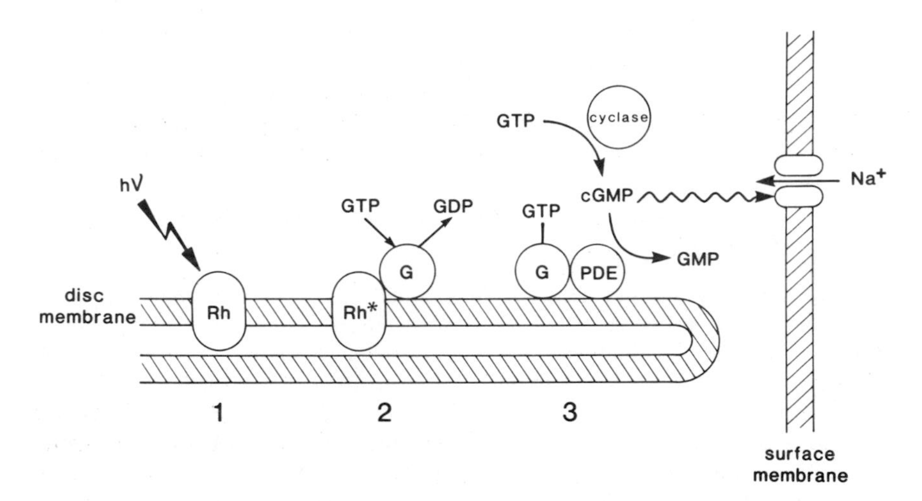

Fig. 12.2. Diagram of the light-sensitive cascade. In step 1 the 11-*cis* retinal chromophore of a rhodopsin molecule, Rh, absorbs a photon of light, converting it into the active form, Rh*. In step 2 an Rh* interacts with a molecule of G-protein (also called transducin), converting it to the active form by catalysing the release of a bound GDP and the uptake of GTP. Each Rh* is capable of activating a large number of G-proteins (probably about 500). In step 3 a single active molecule of G-protein switches a single phosphodiesterase into an active state, thus catalysing the hydrolysis of cGMP to GMP. When the level of cGMP is lowered the light-sensitive channels close. The diagram does not show the inactivation of the various steps in the pathway, nor does it show the possible interactions of intracellular calcium with the pathway.

12.3 The cellular basis of phototransduction

An outline of our current state of knowledge of the cellular basis of phototransduction may be helpful in understanding the experiments which are to be described. The machinery of phototransduction in vertebrate rods is now fairly well understood, as a result of a series of advances in recent years on both biochemical and biophysical fronts (for reviews see McNaughton and Cervetto 1986; Pugh and Cobbs 1986; Stryer 1986). The following account applies to vertebrate rods; there is good evidence that vertebrate cones transduce light in essentially the same way, but it is still unclear how much of this story applies to invertebrates.

A single photon of light activates a single rhodopsin molecule which in turn activates an intermediate G-protein (also called transducin) by mediating an exchange of GDP for GTP (see Fig. 12.2). Each rhodopsin can activate about 500 G-protein molecules and thus part of the large amplification necessary to convert the energy of a single photon into a light response occurs at this first stage. Each active G-protein then activates a phosphodiesterase by displacing an inhibitory subunit, and the phosphodiesterase in its active state breaks down 3′,5′-cyclic GMP to GMP. The resulting drop in the concentration of cGMP closes the light-sensitive channels, which are held open in darkness by direct combination with at least 2 (and probably 3) molecules of cGMP (Fesenko, Kolesnikov, and Lyubarsky 1985; Haynes, Kay, and Yau 1986; Zimmerman and Baylor 1986), thereby shutting off the entry of Na ions and allowing the cell to hyperpolarize.

Calcium ions play an important role in modulating this light-sensitive pathway, although it has now been established that calcium ions do not participate as an essential link in the chain of events leading to a light response. A reduction in internal Ca causes a dramatic increase in the light-sensitive current, and changes in internal $[Ca]_i$ have effects on the form of the light response similar to those of light adaptation. At present it seems likely that Ca acts as the internal messenger for light adaptation, perhaps through an inhibitory action on the guanylate cyclase which synthesizes cGMP from GTP. These topics are discussed further in Section 12.11.

12.4 Methods for studying Na–Ca exchange in photoreceptors

It has recently become clear that the vertebrate photoreceptor outer segment is a surprisingly favourable preparation in which to study the mechanism of Na–Ca exchange. The outer segment can be loaded with a known amount of Ca by exposure to a solution containing only Ca ions; the Ca flows through the light-sensitive channel, and the load can be calculated simply by integrating the light-sensitive current (Hodgkin, McNaughton, and Nunn 1985, 1987; Yau and Nakatani 1984b). The intracellular free [Ca] can be measured with aequorin (Cervetto, McNaughton, and Nunn 1986; McNaughton, Cervetto, and Nunn 1986) or with the Ca-sensitive fluorophores quin-2 (Miller and Korenbrot 1986) and fura-2 (Ratto *et al.* 1988). Finally, the membrane current produced by the operation of the electrogenic Na–Ca exchange can be recorded while the Ca load is extruded. The simplicity of this preparation results in part from the simple intracellular buffering of Ca, with the uptake of Ca being time-independent and approximately linearly related to the free [Ca] (McNaughton, Cervetto, and Nunn 1986; Cervetto, Lagnado, and McNaughton 1987), and in part from the fact that the only significant mechanisms contributing to the calcium flux across the outer segment membrane are the light-sensitive channels, which can be used to load the cell with Ca, and the electrogenic Na–Ca exchange itself.

An even simpler preparation which has been developed in the past year is the outer segment isolated from the inner segment and the rest of the cell (Hestrin and Korenbrot 1987; Lagnado, and McNaughton 1987a,b; Sather and Detwiler 1987). Light responses and Na–Ca exchange currents are normal in these isolated outer segments provided the ionic gradients and high-energy metabolites are maintained by diffusion from a whole-cell pipette. In this preparation the Na–Ca exchange current can be recorded under voltage-clamp conditions, and the ionic environment on both sides of the membrane is under the control of the experimenter. Some early results obtained with this preparation are presented in this review, and it seems likely that many more features of the Na–Ca exchange will be elucidated by subsequent work.

Finally, a number of interesting results have come from studies of calcium fluxes across the outer segment membrane using Ca-sensitive electrodes or dyes to detect the efflux of Ca into the extracellular space. A flash of light causes a rise in the extracellular [Ca] (Gold and Korenbrot 1980; Yoshikami, George, and Hagins 1980; Miller and Korenbrot 1987), which was originally interpreted as support for the 'calcium hypothesis' of phototransduction, in which the closure of light-sensitive channels was proposed to be caused by the release of Ca within the cell (Hagin 1972). It is now clear that the action of light is to suppress the Ca influx, and the rise in $[Ca]_o$ occurs because Ca extrusion via a Na–Ca mechanism is light-insensitive and continues for a time after a bright flash. Measurements of changes in the extracellular [Ca] in the presence of a bright light can therefore be used to monitor Na–Ca exchange activity (Schnetkamp 1986; Schnetkamp and Bownds 1987; see also the measurements on a vesicular preparation made by Bauer 1988).

12.5 Observations implicating a Na–Ca exchange in photoreceptor function

It has been clear for some years that a Na–Ca exchange mechanism is important in the function of both vertebrate and invertebrate photoreceptors. Removing extracellular Na abolishes the light response, but this observation cannot be taken as proof that light-sensitive channels are exclusively permeable to Na because if $[Ca]_o$ is also reduced the light response is maintained (Bastian and Fain 1982; Fain and Lisman 1981; Yau, McNaughton and Hodgkin 1981). A reduction in the $[Na]_o$ bathing a vertebrate rod outer segment reduces the amplitude of the light-sensitive current in proportion to $[Na]_o^2$, and the current can be maintained at a constant level if $[Na]_o$ and $[Ca]_o$ are altered in the ratio $[Na]^{2.5}/[Ca]$ (Hodgkin *et al* 1984). Elevations in internal Na also strongly suppress the light-sensitive current. These observations suggest that Na does not simply participate as a current carrier through the light-sensitive channel, but also acts by activating a Na–Ca exchange. We now think that the main reason why reducing the Na gradient reduces the light-sensitive current is not because of the removal of current carriers, which accounts for only an approximately linear dependence of current on $[Na]_o$ (Hodgkin, McNaughton, and Nunn 1985), but because the reduction in Na–Ca exchange activity causes a rise in $[Ca]_i$ and inhibits the light-sensitive current by a powerful mechanism involving a delay. One strong possibility is that Ca may inhibit the guanylate cyclase responsible for synthesizing cyclic GMP (Hodgkin, McNaughton, and Nunn 1985; Koch and Stryer 1988; Lolley and Racz 1982; Pepe, Panfoli, and Cugnoli 1986; Troyer, Hall and Ferrendelli 1987—see Fig. 12.13).

The dual action of external ions was confirmed in experiments in which the solution bathing a rod outer segment could be changed rapidly. The change in light-sensitive current, after (for instance) substituting another monovalent cation for Na, consisted of a rapid component attributed to the change in a current carried through open light-sensitive channels, followed by a delayed relaxation of current attributed to a change in the number of open light-sensitive channels (see Fig. 12.3). The rapid change in current would be explained if monovalent cations carried current through the light-sensitive channel in the proportions Li:Na:K:Rb:Cs = 1.4:1:0.8:0.6:0.15 (Hodgkin, McNaughton, and Nunn 1985), while the delayed component could be explained if current was regulated by changes in internal Ca, with Ca in turn being regulated by a Na–Ca exhange mechanism. External Na ions strongly promoted Ca efflux through the exchange, while K and Rb were inhibitory, and Li, Cs, and choline had little or no direct action (Hodgkin, McNaughton, and Nunn 1985).

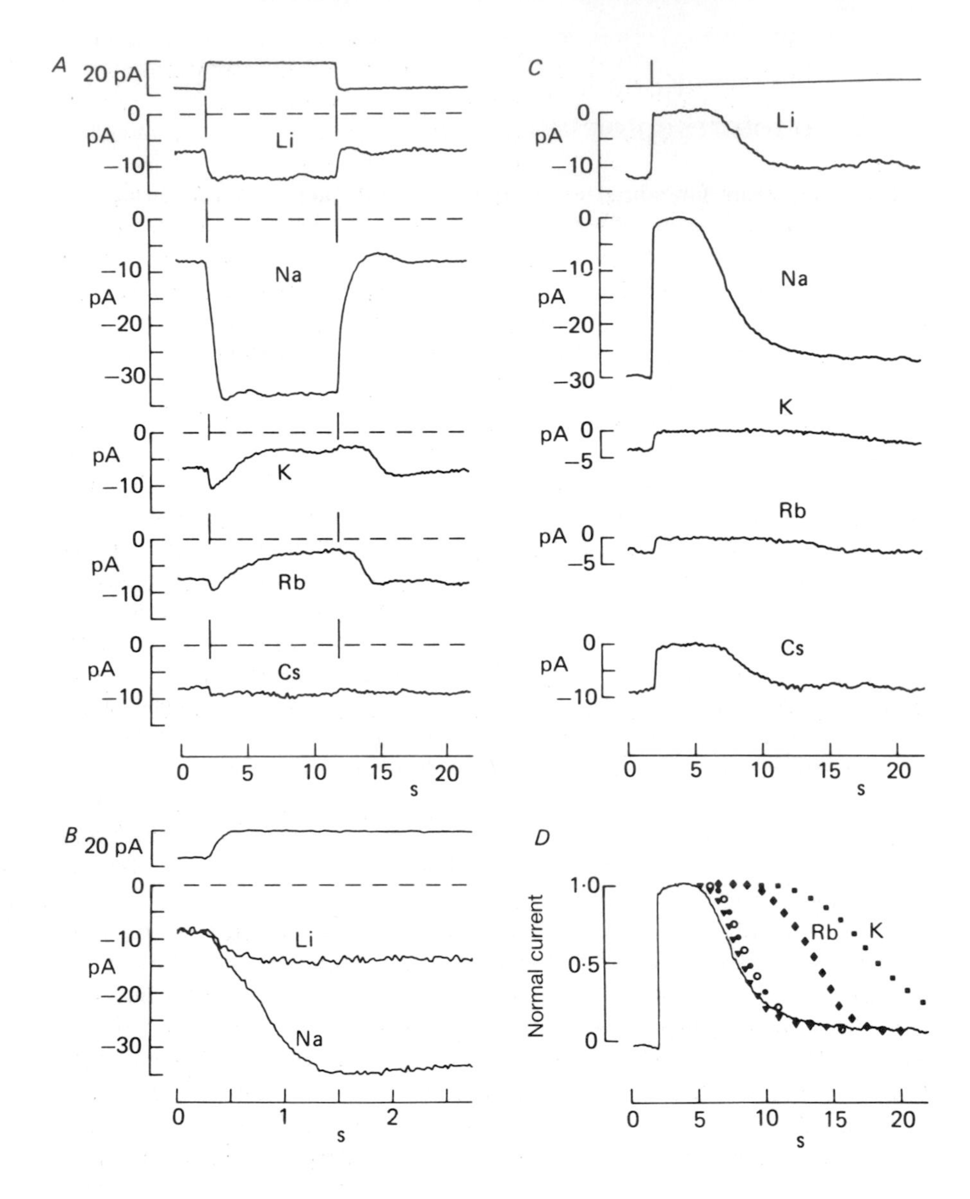

Fig. 12.3. Changes in the light-sensitive current (*A*,*B*) and in the form of the light response (*C*,*D*) when the choline in a Ringer containing half the normal $[Na]_o$ (composition 55 mM Na, 55 mM choline, 2.5 mM KC1, 1 mM $CaC1_2$, 1.6 mM $MgC1_2$, 10 mM HEPES, 3 mM glucose) was replaced by various other monovalent cations.
A. Top trace shows time course of solution change. Other traces show the effect on the light-sensitive current of replacing 55 mM choline with Li, Na, K, Rb, and Cs.
B. Traces from A on an expanded time scale to show that the increased current in Na consists of 2 phases: a rapid increase caused by the increase in concentration of

Most of the work suggestive of a Na–Ca exchange in invertebrate photoreceptors has been carried out on the ventral photoreceptors of *Limulus*, which are large and can be readily impaled, voltage-clamped, and injected with aequorin. Lisman and Brown (1972) found that injections of Na caused a loss of sensitivity which depended on Ca_o; Brown and Mote (1974) and O'Day and Keller (1986) observed a similar effect after Na_o removal; and Fein and Charlton (1977a,b) further showed that the desensitization caused by injecting Ca was localized to the site of injection, while that caused by injecting Na spread throughout the cell. This last result would be expected if Ca diffusion is limited, as in most cells, while Na is free to diffuse throughout the cell and can cause a widespread elevation of $[Ca]_i$ by inhibiting or reversing a Na–Ca exchange. Waloga, Brown, and Pinto (1975) confirmed that changes in external or internal Na have their effects through changes in $[Ca]_i$ by showing that rise in $[Ca]_i$ could be detected using aequorin when the Na gradient was reduced. More recently a current due to the operation of the Na–Ca exchange has been recorded (O'Day and Keller 1987). All of these results show that a Na–Ca exchange operates in the surface membrane of invertebrate photoreceptors, and suggest that elevations in $[Ca]_i$ desensitize the cell's response to light (Brown and Lisman 1975; Levy and Fein 1985). An important difference from the vertebrate is that in vertebrates a fall in $[Ca]_i$ seems to signal light adaptation, while in the invertebrate the signal is a rise in $[Ca]_i$.

12.6 Evidence for a membrane current associated with Na–Ca exchange activity

The first clear indications of a membrane current associated with the operations of a Na–Ca exchange came from experiments in which rod outer segments were loaded with Ca by exposure to a high-Ca solution and then, with the light-sensitive channels closed by a bright light, were returned to a Na-containing solution (Hodgkin, McNaughton, and Nunn 1987; McNaughton, Cervetto, and Nunn 1986; Yau and Nakatani 1984b—see Fig. 12.4). The inward current which is recorded on restoration of Na is transient, and its duration depends on the magnitude of the preceding Ca load; it is reduced

current carriers, followed by a slower increase due to the activation of the Na–Ca exchange. The increased current in Li consists only of the first phase.

C. Responses to bright flashes (2.3×10^4 isomerizations per rod) in the various solutions shown in *A*.

D. Comparison of the time courses of the flash responses shown in C. Symbols correpond to 55 mM Na plus 55 mM of the following cations: Na (continuous line); choline (▼); Li (○); Rb (◆): K (■); Cs (●). Reproduced with permission from Hodgkin, McNaughton and Nunn (1985).

by a decrease in external Na and is increased by removal of external Ca; and lastly, it is abolished by the application of low concentrations of La^{3+}, which is known to be a potent blocker of the Na–Ca exchange (Baker and McNaughton 1978; Van Breeman and de Weer 1970). A similar current is recorded on return to normal Na after a period in low Na (Cook *et al* 1984; McNaughton, Nunn, and Hodgkin 1986). All these properties are those expected of a current carried by the Na–Ca exchange.

An important piece of evidence in support of the idea that the current shown in Fig. 12.4 is actually due to the operation of a Na–Ca exchange came from experiments in which $[Ca]_i$ was measured directly, and in which the operation of the exchange could therefore be linked to the fall in $[Ca]_i$ which it causes (Cervetto, McNaughton, and Nunn 1985; McNaughton, Cervetto and Nunn 1986). In this work $[Ca]_i$ was measured using the Ca-sensitive photoprotein aequorin, which was incorporated into the outer segment of an isolated rod by diffusion through the mouth of a whole-cell pipette sealed against the outer segment membrane. At the end of the loading period the pipette was gently withdrawn, allowing the membrane to reseal. The outer segment membrane current of the aequorin-loaded rod can then be recorded in the usual way with a suction pipette, and both light responses and Na–Ca exchange currents are observed to be normal in form and in amplitude.

Figure 12.4 shows a simultaneous recording of membrane current and free $[Ca]_i$ calculated from the aequorin light output. The rod was loaded with Ca by exposure to a solution in which the only permeable ion was Ca^{2+}, and the Ca load was calculated from the integral of the light-sensitive current. The phosphodiesterase inhibitor IBMX was added to prevent the rapid closure of light-sensitive channels which would normally occur after a rise in internal Ca (See Section 12.11); other experiments showed that the Na–Ca exchange was unaffected by IBMX (see Fig. 12.11). During the loading period the free $[Ca]_i$ rose in approximate proportion to the integral of the Ca influx (solid trace in Fig. 12.4B), showing that a constant fraction (between 5 per cent and 10 per cent) of the Ca influx remains free within the cell, the rest being bound rapidly and reversibly to a buffer of high capacity and low affinity. After the Ca influx had been terminated by closure of the light-sensitive channels the free $[Ca]_i$ remained at a stable high level, showing that Ca is not removed from the cell either by a slow internal uptake mechanism or by Na-independent transport across the outer segment membrane. On restoration of external Na the exchange current was activated and the free $[Ca]_i$ declined rapidly. The rate of decline of free $[Ca]_i$ could be predicted directly from the integral of the exchange current (smooth trace in Fig. 12.4B). This experiment demonstrates that the Na–Ca exchange is the principal mechanism for extrusion of Ca from the outer segment, at least at high $[Ca]_i$, since the Na–Ca exchange working at its maximum rate caused a rate of decline of $[Ca]_i$ of $30\ \mu M\ s^{-1}$, which was at least 50 times greater than in the absence of external Na.

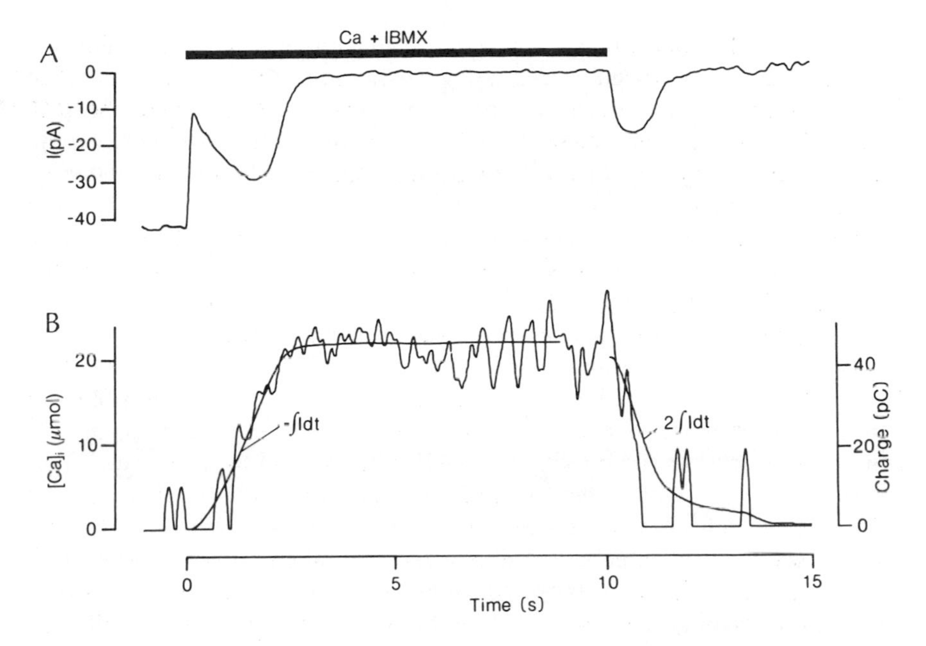

Fig. 12.4. Measurements of Ca influx, Na–Ca exchange current and intracellular free [Ca] in a single rod outer segment.

A. Outer segment membrane current. The duration of exposure to isotonic $CaCl_2$, containing IBMX to keep the light-sensitive channels open, is shown by the horizontal bar. The Ca^{2+} current flowing through light-sensitive channels during this period was terminated by the emission of aequorin light, and the total charge flowing was 46 pC. On return to Ringer's solution a transient light-insensitive current was activated. The charge transferred by this current was 20.6 pC, consistent with an influx of one positive charge per calcium ion extruded.

B. Intracellular free [Ca] measured with aequorin (noisy trace). No decline in free $[Ca]_i$ was observed in the absence of external Na, and the rate of decline on restoration of Na_o was 30 μM s^{-1}. The smooth traces show the integral of the calcium influx and twice the integral of the Na–Ca exchange current. The close correspondence between the form of the free $[Ca]_i$ and the total $[Ca]_i$ obtained from these integrals shows that intracellular buffering of Ca is rapid and reversible. About 5 per cent of the total $[Ca]_i$ is free. Modified from McNaughton, Cervetto, and Nunn 1986.

12.7 Stoichiometry of the Na^+–Ca^{2+} exchange

When an outer segment is exposed to an isotonic calcium chloride solution the membrane current is carried exclusively by calcium ions passing through the light-sensitive channels, because the current is found to be completely suppressed either by reducing external Ca to 1 μM or by turning on a bright light (Hodgkin, McNaughton, and Nunn 1985). This method can therefore be used to load a known amount of calcium into the rod outer segment. On restoration of a normal Na gradient, the Na–Ca exchange current is activated (see Fig. 12.4) and the charge moved per calcium ion expelled by the exchange can be measured comparing the amount of charge flowing during the loading period and during the period of operation of the exchange. A correction, which is usually small, must be made for the amount of calcium present in the cell before the loading period, but this existing load can be calculated from the integral of the small Na–Ca exchange current observed after applying a bright flash of light while in Ringer's solution.

Measurements were first made with intact rods, in which the current is measured with a suction pipette (Yau and Nakatani 1984b; Hodgkin, McNaughton, and Nunn 1987). In these experiments the influx and efflux of Ca must be calculated from the difference between two records, one in darkness (apart from the bright flash used to shut the light-sensitive channels at the end of the loading period) and a second in bright light, because the different mobilities of the ions in the various solutions produce a junction current through the imperfect seal between the rod and the suction pipette. Since this junction current is present in both the dark and the light record it can be removed by subtraction. While there appears to be no problem with this procedure, it does increase the variability in the measurements, as well as introducing an element of experimental uncertainty. A more direct method is to measure the outer segment membrane current using a whole-cell pipette (Lagnado and McNaughton 1987a,b) as shown in the inset of Fig. 12.5. The high resistance of the seal between the pipette mouth and the outer segment membrane ensures that the junction currents are negligibly small, while the preparation offers the additional advantage that the outer segment membrane is under voltage control. When the light-sensitive channels have been closed by a bright flash and when the Na–Ca exchange is inactive the input impedance of the preparation can be as high as 40 GΩ, and even this high figure may be an underestimate of the outer segment membrane impedance in light since the seal resistance is presumably finite. This observation shows that no ionic channels apart from light-sensitive channels exist in the outer segment membrane (see also Baylor and Lamb 1982; Baylor and Nunn 1986), reinforcing the conclusion that the only path for Ca influx is through the light-sensitive channel itself.

These experiments have demonstrated that a single charge enters the cell for every Ca^{2+} ion extruded. The value of r, the ratio of the charge flowing into the outer segment during the period of Ca loading to the charge moved in during the operation of the exchange in extruding the load, was found to be 2.13 by Yau and Nakatani (1984b) and 2.70 by Hodgkin, McNaugton, and Nunn (1987), both estimates made using the suction pipette method with intact rods. The corresponding values of q, the charge transported in exchange for one calcium ion, are 0.94 and 0.74. Measurements of the charge moved during the extrusion of a known Ca load have also been obtained using the voltage-clamped isolated outer segment (see Fig. 12.5). These experiments gave what is probably a more accurate value for r of 1.99 ± 0.02 (mean $\pm$ S.E.M., $n = 84$), corresponding to $q = 1.005 \pm 0.01$, and have also shown that q is independent of external Na (220 mM–35 mM), of membrane potential (-64 mV to $+16$ mV), and of the presence or absence of external K (2.5 mM) or Mg (1.6 mM) (Lagnado, Cervetto, and McNaughton 1988; Lagnado, and McNaughton 1988). It is clear, therefore, that the carrier transports one net positive charge in exchange for one Ca^{2+}, and that the movements of ions through the exchange mechanism are tightly coupled according to a stoichiometry which is fixed over a wide range of conditions.

These results were initially interpreted to mean that the exchange stoichiometry was $3Na^+$–$1Ca^{2+}$. The only other possibility consistent with these experiments would be a mechanism such as $4Na^+$–$1Ca^{2+}$, $1K^+$, in which Ca^{2+} leaves the cell in association with K^+, because all ions other than Na^+ can be removed from the external solution without abolishing the exchange, and the only small internal ions, in the experiments with isolated, perfused outer segments, were K^+ and Ca^{2+}. Operating on the principle that the simplest hypothesis consistent with the data is the one to be preferred, this latter possibility would be considered a long shot. Recent evidence has, however, overturned this assumption. The reversed mode of the exchange—in which Ca^{2+} enters the cell in exchange for internal Na^+—has been found to depend on the presence of external K^+ (Cervetto *et al.* 1989). There are two possible explanations for this dependence: K^+ could be a cotransported with calcium, or it could be a necessary cofactor (for instance, acting to increase the affinity of the calcium binding site for Ca^{2+}). In the former case, the K^+ gradient contributes energy to the exchange, and the equilibrium level of $[Ca^{2+}]_i$ would be altered by changes in the transmembrane K^+ gradient, while in the latter K^+ acts as a catalyst, and the forward and backwards rates of the exchange would be equally affected by changes in $[K^+]$, leaving the equilibrium level of $[Ca^{2+}]_i$ unchanged. In an experiment to distinguish between the two possibilities the exchange was allowed to reach equilibrium in an isolated outer segment, under whole-cell voltage clamp, in which the internal $[Na^+]$ had been increased to facilitate operation of the reversed mode of the exchange. When $[K^+]_o$ was reduced there was an activation of inward current as Ca^{2+} was extruded from the cell, followed by an equal and opposite outward movement of charge on restoration of the previous K^+ gradient (Cervetto *et al.* 1989). We conclude that K^+ is indeed transported by the exchange.

The stoichiometry of the exchange can also be determined directly by a similar method. If n Na^+ ions exchange for one K^+, for instance, then the equilibrium level of $[Ca^{2+}]_i$ will be maintained if $[Na^+]_o$ and $[K^+]_o$ are simultaneously altered so as to preserve the ratio $\{[Na^+]_o\}^n/[K^+]_o$ at a constant value, and n can be determined by testing various levels of $[Na^+]_o$ until the value required to null off a given change in $[K^+]_o$ is obtained. Using this method it was found that $4Na^+$ exchange for one K^+, that $4Na^+$ exchange for one Ca^{2+}, that one K^+ is cotransported with one Ca^{2+}, and finally, by altering the membrane potential and the K^+ gradient simultaneously, that one charge is countertransported in exchange for one K^+. The only stoichiometry consistent with all these measurements is $4Na^+$–$1Ca^{2+}$, $1K^+$, where the hyphen indicates countertransport and the comma cotransport. This stoichiometry is, of course, also consistent with the known countertransport of one Ca^{2+} in exchange for one positive charge.

A stoichiometry of $4Na^+$–$1Ca^{2+}$, $1K^+$ will enable the exchange to attain a much lower level of $[Ca^{2+}]_i$ than previously supposed. If Na and K do not compete for each other's binding sites then the equilibrium level of $[Ca^{2+}]_i$ is given by:

$$[Ca^{2+}]_i = [Ca^{2+}]_o \frac{[Na^+]_i^4}{[Na^+]_o^4} \frac{[K^+]_o}{[K^+]_i} \exp(V_m F/RT) \qquad (1)$$

In the rod outer segment in bright light $V_m \approx -55\,mV$, and taking $[Na^+]_i = 10$ mM and $[K^+]_i = 110$ mM we obtain an equilibrium level in Ringer of $[Ca^{2+}]_i = 1.8 \times 10^{-10}$M, nearly 500 times lower than the value of 8.5×10^{-8}M for a $3Na^+$–$1Ca^{2+}$ exchange. While it seems unlikely that this equilibrium level would be attained in practice, an exchange stoichiometry in which K^+ is transported enables the exchange to continue extruding Ca^{2+} down to much lower levels of $[Ca^{2+}]_i$ than was previously supposed. In the rod outer segment, where $[Ca^{2+}]_i$ inhibits the guanylate cyclase at $[Ca^{2+}]_i \leqslant 10^{-7}$M (Lolley and Racz 1982; Koch and Stryer 1988) the ability to attain a low $[Ca^{2+}]_i$ may be particularly important in the control of the light-sensitive current (Hodgkin, McNaughton, and Nunn 1985).

In other cells there is also evidence of an interaction of K^+ ions with the exchange. In squid axon (Baker *et al.* 1969), bovine rod outer segments (Schnetkamp 1986) and ventricular myocytes (Gadsby *et al.* 1988) the reversed exchange depends on the presence of an alkali metal ion in the external solution, with K being considerably more effective than Li (Allen and Baker 1986). The forward mode of the exchange in squid axon (DiPolo and Rojas 1984) and in bovine rod outer segments (Schnetkamp *et al.* 1988) depends on internal K^+. We conclude that an exchange of Na^+ for Ca^{2+} and K^+ is a general phenomenon, although in the rod outer segment the K^+ binding site may have evolved a higher degree of selectivity in order to maintain Ca extrusion at low $[Ca^{2+}]_i$ without the assistance of an ATP-dependent Ca pump.

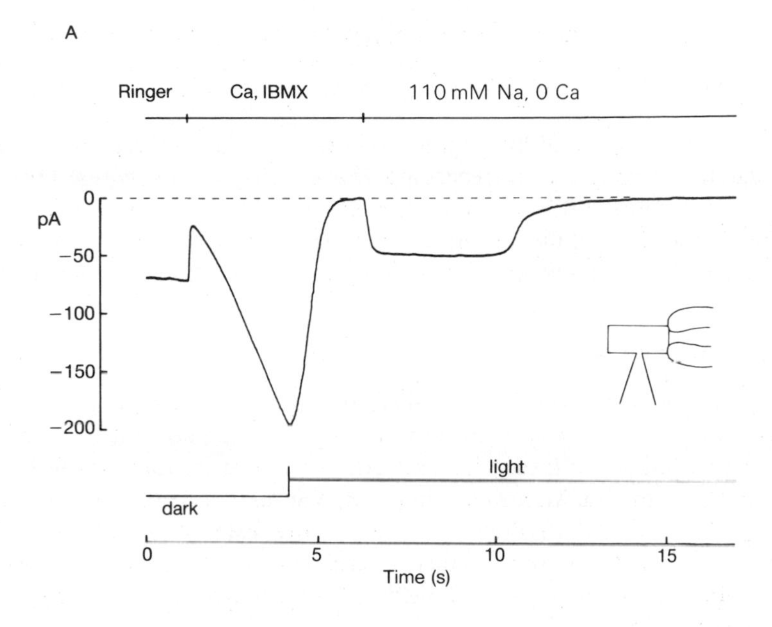

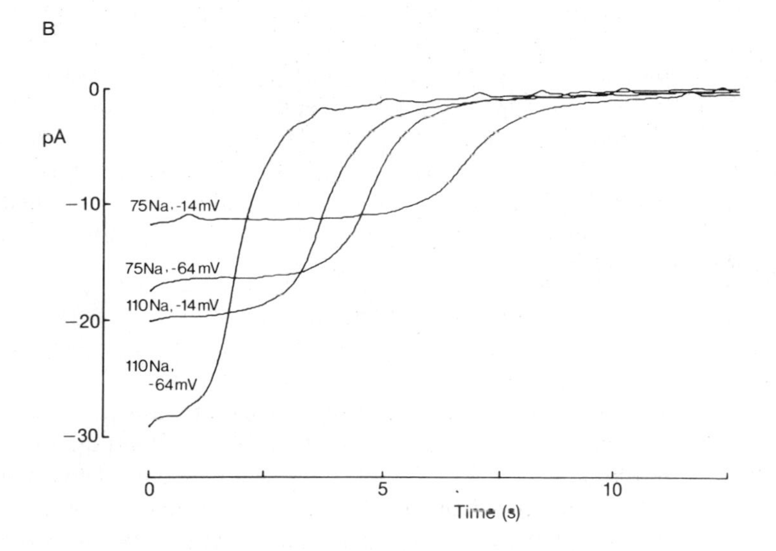

Fig.12.5. A. Measurement of Na–Ca exchange stoichiometry in an isolated rod outer segment under whole-cell voltage clamp. Rod was loaded with Ca and the exchange current activated as in Fig. 12.5. Charge transferred during loading was 417 pC and during exchange 220 pC, so charge ratio *r* is 1.90.
B. Effect of polarization and $[Na]_o$ on the form of the exchange current. Measured charge ratio *r* was (from bottom) 2.02, 2.00, 1.90, and 2.08.

12.8 Activation of the exchange by external and internal ions

The effects of external and internal ions on the Na–Ca exchange have been investigated in two ways: by recording the electrogenic exchange current (Hodgkin and Nunn 1987; Lagnado, Cervetto, and McNaughton 1988) or by monitoring the efflux of Ca^{2+} from the cell using a Ca^{2+}-sensitive dye in the external solution (Schnetkamp 1986; Schnetkamp and Bownds 1987).

12.8.1 External Na

The exchange is, so far as is known, absolutely dependent on the presence of external Na for its forward operation, as no exchange current is observed when Li or Choline substitute for external Na (Hodgkin and Nunn 1987; Lagnado, Cervetto, and McNaughton 1988; Yau and Nakatani 1984b) and the free internal [Ca] declines at a very low rate, if at all, in these circumstances (McNaughton, Cervetto, and Nunn 1986). Similar results were obtained by monitoring the Ca efflux from isolated rod outer segments (Schnetkamp 1986).

When the external [Na] is increased both the exchange current and the Ca efflux rate from the outer segment increase in a sigmoidal manner. Hodgkin and Nunn found that the foot of the curve could be described by the relation of $j \propto [Na]_o^{2.4}$, where j is the exchange current, while Schnetkamp (1986) notes that a reasonably satisfactory fit of the Ca efflux rate from isolated rod outer segments could be obtained with $n=2$ in the presence of external K.

Fig. 12.6, curve, shows the more complete activation curve obtained by Lagnado, Cervetto, and McNaughton (1988). The activation curve in these experiments was well fitted by the Hill equation:

$$\frac{j}{j_{max}} = \frac{[Na]_o^h}{[Na]_o^h + [K_{\frac{1}{2}}]_o^h}$$

At a membrane value of -14 mV in the absence of competing external cations values for the Hill coefficient h, of 2.26 and for the $[Na]_o$ giving half-activation of the exchange current, $K_{\frac{1}{2}}$, of 93 mM were obtained. Hyperpolarization or the addition of competing cations (curves 2 and 3 in Fig. 12.6) affected the value of $K_{\frac{1}{2}}$ but did not significantly alter either the value of h or the saturated exchange current at high $[Na]_o$ (see below).

The Hill equation is derived from the extreme assumption that the co-operativity between the Na binding site is infinite, so that the binding of the first Na ion increases the affinity of the exchange for the subsequent ions to such an extent that binding is inevitable and we can neglect the concentrations of intermediate states such as E Na and E Na_2 (where E is the exchange

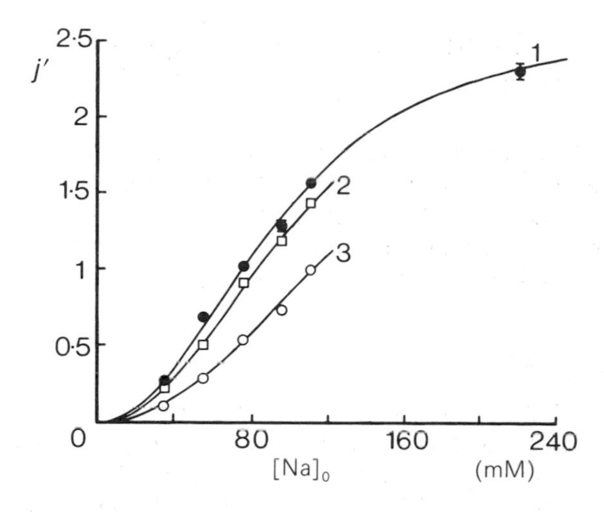

Fig. 12.6. Activation of the Na–Ca exchange current in an isolated rod outer segment by external [Na]. Vertical axis shows exchange current at a saturating level of $[Ca]_i$, expressed relative to its value in 110 mM Na, 1.6 mM Mg, 2.5 mM K at $V_m = 14$ mV. Curve 1 was obtained with $V_m = -14$ mV and in the absence of all external cations apart from Na and Li, which substituted for Na. Curves 2 and 3 were obtained at $V_m = -64$ mV and -14 mV, respectively, in the presence of 1.6 mM Mg and 2.5 mM K. All data could be fitted by a Hill equation (see eqn. 1) with $h = 2.26$ and $j_{max} = 2.66$. Values of $K_{\frac{1}{2}}$ were 93 mM (curve 1), 103 mM (curve 2) and 139 mM (Curve 3).

molecule). An alternative extreme assumption, that the co-operativity is zero and that all Na binding sites have equal affinity, generates a relation which is a power of the simple Michaelis relation:

$$\frac{j}{j_{max}} = \frac{[Na]_o^h}{\{[Na]_o + K_{Na}\}^h} \tag{3}$$

This equation produces a substantially worse fit to the data in Fig. 12.6 than does the Hill equation, and it seems clear therefore that some degree of co-operativity is involved in the binding of Na ions to the exchange.

All equations of the same type as (2) and (3) reduce to a simple power law $j \propto [Na]_o^h$ at low $[Na]_o$, and we might expect this to be true of the Na–Ca exchange, with the value of h being equal to 4. The measurements on the rod outer segment Na–Ca exchange agree with others from the squid giant axon and other preparations in showing that the optimal value of h in the Hill equation used to describe the activation of efflux by external Na is less than the value of 4 which might be expected if four Na ions must bind to the pump before the exchange can take place. If the four Na binding sites are very unequal in their affinities for Na, though, the fourth-power dependence on

$[Na]_o$ might not be observed until $[Na]_o$ has been reduced below the level at which reliable estimates of exchange currents can be made.

In conclusion, the form of the activation of the Na–Ca exchange by external Na supports the idea that there is some degree of co-operativity between the binding of the Na ions necessary to activate the exchange. A good empirical fit to the activation can be obtained by using the Hill equation with a value of $h=2.26$, but the non-integral value of h and the strong independent evidence that the exchange stoichiometry is $4Na^+–1Ca^{2+}$, $1K^+$ (see above) implies that a more complex model employing a lower degree of co-operativity, and probably binding sites of unequal affinities for Na, is required to describe the activation of the exchange by external Na.

12.8.2 Internal Ca

The activation of the Na–Ca exchange current by internal Ca has been measured directly in intact rods loaded with aequorin, in which the exchange current can be compared directly with the free intracellular [Ca]. In the experiment of Fig. 12.7A the rod outer segment was loaded with Ca to varying degrees in different trials and the decline of both free $[Ca]_i$ and the exchange current were recorded as the Ca load was extruded. The relation between free $[Ca]_i$ and exchange current, shown in Fig. 12.7C, is described by a Michaelis relation with $K_m=2.3$ μM (Cervotto, Lagnado, and McNaughton 1987). This simple result shows that one calcium ion interacts with the internal binding site.

These measurements, while direct and unequivocal, are difficult to carry out and for many purposes an alternative method, in which changes in the affinity for internal Ca can be measured, is a more practical proposition. This method depends on expressing the exchange current, not as a function of free $[Ca]_i$, but rather as a function of the total exchangeable Ca load within the cell, which can be obtained by integrating the exchange current. The relation between these two quantities is not well-described by a Michaelis relation (Hodgkin, McNaughton, and Nunn 1987; Nakatani and Yau 1988a) because of the existence within the cell of saturable Ca buffer systems which are present to variable extents in different preparations (Hodgkin, McNaughton, and Nunn 1987). The buffering is sufficiently rapid to be considered time-independent, at least on the time scale of a few seconds over which a typical exchange current is recorded (McNaughton, Cervetto, and Nunn 1986), and so the relation between the total exchangeable Ca and the free $[Ca]_i$ is also a time-independent function. This simplification means that the changes in the form of the relation between total [Ca] and the exchange current – as opposed to a simple vertical scaling on the exchange current axis – must be due to changes in the affinity for internal Ca.

An example of this approach is shown in Fig. 12.8, taken from Hodgkin

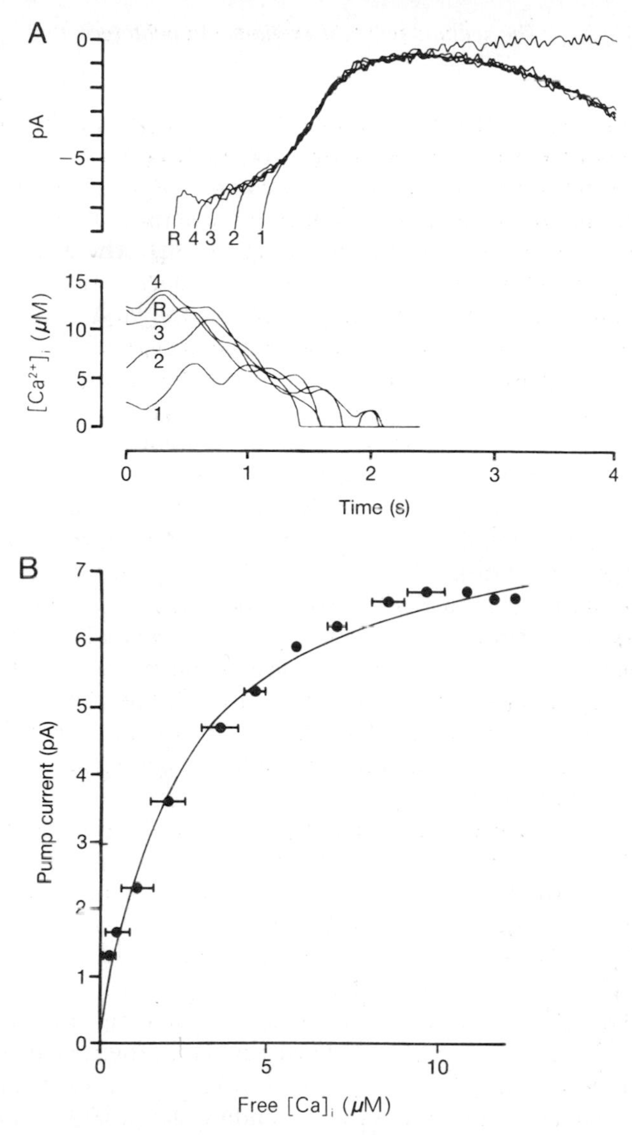

Fig. 12.7. Dependence of Na–Ca exchange current on $[Ca]_i$ in an intact rod loaded with aequorin.

A. top panel: Na–Ca exchange current recorded after closing light-sensitive channels with a bright flash; the different degrees of Ca load (traces 1–4) were obtained by different durations of exposure to the PDE inhibitor, IBMX. In the trace labelled R the rod was returned to Ringer's solution as the flash was given, and the similar form of the Na–Ca exchange current shows that IBMX has no direct effect on the exchange.

A. lower panel: Decline of $[Ca]_i$ calculated from aequorin light output.

B. Na–Ca exchange current (ordinate), as a function of mean $[Ca]_i$, obtained from traces in A. Error bars show ±S.E.M. Smooth curve shows a Michaelis function with $K_m = 2.3$ μM.

(From data of Cervetto, Lagnado and McNaughton 1987.)

and Nunn (1987). Here the Na–Ca exchange current in normal condition (A and C) is compared with that after the rod has been loaded internally with Na; the effect of the Na load is to reduce the apparent affinity of the exchange for Ca at the internal binding site, producing the horizontal shift in the charge *vs.* exchange current curve shown in Fig. 12.8D. The shift in Fig. 12.8D corresponds to an approximately 10-fold reduction in the affinity of the exchange for internal Ca in the presence of the large Na load (which was approximately 28 mM). This change in the apparent affinity was shown by Hodgkin and Nunn to occur without a change in the stoichiometry of the exchange.

An extension of this method, introduced by Lagnado, Cervetto, and McNaughton (1988) is to assume that the Michaelis activation curve shown in Fig. 12.7 is valid for one particular set of conditions. This assumption allows the intracellular free Ca to be deduced from the magnitude of the exchange current using the relation in Fig. 12.7, and, since the Ca load can also be measured from the integral of the exchange currents as explained above, the relation between the free $[Ca]_i$ and the Ca load – in other words, the buffering power of the cell cytoplasm – can thereby be deduced. On the reasonable assumption that this relation is fixed, the relation between Ca load and exchange current obtained under a different set of conditions can be transformed into the more straightforward relation between exchange current and free $[Ca]_i$.

Figure 12.9 shows this method applied to the effects of changes in external Na and in transmembrane voltage on the activation of the Na^+Ca^{2+} exchange current by free $[Ca]_i$. No change in the apparent affinity for internal Ca can be detected using this method, as the same Michaelis curve which was assumed for the condition $[Na]_o = 110$ mM, $V_m = -14$ mV (continuous curve without points) also provides a good fit to the activation curves obtained at low [Na] (filled triangles) and when the exchange rate is increased by hyperpolarization to -64 mV (filled squares). The conclusion of this approach that changes in the activity of the exchange, caused either by changes in the occupancy of the external Na-binding site (Hodgkin and Nunn 1987) or by hyperpolarization (Lagnado, Cervetto, and McNaughton 1988) do not change the affinity of the internal Ca-binding site. Similar observations have been made in squid axon (Baker and McNaughton 1976a; Blaustein 1977) and it is difficult to reconcile these results with a simple model of serial binding, in which the binding of ions occurs in sequence, and in which the availability of the internal Ca-binding site therefore depends on the binding of Na at the external membrane surface. Before coming to a definite conclusion, though, it is clearly important to measure the calcium affinity of the exchange by more direct methods, perhaps by controlling the [Ca] bathing the cytoplasmic face of an isolated patch of rod outer segment membrane.

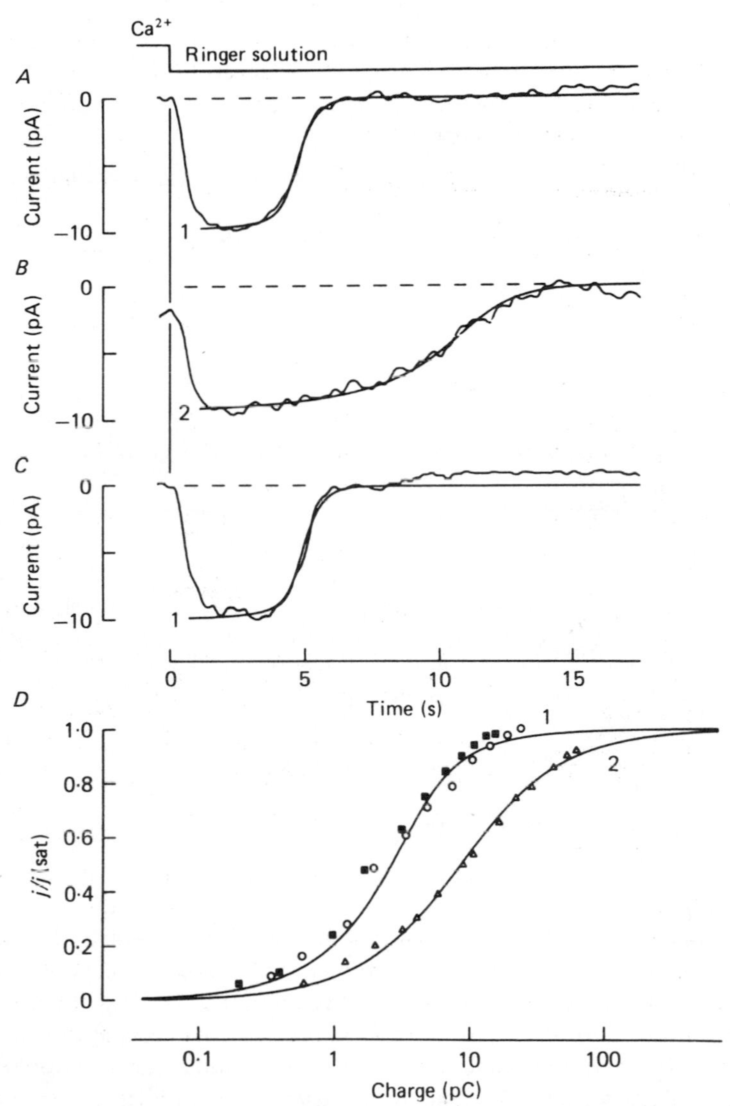

Fig. 12.8. Effect of internal Na on Na–Ca exchange current. Traces in *A* and *C* are normal exchange current activated on return to Ringer's solution after a Ca load; in *B* the rod was preloaded with *c*.28 mM Na by exposing the rod to 110 mM Na, 1 μM Ca for 50 s.

D shows the relations between normal pump current and the change remaining to be transferred by the pump; the latter quantity is proportional to the total Ca within the cell. Curve 2 plotted with a 3.27-fold greater value of $q_{\frac{1}{2}}$, the charge necessary to half-activate the exchange current. For details of other parameters used to fit the points in *D* see Hodgkin and Nunn (1987). Reproduced with permission from Hodgkin and Nunn (1987).

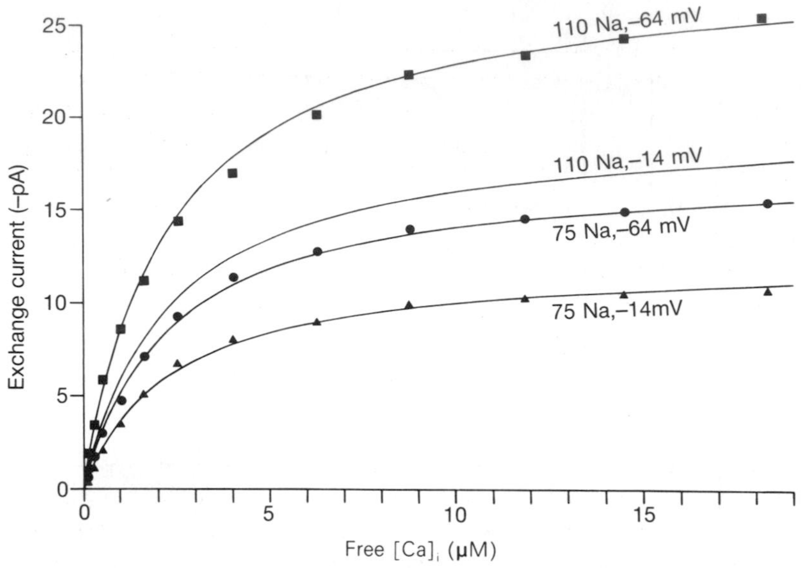

Fig. 12.9. Dependence of Na–Ca exchange current on free $[Ca]_i$, at different values of V_m and $[Na]_o$. The exchange current is plotted as a function of free $[Ca]_i$, obtained as outlined in the text. Data taken from experimental traces in Fig. 12.6B. Reproduced with permission from Lagnado, Cervetto, and McNaughton (1988).

12.8.3 Other external ions

External Ca^{2+} and Mg^{2+} inhibit the forward mode of the Na–Ca exchange, probably by competing with Na^+ for the external binding site. Inhibitory constants of 1 mM for Ca^{2+} and 4 mM for Mg^{2+} were measured by Hodgkin and Nunn (1987). The binding of external Ca^{2+} is presumably one step in the activation of reversed Na–Ca exchange. Whether the binding of Mg^{2+} leads to reversed Na–Ca exchange is not known, but it seems unlikely since a $4Na^+$–$1Mg^{2+}$, $1K^+$ exchange would, if operating near equilibrium, produce intracellular levels of ionized Mg^{2+} far below those measured in other cells. This consideration, together with the observation by Yau and Nakatani (1984b) that a Mg^{2+} load is not extruded from the cell by an electrogenic mechanism, implies that Mg^{2+} is probably transported by a different route.

External potassium and rubidium ions inhibit calcium extrusion by the Na–Ca exchange in rods from lower vertebrates (Hodgkin *et al.* 1985; Hodgkin and Nunn 1987; Schnetkamp and Bownds 1987). A similar inhibition was not observed in bovine rod outer segments (Schnetkamp 1986), but a recent report suggests that internal K may promote the exchange in this preparation (Schnetkamp, Szerencsei, and Basu 1988). Both the inhibition of the forward mode of the exchange by external K^+ and its promotion by internal K^+ are consistent with the recently discovered cotransport of Ca and K by the exchange. The inhibition of calcium extrusion by K^+, together with the effect of internal Ca on the light-sensitive current (see above) neatly explains the strong inhibition of the light-sensitive current by potassium ions (Hodgkin, McNaughton, and Nunn 1985; Hodgkin *et al.* 1984; see Fig. 12.3).

Hydrogen ions also inhibit the exchange (Hodgkin and Nunn 1987), and the form of the inhibition as a function of pH suggests that one inhibitory proton combines between pH 8 and 10 and a pair between pH 6 and 7. It seems a reasonable (though unproven) hypothesis that H^+ is able to occupy the K^+-binding site on the exchange.

12.8.4 Blockers

Everyone who has worked on the Na–Ca exchange has wished for a potent and specific blocker, the 'ouabain' of the Na–Ca exchange. Lanthanum is certainly very effective (Cervetto and McNaughton 1983, 1986; Yau and Nakatani 1984b), although it is also a potent blocker of the light-sensitive channel (Cervetto and McNaughton 1986). Quinidine is a somewhat less potent blocker, inhibiting the Na–Ca exchange current with a K_m of 290 μM, but the inhibition is only 50 per cent complete even at saturating doses (Lagnado and McNaughton 1987c). The inhibition does not appear to occur by competition with Na ions for the binding site, and our current guess is that some non-specific effects on the properties of the membrane are responsible. Dichlorobenzamil, a blocker of the Na–Ca exchange in other preparations, is a potent blocker of the light-sensitive channel in frog rods (Nicol *et al.* 1987), and its usefulness in studying the photoreceptor Na–Ca exchange is clearly limited.

12.9 Voltage sensitivity of the Na–Ca exchange

The voltage-sensitivity of the Na–Ca exchange has been measured by recording the exchange current from isolated outer segments voltage-clamped in the whole-cell mode (Lagnado, Cervetto, and McNaughton 1988). The experiment is made possible by the high resistance of the seal which can be obtained between the whole-cell pipette and the surface membrane of the outer segment, and by the intrinsically high resistance of the outer segment membrane when light-sensitive channels are closed. These advantages mean that a pure exchange current can be recorded, free of ionic currents passing through the membrane channels.

Collected results from a large number of experiments are shown in Fig. 12.10. The experimental points are well-fitted by the relation:

$$j(V_2) = j(V_1)\exp\{(V_1 - V_2)/70\} \qquad (4)$$

This relation is valid not only for the saturating levels of internal Ca at which the measurement was performed, but for all $[Ca]_i$ since the binding of internal Ca is unaffected by membrane potential (see Fig. 12.9). This relation therefore provides a useful definition of the voltage-sensitivity of the exchange over the normal physiological range of membrane potentials and ionic conditions.

The most intuitively obvious explanation of the voltage-sensitivity is that a charged entity, perhaps the exchange molecule loaded with ions, must cross a rate-limiting energy barrier located partway across the membrane electrical field (Baker and McNaughton 1976b; Allen and Baker 1986; Hodgkin and Nunn 1987). A single charge crossing an energy barrier which senses a fraction $\gamma = 0.37$ of the membrane electric field would account for the observed sensitivity to membrane potential (Lagnado and McNaughton 1987b). In this model the binding of ions at the external and internal membrane faces is assumed to be voltage-independent, since all of the voltage dependence is accounted for by the translocation step. Figure 12.6 shows that this assumption is unlikely to be correct, because the apparent affinity for external Na ions is greater when the rod is hyperpolarized (curve 2) than when it depolarized (curve 3).

Another prediction of the simple single-energy barrier model is that the voltage-sensitivity of the exchange should be constant, independent of the ionic conditions at the external and internal face of the membrane. While the voltage-sensitivity does appear to be independent of the level of internal Ca (see Fig. 12.9), the same is not true for external Na, as shown in Fig. 12.11A. As $[Na]_o$ is elevated the voltage-sensitivity of the exchange is reduced, and with the external binding site close to saturation little or no voltage-sensitivity is observed.

A simple model which may account for these results is shown in Fig. 12.11B. The external Na-binding site is assumed to be located at the foot of a short channel, so that it senses a fraction of the transmembrane potential, while the Ca-binding site is outside the membrane electric field. Hyperpolarization therefore increases the apparent affinity of the Na-binding site by increasing the effective concentration of Na ions at the site (Läuger 1987).

Because the exchange is electrogenic a charged species must cross the membrane at some stage in the exchange cycle, but the observation that the exchange rate is voltage-independent at high $[Na]_o$ implies that this charge movement does not limit the overall exchange rate, at least when the exchange is driven exclusively in the forward direction as in the experiments

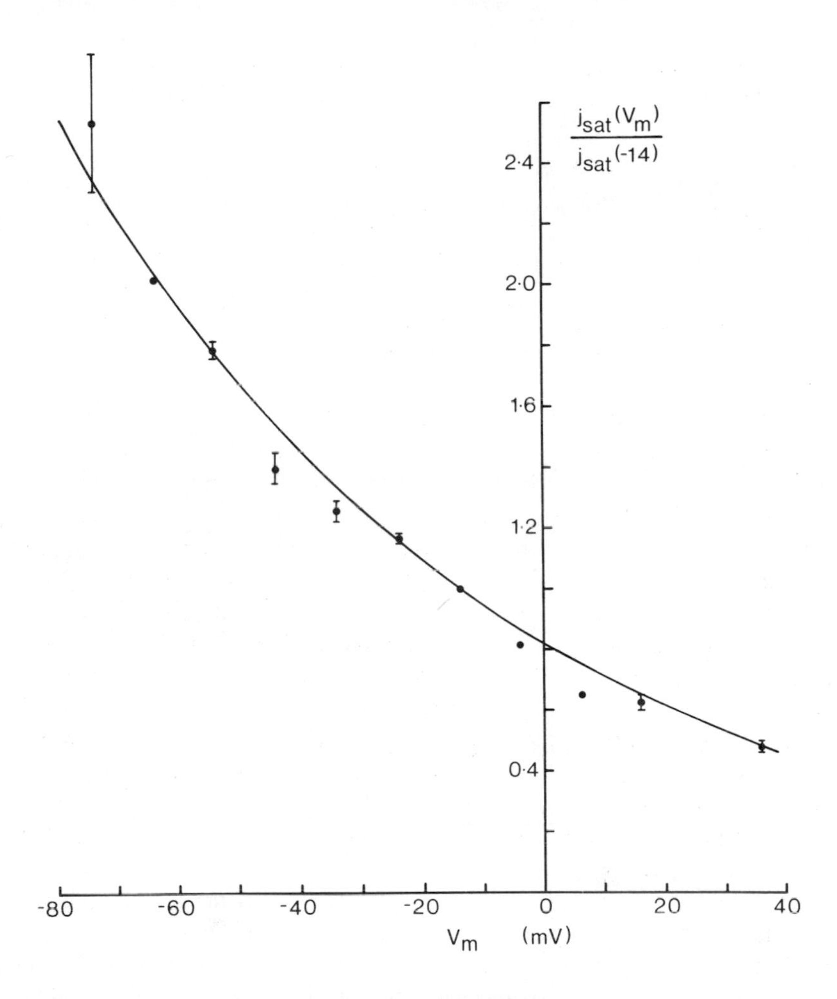

Fig. 12.10. Voltage dependence of the Na–Ca exchange in normal Ringer's solution containing 1 mM Ca. The exchange current j_{sat}, measured after an internal Ca load sufficient to saturate the internal Ca binding site, is expressed relative to the value observed at $V_m = -14$ mV in each experiment. Continuous line is:

$$j_{sat}(V_m)/j_{sat}(-14\ \text{mV}) = \exp\{-(V_m + 14)/70\}.$$

Reproduced with permission from Lagnado, Cervetto, and McNaughton (1988).

described here. Voltage-independence of the actual exchange itself, as opposed to the binding of ions, could arise if the exchange process consisted of a sequence of steps, with the rate-limiting step being different from the step or steps in which ions are moved through the membrane electric field. An analogy can be drawn with models of ion permeation through membrane channels which involve a single energy barrier close to the external face of

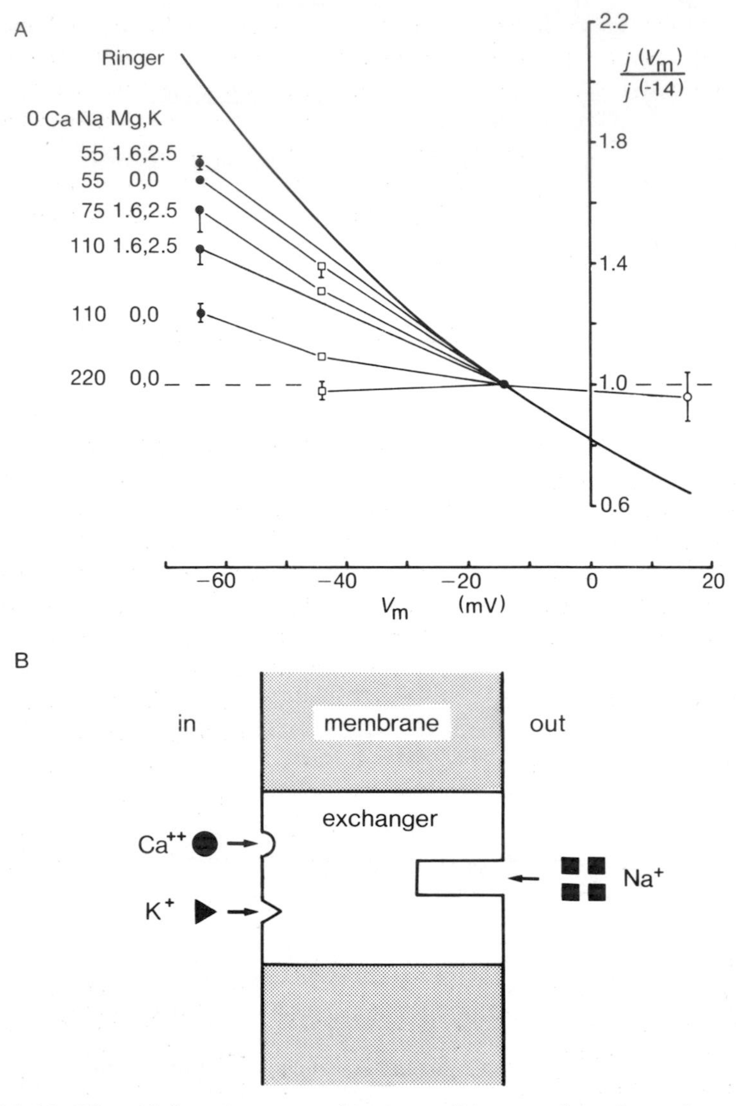

Fig. 12.11. Effect of changing external ionic conditions on the voltage-dependence of the Na–Ca exchange. A shows the voltage-dependence of the exchange current, normalized to its value at −14 mV under a given set of conditions. Note that the steepness of the curve depends on the degree of saturation of the external Na-binding site. B shows a possible arrangement of the ion binding sites of the Na–Ca exchange. The internal Ca binding site is outside the membrane electrical field, while the Na binding site senses a fraction of the transmembrane potential. The model is intended to be schematic, and does not show the possibility that the Na-binding sites may not be equivalent. No information is at present available concerning the location of the K-binding site.

the membrane (Jack, Noble and Tsien 1975; Baylor and Nunn 1986); here the voltage-sensitivity vanishes at extreme hyperpolarizations because the rate-limiting step is the rate at which ions from the external solution reach the top of the energy barrier, the height of which is voltage-independent because of its location at the external surface. The sequence of events involved in the exchange cycle is undoubtedly more complex, and there is clearly no *a priori* reason to assume that the ion-translocation step will be rate-limiting in the exchange cycle.

12.10 Calcium homeostasis in the rod outer segment

The Na–Ca exchange mechanism contributes in at least two respects to the form of the light response: it is the principal (and perhaps the only) mechanism for pumping Ca from photoreceptors, and therefore is important in regulating both the amplitude of the light-sensitive current and the form of the light response; and it carries a fraction of the outer segment membrane current. The latter role is perhaps not as significant as the first, as in darkness the exchange current in normal Ringer's solution in a rod from the lower vertebrate is only about 1 pA in amplitude, or about 5 per cent of the current carried through light-sensitive channels (Cervetto and McNaughton 1986; Hodgkin, McNaughton, and Nunn 1987; Nakatani and Yau 1988a; Yau and Nakatani 1985). The amplitude of the exchange current can be used to estimate the Ca influx through light-sensitive channels, since with a $4Na^+$–$1Ca^{2+}$, $1K^+$ stoichiometry the Ca current must in the steady-state be equal to twice the exchange current. This apparently simple result is subject to a number of provisos, but it seems clear that a substantial fraction, probably about 10 per cent, of the light-sensitive current is carried by Ca^{2+}.

In darkness the Ca influx is relatively large, therefore, and an efficient Ca extrusion mechanism is required. The Na–Ca exchange mechanism fulfils this requirement, as the normal Ca efflux extruded by the exchange across a salamander rod outer segment membrane is about 2.8 pmol cm^{-2} s^{-1}, increasing to about 50 pmol cm^{-2} s^{-1} when internal Ca is high. The maximum Ca efflux generated by the exchange under optimal conditions when the external Na binding site is saturated, is five times this last figure. In Ringer's solution the exchange is capable of lowering the free $[Ca]_i$ at a rate of 30 $\mu M\,s^{-1}$ and the total $[Ca]_i$ at 0.6 $mM\,s^{-1}$ (McNaughton, Cervetto, and Nunn 1986).

The free intracellular Ca in darkness has not been determined with any certainty, mainly because all the methods applied so far depend on the use of optical probes which either emit light or require an exciting beam of light. Measurement using aequorin have set an upper limit of 0.6 μM on the free

$[Ca]_i$ in darkness (McNaughton, Cervetto and Nunn 1986). The fluorophore quin-2 was used to obtain a value of 0.22 μM, claimed to be in darkness although some UV light must have been used to carry out the measurement (Miller and Korenbrot 1986). More recently Ratto *et al.* (1988) using fura–2, observed an exponential decline in free $[Ca]_i$, to a steady value of 0.14 μM, after switching on the UV light used to perform the measurement. By extrapolating to the time origin a value of 0.22 μM was obtained for the free $[Ca]_i$ in darkness.

An independent method of determining the free $[Ca]_i$ in darkness depends on using the Na–Ca exchange current itself. The mean dark exchange current in salamander rods is 1.1 pA, and the maximum is 20 pA (Hodgkin, McNaughton and Nunn 1987). The exchange current has been shown to be activated in a Michaelis fashion by Ca_i with a K_M of 2.3 μM (Cervetto, Lagnado and McNaughton 1987), and from this function a free $[Ca]_i$ of 0.13 μM above the basal level in bright light can be calculated. The level in bright light can in turn be predicted from the stoichiometry of the exchange using the assumption that it will come into equilibrium when the Ca influx through the light-sensitive channels is shut down. The calculation outlined in Section 12.7 (eqn 1) yields an estimate in the subnanomolar range, which is considerably lower than the estimate of 0.14 μM measured in the presence of bright light using fura-2 (Ratto *et al.* 1988).

Implicit in the above discussion is the assumption that light does not itself release Ca within the rod outer segment. The central tenet of the 'calcium hypothesis' (Hagins 1972) was that light released Ca within the outer segment, and that this Ca closed light-sensitive channels. A variety of approaches (reviewed by McNaughton and Cervetto 1986) showed that a Ca release was not essential to a normal light response, and that the well-established effect of internal Ca in suppressing light-sensitive current was due to an indirect action and not to a direct combination of Ca ions with channels. Measurements of free $[Ca]_i$ using aequorin showed clearly that a flash of light bright enough to suppress all the light-sensitive current does not release Ca within the cell, but instead causes a decline in free $[Ca]_i$ due to the suppression of Ca influx through channels (McNaughton, Cevetto, and Nunn 1986).

Our current picture of Ca transport and buffering in the rod outer segment is in essence a simple one. Calcium enters the cell through the light-sensitive channels. No other ionic channels are present in the outer segment membrane, because the membrane resistance exceeds 40 GΩ in bright light (Baylor and Lamb 1982; Baylor and Nunn 1986; Lagnado and McNaughton 1987a). Calcium buffering within the cell is also simple, with the relationship between free and bound Ca being approximately linear and effectively instantaneous on the time scale of the light response (McNaughton, and Nunn 1986; Cervetto, Lagnado, and McNaughton 1987), though there are some indications that a high-affinity buffer of low capacity may be important at low $[Ca]_i$ (Cervetto, Lagnado, and McNaughton; Hodgkin, McNaughton, and Nunn 1987), and that a slow uptake and release of Ca, perhaps into the discs, may determine the total Ca content of the cell on a longer time scale (Fain and Schröder 1985, 1987; Schnetkamp 1979). Finally, calcium transport from the cell appears to be dominated by the Na–Ca exchange, perhaps because the absence of mitochondria in the outer segment rules out an ATP-driven pump. The maximum rate of decline of free $[Ca]_i$ is at least 50 times greater in the presence of external Na than in its absence, showing that the maximum rate of Ca efflux through the exchange must be at least 50 times greater than the efflux through other Ca transport systems such as the Ca ATPase.

The membranes of the discs are formed by invagination of the external membrane, and one might expect that a Na–Ca exchange would be active in transporting Ca into the disc lumens (Schnetkamp, Daemen, and Bonting 1977). If such an exchange existed it would substantially increase Ca buffering in the outer segment. Current evidence does not favour this idea because isolated vesicles from bovine rod outer segments seem to consist of two populations: those which can be loaded with Ca in low Na or by keeping light-sensitive channels open with cGMP, and from which Ca can be extruded by an elevation in [Na]; and a second population which can be loaded with Ca only by passive diffusion, and from which the Ca is not released either by cGMP or by elevations in [Na] (Bauer 1988). The first pool is proposed to consist of vesicles formed from outer segment surface membane and the second of isolated discs, and Bauer (1988) concludes that the Na–Ca exchange and the light-sensitive channels are located principally or exclusively in the surface membrane. A second piece of evidence pointing to the same conclusion is the observation that a Na load sufficient to cause a large reduction in the Ca affinity of the surface membrane Na–Ca exchange in an intact rod has no detectable effect on the intracellular buffering of Ca (Cervetto, Lagnado, and McNaughton 1988), implying that a Na–Ca exchange across the disc membranes does not contribute significantly to the sequestration of internal Ca.

12.11 Functional role of the Na–Ca exchange in vertebrate rods

Changes in the Na–Ca exchange activity affect both the form of the light response and the amplitude of the light-sensitive current. Inhibiting the Na–Ca exchange by reducing the external Na concentration prolongs the response to both strong and weak flashes, and makes the cell more sensitive in the sense that the fraction of the maximum current suppressed by each absorbed photon is increased. The correlation between the time to peak and amplitude of the dim flash response is reminiscent of the adapting effects of background light (Yau, McNaughton, and Hodgkin 1981). This action can be seen in Fig. 12.3C and D, which show that increasing external [Na] speeds the recovery from a flash, while inhibiting the exchange by increasing external K or Rb prolongs the recovery. Similar effects are seen if external Ca is elevated, and this action of Ca appears to be most prominent in an early stage of phototransduction. A brief increase in the external [Ca] just before a bright flash prolongs the plateau of the saturating light response, while brief changes during the recovery phase do not alter its subsequent time-course (Hodgkin, McNaughton, and Nunn 1986; Hodgkin and Nunn 1988). Conversely, reducing the external [Ca] (and therefore the internal [Ca]) causes a small reduction in the duration of the light response (Hodgkin, McNaughton, and Nunn 1985). Taken together, these results suggest that the light-induced fall in [Ca] mediated by the Na–Ca exchange is involved in signalling the decreased sensitivity and faster response kinetics characteristic of light adaptation (Cervetto and McNaughton 1986; Cervetto *et al.* 1985; Hodgkin, McNaughton, and Nunn 1986).

Independent evidence for this idea comes from experiments in which the calcium buffer BAPTA was introduced into a rod so as to slow down changes in the internal [Ca]. Buffering $[Ca]_i$ was found to cause an increase in the rod's sensitivity, a prolongation of the light response and a slowing of the rate of onset of light adaptation (Torre, Matthews and Lamb 1986). The role of the Na–Ca exchange in light adaptation has also been investigated by recent experiments aimed at preventing changes in [Ca] within the outer segment. Matthews *et al.* (1988) and Nakatani and Yau (1988b) minimized the Ca influx into the outer segment by reducing the external [Ca] to about 1 μM, and simultaneously blocked Ca efflux through the exchange by replacing external Na with the permeant cation guanidinium, which does not exchange with Ca. In both rods and comes from the salamander, this treatment abolished the adaptation caused by steady light.

One important question which still remains to some extent open is how changes in internal [Ca] modulate the light-sensitive current and the form of the light response. Electrophysiological experiments suggest that Ca may act on the guanylate cyclase that synthesizes cGMP from GTP. When the Na–Ca exchange is suddenly blocked by rapidly replacing external Na with Li there is an increase in the [Ca] within the outer segment (L. Cervetto, L. Lagnado, and P. A. McNaughton, unpublished results) and a consequent exponential decline in the light-sensitive current. Hodgkin, McNaughton, and Nunn (1985) found the rate of decline of current to be accelerated by weak background lights and slowed by the PDE inhibitor IBMX. They therefore proposed that the rise in [Ca] reduces the level of cGMP, by inhibiting its production by the cyclase while its hydrolysis by the PDE continues (Hodgkin, McNaughton, and Nunn 1985; McNaughton, Nunn and Hodgkin 1986).

The effects of light adaptation and IBMX can then be explained by their respective stimulatory and inhibitory actions on the PDE.

Direct biochemical evidence for an effect of Ca on the guanylate cyclase has now been provided by several groups (Fleischmann and Denisevich 1979; Lolley and Racz 1982; Troyer, Hall, and Ferendelli 1978). Koch and Stryer (1988) found the activity of the enzyme to increase up to 20–fold when the free [Ca] was reduced from 200 nM to 50 nM. Can the rod outer segment achieve such low levels of internal Ca? As was discussed in Section 12.7, a $4Na^+:1Ca^{2+}, 1K^+$ exchange would come into equilibrium at subnanomolar levels of internal Ca, and while it does not seem likely that this equilibrium level would actually be attained, an exchange in which K is cotransported with Ca will be able to continue extruding Ca to much lower levels of $[Ca]_i$ than would be expected of a simpler mechanism with a stoichiometry of $3Na^+:1Ca^{2+}$. The $4Na^+:1Ca^{2+}, 1K^+$ stoichiometry of the exchange may therefore be of particular importance in the rod outer segment, where the ability to reach a low $[Ca]_i$ will, via the activation of the cyclase, control the light-sensitive current. Koch and Stryer (1988) found this effect to be highly co-operative, the Hill coefficient of activation by lowered [Ca] being 4. Interestingly, a soluble component (which is not calmodulin) is required for the Ca-regulation of cyclase activity (Koch and Stryer 1988).

The model proposed by Hodgkin, McNaughton, and Nunn (1985) is outlined in Fig. 12.2, in which the light-dependent regulation of the PDE activity occurs through the pathway shown in Fig. 12.2 and where the guanylate cyclase is subject to inhibition by Ca. The light-sensitive current is therefore regulated by a system involving negative feedback with two delays: a decrease in light-sensitive current causes a decrease in $[Ca]_i$ with a time constant of about 0.5 seconds, and this change in $[Ca]_i$, in turn, causes an increase in [cGMP], with a delay determined by the turnover time of cGMP, by releasing the guanylate cyclase from inhibition. A system of this kind involving negative feedback with delays is potentially oscillatory, and oscillations in the light-sensitive current are indeed observed in rods after changes in Na–Ca exchange activity caused by changes in external ionic conditions (Hodgkin, McNaughton, and Nunn 1985), in membrane potential (Baylor and Nunn 1986) or in the buffering of internal Ca (Lamb, Matthews, and Torre 1987; Matthews, Torre, and Lamb 1985). Oscillations which may have a similar origin are also observed in the normal flash response of cones (Baylor, Nunn, and Schnapf 1987; Craig and Perry 1988). An interesting feature of the system outlined in Fig. 12.12 is its sensitivity to intracellular calcium: there is a high degree of co-operativity in the action of Ca on the cyclase (Koch and Stryer 1988), and in the action of cGMP on the light-sensitive channel (Fesenko, Kolenikov, and Lyubarsky 1985; Zimmerman and Baylor 1986). This may explain the large changes in the amplitude of the light-sensitive current on altering the external [Ca] (Hodgkin, McNaughton, and Nunn 1985).

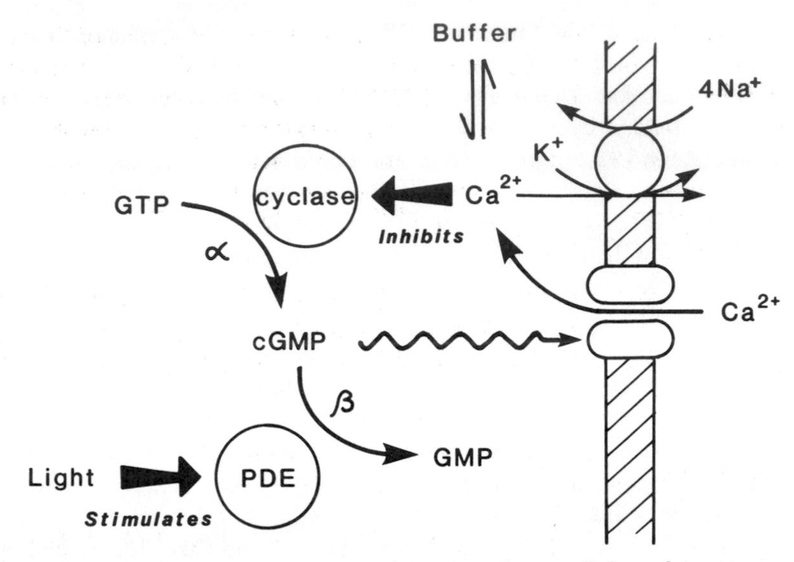

Fig. 12.12. Possible mechanisms for the interaction of intracellular calcium ions with the light-sensitive pathway. The rate of formation of cGMP from GTP, α, is increased by a fall in intracellular [Ca] while the rate of hydrolysis of cGMP, β, is increased by light means of the pathway shown in Fig. 12.2.

Torre, Matthews, and Lamb (1986) has proposed a second action of Ca within the outer segment, namely a stimulation of the PDE that hydrolyses cGMP. They observed a prolongation of the plateau of the light response after a Ca buffer had been introduced into a rod, and they therefore suggest that the slower fall in [Ca] increases the overall gain of the pathway linking the photosomerization of a rhodopsin molecule to the activation of the PDE. This idea is supported by the prolongation of the flash response observed when a rod is loaded with Ca, but the cyclase is by-passed by the continuous infusion of cGMP via a patch pipette (Matthews, Torre, and Lamb 1985; Cobbs, Barkdoll, and Pugh 1985; Cobbs, Pugh, and Matthews 1985). However, the early rising phase of the flash response is not altered by the presence of a Ca buffer, indicating that if Ca stimulates the activation of the PDE, this effect is not the absolute activity of the PDE, but on its life-time of activation (Matthews *et al.* 1985; Lamb *et al.* 1986; Torre, Matthews, and Lamb 1986). The site of action of Ca^{2+} in these experiments is therefore unclear; a prolonged life-time of the PDE may be a direct effect, or due to an increase in the life-time of isomerized rhodopsin or of activated G-protein. Some biochemical evidence suggesting that Ca may alter PDE activity has been obtained by Robinson *et al.* (1980).

Whether there are two functionally important effects of internal Ca, one on the cyclase and one on the chain of events controlling the phosphodiesterase, has still to be resolved. It is also unclear whether the action of Ca on the magnitude of the light-sensitive current is distinct from its action on the adaptive state of the photoreceptor. Whatever the final outcome, the importance of a Na–Ca exchange in modulating the machinery of phototransduction has now been established beyond doubt.

Acknowledgements

We thank L. Cervetto, A. L. Hodgkin, and R. J. Perry for helpful discussions and G. L. Fain, T. D. Lamb, P. P. M. Schnetkamp, and K.-W. Yau for their comments on the manuscript. Supported by the MRC and by the Wellcome Trust.

References

Allen, T. J. A. and Baker, P. F. (1986). Influence of membrane potential on calcium efflux from giant axons of *Loligo*. *J. Physiol.* **378**, 77–96.

Bastian, B. L. and Fain, G. L. (1982). The effects of sodium replacement on the responses of toad rods. *J. Physiol.* **330**, 331–47.

Baker, P. F., Blaustein, M. P., Hodgkin, A. L., and Steinhardt, R. A. (1969). The influence of calcium on sodium efflux in squid axons. *J. Physiol.* **200**, 431–58.

Baker, P. F. and McNaughton, P. A. (1976a). Kinetics and energetics of calcium efflux from intact squid axons. *J. Physiol.* **259**, 103–44.

Baker, P. F. and McNaughton, P. A. (1976b). The effect of membrane potential on the calcium transport systems in squid axons. *J. Physiol.* **260**, 24P.

Baker, P. F. and McNaughton, P. A. (1978). the influence of extracellular calcium binding on the calcium efflux from squid axons. *J. Physiol.* **276**, 127–50.

Bastian, B. L. and Fain, G. L. (1982). The effects of sodium replacement on the responses of toad rods. *J. Physiol.* **330**, 331–47.

Bauer, P. J. (1988). Evidence for two functionally different membrane fractions in bovine retinal rod outer segments. *J. Physiol.* **401**, 309–27.

Baylor, D. A. and Lamb, T. D. (1982). Local effects of bleaching in retinal rods of the toad. *J. Physiol.* **328**, 40–71.

Baylor, D. A. and Nunn, B. J. (1986). Electrical properties of the light-sensitive conductance of rods of the salamander *Ambystoma tigrinum*. *J. Physiol.* **371**, 115–45.

Baylor, D. A., Nunn, B. J., and Schnapf, J. L. (1987). Spectral sensitivity of cones of the money *Macaca fasicularis*. *J. Physiol.* **390**, 145–60.

Blaustein, M. P. (1977). Effects of internal and external cations and of ATP on sodium-calcium and calcium-calcium exchange in squid axons. *Biophys. J.* **20**, 79–110.

Blaustein, M. P. and Hodgkin, A. L. (1969). The effect of cyanide on the efflux of calcium from squid axons. *J. Physiol.* **200**, 497–527.

Brown, J. E. and Blinks, J. R. (1974). Changes in intracellular free calcium concentration during illumination of invertebrate photoreceptors: detection with aequorin. *J. Gen. Physiol.* **64**, 643–65.

Brown, J. E. and Lisman, J. E. (1975). Intracellular calcium modulates sensitivity and time scale in *Limulus* ventral photoreceptors. *Nature* **258**, 252–4.
Brown, J. E. and Mote, M. K. (1974). Ionic dependence of reversal voltage of the light response in Limulus ventral photoreceptors. *J. Gen. Physiol.* **63**, 337–50.
Cervetto, L., Lagnado, L., and McNaughton, P. A. (1987). Activation of the Na–Ca exchange in salamander rods by intracellular Ca. *J. Physiol.* **382**, 135P.
Cervetto, L., Lagnado, L., and McNaughton, P. A. (1988). The effects of internal Na on the activation of the Na–Ca exchange in salamander rods. *J. Physiol.* **407**, 81P.
Cervetto, L, and McNaughton, P. A. (1983). Inhibition of the light-sensitive current in vertebrate rods by La^{3+} ions. *J. Physiol.* **341**, 75P.
Cervetto, L and McNaughton. P. A. (1986). The effects of phosphodiesterase inhibitors and lanthanum ions on the light-sensitive current of toad retinal rods. *J. Physiol.* **370**, 91–109.
Cervetto, L., McNaughton, P. A., and Nunn, B. J. (1986). Aequorin signals from isolated salamander rods. *J. Physiol.* **371**, 36P.
Cervetto, L., Torre, V., Rispoli, G., and Marroni, P. (1985). Mechanisms of light adaptation in toad rods. *Exp. Biol.* **44**, 147–57.
Cervetto, L., Lagnado, L., Perry, R. J., Robinson, D. W., and McNaughton, P. A. (1989). Extrusion of calcium from rod outer segments is driven by both sodium and potassium gradients. *Nature* **337**, 740–3.
Cobbs, W. H., Barkdoll, A. E. III, and Pugh, E. N. Jr. (1985). Cyclic GMP increases photocurrent and light-sensitivity of rentinal cones. *Nature* **317**, 64–66.
Cobbs, W. H. and Pugh, E. N. Jr. (1985). Cyclic GMP can increase rod outer segment light-sensitive current without delay of excitation. *Nature* **313**, 585–7.
Cook, R. H., Hodgkin, A. L., McNaughton, P. A., and Nunn, B. J. (1984). Rapid change of solutions bathing a rod outer segment. *J. Physiol.* **357**, 2P.
Craig, A. J. and Perry, R. J. (1988). Differences in absolute sensitivity and response kinetics between red- and blue-sensitive salamander cones. *J. Physiol.* **407**, 83P.
DiPolo, R. and Rojas, H. (1984). Effect of external and internal K^+ on Na–Ca exchange in dialysed squid axons under voltage-damp conditions. *Biochim. Biophys. Acta* **776**, 313–6.
Fain, G. L. and Lisman, J. E. (1981). Membrane conductances of photoreceptors. *Prog. Biophys. Mol. Biol.* **37**, 91–147.
Fain, G. L. and Schröder, W. H. (1985). Calcium content and calcium exchange in dark adapted toad rods. *J. Physiol.* **368**, 641–55.
Fain, G. L. and Schröder, W. H. (1987). Calcium in dark-adapted toad rods: evidence for pooling and guanosine-3′-5′-monophosphate-dependent release. *J. Physiol.* **389**, 361–84.
Fein, A. and Charlton, J. S. (1977a). A quantitative comparison of the effects of intracellular calcium injection and light adaptation on the light response of *Limulus* photoreceptors. *J. Gen. Physiol.* **70**, 591–600.
Fein, A. and Charlton, J. S. (1977b). Increased intracellular sodium mimics some but not all aspects of photoreceptor adaptation in the ventral eye of *Limulus. J. Gen. Physiol.* **70**, 601–20.
Fesenko, E. E., Kolenikov, S. S., and Lyubarsky, A. L. (1985). Induction by cyclic GMP of cationic conductance in plasma membrane of retinal rod outer segment. *Nature* **313**, 310–13.

Fleischman, D. and Denisevich, M. (1979). Guanylate cyclase of isolated bovine retinal rod axonemes. *Biochemistry* **18**, 5060–6.

Gadsby, D. C., Nakao, M., Noda, M. & Shepherd, R. N. (1988). Na dependence of outward Na–Ca exchange current in guinea-pig ventricular myocytes. *J. Physiol.* **407**, 135P.

Gold, G. H. and Korenbrot, J. I. (1980). Light-included Ca release by intact retinal rods. *Proc. Natl Acad. USA.* **77**, 5557–61.

Hagins, W. A. (1972). The visual process: excitatory mechanisms in the primary receptor cells. *Ann. Rev. Biophys. Bioeng.* **1**, 131–58.

Hagins, W. A. and Yoshikami, S. (1975). Ionic mechanisms in excitation of photoreceptors. *Ann. NY Acad. Sci.* **264**, 314–25.

Haynes, L. W., Kay, A. R., and Yau, I.-W. (1986). Single cyclic GMP-activated channel activity in excised patches of rod outer segment membrane. *Nature* **321**, 66–70.

Hestrin, S. and Korenbrot, J. I. (1987). Effects of cyclic GMP on the kinetics of the photocurrent and detached rod outer segments. *J. Gen. Physiol.* **90**, 527–51.

Hodgkin, A. L., McNaughton, P. A., Nunn, B. J., and Yau, K. W. (1984). Effects of ions on retinal rods from *Bufo marinus*. *J. Physiol.* **350**, 649–80.

Hodgkin, A. L., McNaughton, P. A., and Nunn, B. J. (1985). The ionic selectivity and calcium dependence of the light-sensitive pathway in toad rods. *J. Physiol.* **358**, 447–68.

Hodgkin, A. L., McNaughton, P. A., and Nunn, B. J. (1986). Effects of changing Ca before and after light flashes in salamander rods. *J. Physiol.* **372**, 54P.

Hodgkin, A. L., McNaughton, P. A., and Nunn, B. J. (1987). Measurement of sodium–calcium exchange in salamander rods. *J. Physiol.* **391**, 347–70.

Hodgkin, A. L. and Nunn, B. J. (1987). The effects of ions on sodium–calcium exchange in salamander rods. *J. Physiol.* **391**, 371–98.

Hodgkin, A. L. and Nunn, B. J. (1988). Control of light-sensitive current in salamander rods. *J. Physiol.* **403**, 439–71.

Jack, J. J. B., Noble, D., and Tsien, R. W. (1975). *Electric current flow in excitable cells.* Oxford University Press, Oxford.

Koch, K.-W. and Stryer, L. (1988). High degree of cooperativity in the inhibition of retinal rod guanylate cyclase by calcium. *Nature* **334**, 64–6.

Lagnado, L., Cervetto, C., and McNaughton, P. A. (1988). Ion transport by the Na:Ca exchange in isolated rod outer segments. *Proc. Nat. Acad. Sci. USA.* **85**, 4548–52.

Lagnado, L. and McNaughton, P. A. (1987a). Light responses and Na:Ca exchange in isolated salamander rod outer segments. *J. Physiol.* **390**, 11P.

Lagnado, L. and McNaughton, P. A. (1987b). Voltage dependence of Na:Ca exchange in isolated salamander rod outer segments. *J. Physiol.* **390**, 162P.

Lagnado, L. and McNaughton, P. A. (1987c). Inhibition of Na:Ca exchange in isolated salamander rods by quinidine. *J. Physiol.* **390**, 163P.

Lagnado, L. and McNaughton, P. (1988). The stoichiometry of Na–Ca exchange in isolated salamanda rod outer segments. *J. Physiol.* **407**, 82P.

Lamb, T. D., Matthews, H. R., and Torre, V. (1986). Incorporation of calcium buffers into salamander retinal rods: a rejection of the calcium hypothesis of phototransduction. *J. Physiol.* **372**, 315–49.

Läuger, P. (1987). Voltage dependence of the sodium:calcium exchange: predictions from kinetic models. *J. Membr. Biol.* **99**, 1–11.

Levy, S. and Fein, A. (1985). Relationship between light sensitivity and intracellular free Ca concentration in *Limulus* ventral photoreceptors. *J. Gen. Physiol.* **85**, 805–41.

Lisman, J. E. and Brown, J. E. (1972). The effects of intracellular iontophoretic injection of calcium and sodium ions on the light response of *Limulus* ventral photoreceptors. *J. Gen. Physiol.* **59**, 701–19.

Lolley, R. N. and Racz, E. (1982). Calcium modulation of cyclic GMP synthesis in rat visual cells. *Vision Res.* **22**, 1481–6.

McNaughton, P. A. and Cervetto, L. (1986). The role of calcium in the light response. *Photobiochem. Photobiophys.* **13**, 399–414.

McNaughton, P. A., Cervetto, L., and Nunn, B. J. (1986). Measurement of the intracellular free calcium concentration in salamander rods. *Nature* **322**, 261–3.

McNaughton, P. A., Nunn, B. J., and Hodgkin, A. L. (1986). Evaluation of internal transmitter candidates: Ca. In: *Molecular mechanism of photoreception* (ed. H. Stieve) pp. 79–92. Springer-Verlag, Berlin, Heidelberg, New York, Tokyo.

Mathews, H. R., Torre, V., and Lamb, T. D. (1985). Effects on the photoresponse of calcium buffers and cyclic GMP incorporated into the cytoplasm of retinal rods. *Nature* **313**, 582–5.

Matthews, H. R., Murphy, R. L. W., Fain, G. L., and Lamb, T. D. (1988). Photoreceptor light adaptation is mediated by cytoplasmic calcium concentration. *Nature* **334**, 67–9.

Miller, D. L. and Korenbrot, J. I. (1986). Effects of the intracellular calcium buffer quin 2 on the photocurrent responses of toad rods. *Biophys. J.* **49**, 281a.

Miller, D. L. and Korenbrot, J. I. (1987). Kinetics of light-dependent Ca fluxes across the plasma membrane of rod outer segments. *J. Gen. Physiol.* **90**, 397–425.

Nakatani, K. and Yau, K.-W. (1988a). Calcium and magnesium fluxes across the plasma membrane of the toad rod outer segment. *J. Physiol.* **395**, 675–730.

Nakatani, K. and Yau, K.-W. (1988b). Calcium and light adaptation in retinal rods and cones. *Nature* **334**, 69–71.

Nicol., G. D., Schnetkamp, P. P. M., Saimi, Y., Cragoe, E. J. Jr., and Bownds, M. D. (1987). An amiloride derivative blocks the light and cyclic GMP regulated conductance in frog rod photoreceptors. *Biophys. J.* **51**, 272a.

Nunn, B. J., Hodgkin, A. L., and McNaughton, P. A. (1986). Effects of raised internal sodium on the Na–Ca exchange in isolated retinal rods of the salamander. *Biophys. J.* **49**, 282a.

Owen, W. G. (1987). Ionic conductances in rod photoreceptors. *Ann. Rev. Physiol.* **49**, 743–64.

O'Day, P. M. and Keller, M. P. (1986). Evidence for Na–Ca exchange in 'forward' and 'reverse' directions in *Limulus* ventral photoreceptors. *Biophys. J.* **49**, 282a.

O'Day, P. M. and Keller, M. P. (1987). Evidence for electrogenic Na–Ca exchange in *Limulus* ventral photoreceptors. *Biophys. J.* **51**, 273a.

Pepe, I. M., Panfoli, I., and Cugnoli, C. (1986). Guanylate cyclase in outer segments of the toad retina. *FEBS Lett.* **203**, 73–6.

Pugh, E. N. and Cobbs, W. H. (1986). Visual transduction in vertebrate rods and cones: a tale of two transmitters, calcium and cyclic GMP. *Vision Res.* **26**, 1613–43.

Ratto, G. M., Payne, R. Owen, W. G., and Tsien. R. Y. (1988). The ion concentration of cytosolic free Ca^{2+} in vertebrate rod outer segments measured with fura-2. *J. Neurosci.* **8**, 3240–6.

Reeves, J. P. and Hale, C. C. (1984). The stoichiometry of the cardiac sodium–calcium exchange system. *J. Biol. Chem.* **259**, 7733–9.

Robinson, P. R., Kawamura, S., Abramson, R. and Bownds, M. D. (1980). Control of the cyclic GMP phosphodiesterase of frog photoreceptor membranes. *J. Gen. Physiol.* **76**, 631–45.

Sather, W. A. and Detwiler, P. B. (1987). Intracellular biochemical manipulation of phototransduction in detached rod outer segments. *Proc. Nat. Acad. Sci. USA* **84**, 9290–4.

Schnetkamp, P. P. M. (1986). Sodium–calcium exchange in the outer segments of bovine rod photoreceptors. *J. Physiol.* **373**, 25–45.

Schnetkamp, P. P. M. (1979). Calcium translocation and storage of isolated cattle rod outer segments in darkness. *Biochim. Biophys. Acta* **554**, 441–9.

Schnetkamp, P. P. M. and Bownds, M. D. (1987). Na^+ – and cGMP – induced Ca^{2+} fluxes in frog rod photoreceptors. *J. Gen. Physiol.* **89**, 481–500.

Schnetkamp, P. P. M., Daemen, F. J. M., and Bonting, S. L. (1977). Calcium accumulation in cattle rod outer segments: evidence for a calcium–sodium exchange carrier in the rod sac membrane. *Biochim. Biophys. Acta* **468**, 259–70.

Schnetkamp, P. P. M., Szerencsei, R. T., and Basu, D. K. (1988). Na–Ca exchange in bovine rod outer segments requires potassium. *Biophys. J.* **53**, 389a.

Stryer, L. (1986). Cyclic GMP cascade of vision. *Ann. Rev. Neurosci.* **9**, 87–119.

Tomita, T. (1970). Electrical activity of vertebrate photoreceptors. *Quart. Rev. Biophys.* **3**, 179–222.

Torre, V., Matthews, H. R., and Lamb, T. D. (1986). Role of calcium in regulating the cyclic GMP cascade of phototransduction in retinal rods. *Proc. Natl. Acad. Sci. USA* **83**, 7109–13.

Troyer, E. W., Hall, I. A., and Ferrendelli, J. A. (1978). Guanylate cyclase in CNS: Enzymatic characteristics of soluble and particulate enzymes from mouse cerebellum and retina. *J. Neurochem.* **31**, 825–33.

Van Breeman, C. and de Weer, P. (1970). Lanthanum inhibition of the ^{45}Ca efflux from squid giant axon. *Nature* **226**, 760–1.

Waloga, G., Brown, J. E., and Pinto, L. H. (1975). Detection of changes in $[Ca^{2+}]_i$ from *Limulus* ventral photoreceptors using Arsenazo III. *Biol. Bull.* **149**, 449–50.

Yau, K.-W., McNaughton, P. A., and Hodgkin, A. L. (1981). Effects of ions on the light-sensitive current in retinal rods. *Nature* **292**, 502–5.

Yau, K.-W. and Nakatani, K. (1984a). Cation selectivity of light-sensitive conductance in retinal rods. *Nature* **309**, 352–4.

Yau. K.-W. and Nakatani, K. (1984b). Electrogenic Na–Ca exchange in retinal rod outer segment. *Nature* **311**, 661–3.

Yau, K.-W. and Nakatani, L. (1985). Light-induced reduction of cytoplasmic free calcium in retinal rod outer segment. *Nature* **313**, 579–82.

Yoshikami, S., George, J. S. and Hagins, W. A. (1980). Light-induced calcium fluxes from outer segment layer of vertebrate retina. *Nature* **286**, 395–8.

Zimmerman, A. L. and Baylor, D. A. (1986). Cyclic GMP-sensitive conductance of retinal rods consists of aqueous pores. *Nature* **321**, 70–2.

13 The role of sodium–calcium exchange in sodium-transporting epithelia

Ann Taylor

During the past decade, several lines of investigation have provided evidence that cytosolic calcium ions play a role in the control of transepithelial sodium transport. At present, our understanding of the processes that determine the level of ionized calcium in sodium-absorbing epithelial cells is relatively incomplete. However, a Na–Ca exchange system has been identified in the basolateral membrane of several types of absorptive epithelial cells and has been implicated in the regulation of cytosolic calcium ion activity. The Na–Ca exchanger has also been implicated in an intrinsic negative-feedback system that may operate to modulate the rate of transcellular sodium transport and serve to co-ordinate the ionic events that occur at the two surfaces of the epithelial cells. This review will attempt to evaluate current evidence regarding the role of Na–Ca exchange and cytosolic calcium ions in the regulation of transepithelial sodium absorption.

13.1 Transepithelial sodium transport: basic cellular mechanisms

Epithelial cells are specialized for the vectorial transport of solutes and water, and are characterized by their asymmetry. In traversing the cell, ions cross two plasma membranes in series, and these membranes differ markedly in their functional properties. Whereas the basolateral membrane of most sodium-transporting epithelial cells has characteristics in common with the resting membrane of excitable cells, the apical membrane has specialized sodium entry systems (channels or carriers) adapted to the specific functions of the epithelium (reviewed in Taylor and Palmer 1982). According to the model advanced by Koefoed-Johnsen and Ussing (1958), transcellular sodium transport occurs as a consequence of the functional polarity of the cells: sodium ions enter the epithelial cell, passively, across the sodium-selective apical membrane and are subsequently extruded, actively, by the NaKATPase that is located in and confined to the basolateral cell membrane.

Studies in a variety of epithelia have provided evidence that the apical entry step is normally rate-limiting for the overall transcellular sodium transport process (see e.g. Lewis, Eaton, and Diamond 1976; Spring and Giebisch 1977). The rate of apical sodium entry is believed to limit the availability of sodium ions to the basolateral sodium pump and thereby determine the rate of active transport. The rate of passive sodium entry depends in part on the sodium permeability of the apical membrane and spontaneous changes in the rate of sodium transport are associated with parallel alterations in apical membrane sodium conductance (Lewis, Eaton, and Diamond 1976; see Schultz 1981). At high luminal (external) sodium concentrations, the rates of sodium entry and transepithelial net sodium transport saturate, apparently as a result of an adaptive reduction in apical sodium permeability (Fuchs, Larson, and Lindemann 1977).

13.2 Coupling of sodium entry and exit: negative feedback concept

Sodium-absorbing epithelial cells may be subject to wide variations in solute throughput. Clearly the rate of entry of sodium ions across the apical plasma membrane and their rate of exit across the basolateral membrane must somehow be co-ordinated if cellular ion content and cell volume are to be maintained within reasonable limits (see e.g. Schultz 1981). In tight epithelia, in which sodium ions enter the cell via amiloride-sensitive sodium channels, there is evidence that the rate of apical sodium entry varies with the rate of basolateral sodium extrusion. Studies on frog skin (Biber 1971; Erlij and Smith 1973), toad urinary bladder (Essig and Leaf 1963; Finn 1975), rabbit urinary bladder (Lewis, Eaton, and Diamond 1976), and rabbit colon (Turnheim, Frizzell, and Schultz 1978) have demonstrated that inhibition of the sodium pump results in a decrease in apical sodium entry which is attributable to a reduction in apical membrane sodium permeability. This effect is dependent on the sodium concentration in the mucosal medium and thus presumably on cell sodium (Erlij and Smith 1973; Turnheim, Frizzell, and Schultz 1978). Such findings led to the concept tht a negative feedback operates between the basolateral and apical cell membranes, whereby the rate of entry of sodium ions into the cell is coupled to the rate of basolateral sodium exit (Lewis, Eaton, and Diamond 1976). Changes in intracellular sodium ion activity, secondary to changes in pump rate, were originally implicated in this negative-feedback process; an increase in cell sodium was presumed to elicit a decrease in the sodium conductance of the apical membrane (e.g. Lewis, Eaton, and Diamond 1976). However, electrophysiological measurements in rabbit and toad urinary bladder (Eaton 1981; Palmer 1985), and studies of ^{22}Na uptake into toad bladder apical membrane vesicles (Chase and Al-Awqati 1983; Garty, Asher, and Yeger 1987) suggest that changes in cell sodium *per se* have little or no effect on apical sodium conductance.

In recent years, evidence has emerged indicating that changes in cytosolic calcium ion activity, resulting from alterations in the rate of Na–Ca exchange across the basolateral cell membrane, in turn dependent on cell sodium, may participate in the negative-feedback control of apical membrane sodium permeability. A hypothetical model of a calcium-mediated intrinsic negative-feedback system incorporating a role for a basolateral Na–Ca exchanger is shown in Fig. 13.1. According to this model, as proposed by Taylor and Windhager (1979), a change in the activity of an electrogenic basolateral sodium pump would alter the electrochemical driving force for sodium entry via the Na–Ca exchanger, and hence influence the rate of basolateral calcium efflux. Alterations in intracellular sodium ion activity would thus be accompanied by parallel, but presumably amplified, changes in cytosolic calcium

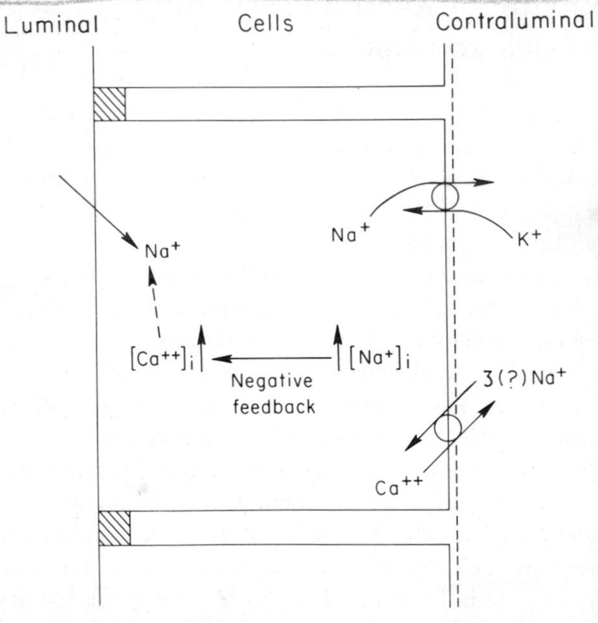

Fig. 13.1. Model of epithelial cell indicating possible role of Na–Ca exchange and cytosolic calcium ions in negative feedback control of apical membrane permeability to sodium, as proposed by Taylor and Windhager (1979) (see text). (Modified from Taylor and Windhager (1979) with permission.)

ion activity (assuming that these are not immediately buffered by other calcium uptake or extrusion systems). The alteration in cytosolic calcium ion levels might then modify (directly or indirectly) apical membrane permeability to sodium in such a way that the rates of apical entry and basolateral exit of sodium are kept in step. In addition, changes in the rate of apical sodium entry, e.g. following alterations in mucosal sodium concentration, might be damped in part through the operation of such a calcium-mediated negative-feedback mechanism. According to this concept, the Na–Ca exchanger and cytosolic calcium ions would constitute links in an intracellular servosystem which would serve to stabilize intracellular sodium ion activity and cell volume, and modulate the rate of transcellular sodium transport, in the face of changes in sodium entry or exit (Taylor 1981).[1]

This model, which was originally based on indirect evidence, has now received some direct experimental support. The results of a number of recent studies suggest that at least some sodium-absorbing epithelial cells possess the cellular machinery that would allow such a calcium-mediated feedback system to operate (Chase 1984). However, some of the tenets of the model as originally proposed (Taylor and Windhager 1979) would appear to require modification in the light of the information that is now available regarding the properties of the basolateral Na–Ca exchanger. Moreover, the relevance of the Na–Ca exchange process, and indeed of the postulated negative-feedback system (see Jensen, Fisher, and Spring 1984), in the fine control of sodium absorption under physiological conditions remains to be evaluated.

13.3 Evidence for modulation of sodium absorption by cytosolic calcium ions

Experiments in amphibian tissues with agents presumed to increase ionized calcium levels provided the first indirect evidence that changes in cytosolic free calcium can modulate the rate of sodium transport in absorptive epithelia. Many studies now indicate that sodium absorption is partially inhibited under experimental conditions in which cytosolic calcium is elevated (reviewed in Taylor and Windhager 1985; Windhager and Taylor, 1983). Thus, quinidine (a drug shown to increase cytosolic calcium ion activity in *Necturus* proximal tubules (Lorenzen, Lee, and Windhager 1984), and the calcium ionophores (A23187, X537A, ionomycin) decrease net sodium transport across the toad urinary bladder (Ludens 1978; Taylor 1975; Taylor *et al.* 1987; Wiesmann, Sinha, and Klahr 1977; Taylor and Barber, unpublished observations). Similar results have been obtained in

[1] Cytosolic calcium ions may also participate in cell volume control and cellular potassium homeostasis during changes in the rate of transepithelial sodium transport by modifying the potassium permeability of the basolateral cell membrane (Schultz 1981; Davis and Finn 1987).

turtle urinary bladder (Arruda and Sabatini 1980), in isolated frog skin epithelium (Erlij *et al.* 1986), and in isolated perfused proximal and cortical collecting tubules from rabbit kidneys (Friedman *et al.* 1981; Frindt and Windhager 1983, 1986). These findings suggest that specific rate-limiting steps in the transcellular sodium transport process are subject to regulation by cytosolic calcium ions in both tight and leaky epithelia.

As discussed in detail elsewhere (Taylor and Windhager 1985), several considerations point to the conclusion that the inhibition of net sodium transport elicited by increased cytosolic calcium is due to a decrease in the sodium permeability of the apical membrane of the transporting epithelial cells. A variety of experimental approaches have provided indirect support for this view in studies on the toad bladder (Chase and Al-Awqati 1981; Garty and Lindemann 1984; Palmer 1985; Taylor *et al.* 1987) and on isolated renal proximal and cortical collecting tubules (see Windhager *et al.* 1986). In tight epithelia, there is now direct evidence that cytosolic calcium ions can induce a reduction in apical membrane sodium permeability. Studies of ^{22}Na uptake into membrane vesicles derived from toad bladder epithelium have shown that amiloride-sensitive sodium uptake via the apical membrane sodium channels is decreased in the presence of submicromolar concentrations of free calcium ($K_{\frac{1}{2}} \sim 500$ nM) (Chase and Al-Awqati 1983). Recently, Palmer and Frindt (1987) have studied the activity of amiloride-sensitive sodium channels in apical membranes of rat cortical collecting tubule cells using the patch-clamp technique. These investigators demonstrated that exposure of isolated tubules to ionomycin led to a marked reduction in the activity of the apical sodium channels in cell-attached (but not detached) patches. These findings are consistent with the results of both electrophysiological and vesicle studies on the toad bladder which suggest that cytosolic calcium ions may modulate sodium channel activity indirectly (Palmer 1985; Garty and Asher 1985). However, there is also evidence for direct, pH-sensitive, Ca^{2+}-sodium channel interaction (Garty, Asher, and Yeger, 1987).

Although the molecular mechanisms whereby calcium ions modify apical membrane permeability to sodium in either tight or leaky epithelia are unclear at present, these studies indicate that cytosolic calcium ions can act (at least potentially) to modulate the rate of sodium entry across the apical surface of the epithelial cells.

13.4 Evidence for basolateral Na–Ca exchange

13.4.1 Functional studies in epithelia

Based on functional studies by Martin and DeLuca (1969) and others, Blaustein (1974) suggested that a Na–Ca exchanger comparable to that found in excitable cells might be present in intestinal and renal epithelial cells, and might mediate the active extrusion of calcium ions across the basolateral cell membrane. Calcium efflux measurements in renal tubules provided preliminary support for this view (Blaustein 1974).

The first evidence that a basolateral Na–Ca exchanger might play a role in the regulation of transepithelial sodium transport was obtained in studies on frog skin. Grinstein and Erlij (1978) recognized that the inhibition of net sodium transport that follows removal of sodium from the medium bathing the serosal (inner) surface of this tissue (Mandel and Curran 1973) might be related to an alteration in Na–Ca exchange across the basolateral cell membrane. These investigators demonstrated that the transport inhibition induced by sodium-free serosal media is dependent on the presence of calcium in the serosal solution, and is associated with increased uptake of ^{45}Ca across the basolateral surface of the epithelium and with diminished tracer efflux from preloaded tissue. They inferred that a Na–Ca exchanger is involved in the regulation of intracellular calcium in the epithelial cells of frog skin, as in excitable cells; they postulated that removal of serosal sodium leads to calcium entry via the reverse mode of operation of the exchanger, and hence to an increase in cytoplasmic calcium which, in turn, results in inhibition of net sodium transport.

Indirect evidence for a role of Na–Ca exchange in the regulation of sodium transport has been obtained in several functional studies on the toad bladder in which the effects of a reduction in the driving force for sodium entry across the basolateral surface of the epithelium have been evaluated. Thus, as illustrated in Fig.13.2, the short circuit current (a measure of net sodium transport) across isolated toad bladders is markedly decreased when sodium is replaced by choline in the serosal bathing solution (cf. Bentley 1960). Consistent with the operation of a Na–Ca exchanger, the inhibition of sodium transport induced by low serosal sodium has been found to be a function of the serosal calcium concentration, being least at low extracellular calcium levels (Fig. 13.2b) (Taylor and Eich 1978; Taylor *et al.* 1987). In other studies, the effects of ouabain on net sodium transport (Chase and Al-Awqati 1981) and on apical membrane sodium conductance (Palmer 1985; Garty and Lindemann 1984) have likewise been shown to vary with the calcium concentration in the serosal bathing solution.

Further evidence for the operation of a basolateral Na–Ca exchanger in the toad bladder has been obtained in experiments in which the water permeability response to vasopressin was used as 'bioassay' system. In these studies, exposure of isolated bladders to either low sodium or potassium-free serosal solutions resulted in a decrease in vasopressin- or cAMP-stimulated water flow: on reduction of extracellular sodium, the degree of inhibition of the response to cAMP varied inversely with the serosal sodium and directly with the serosal calcium concentration; on exposure to potassium-free media,

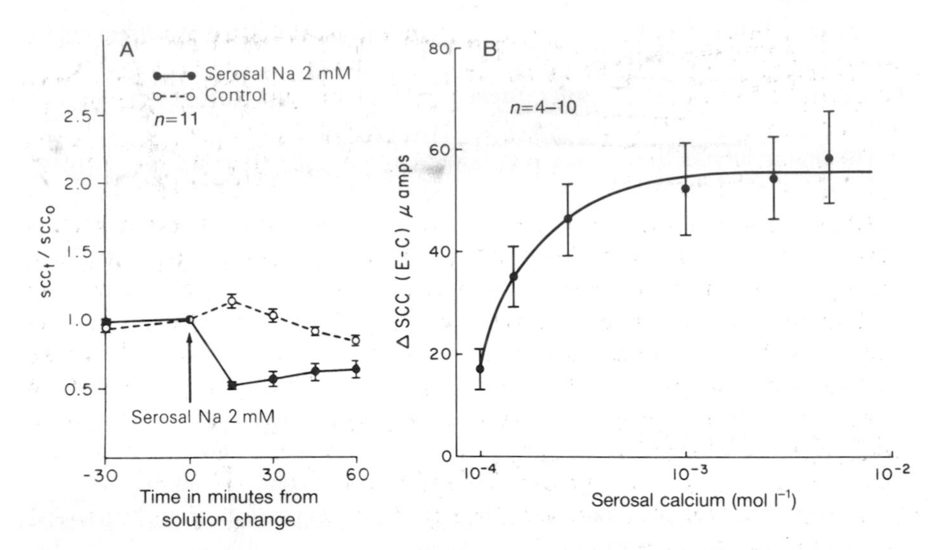

Fig. 13.2. A reduction in serosal sodium concentration inhibits the short circuit current (SCC) in a calcium-dependent manner in isolated toad urinary bladders: (A) Effect of low serosal sodium on SCC. At time 0, the medium bathing the serosal surface of the experimental hemibladders was replaced with an isosmotic solution containing only 2 mM sodium. (B) The absolute fall in SCC induced by exposure to low serosal sodium: dependence on serosal calcium concentration (from Taylor *et al.* 1987, with permission).

the inhibition of cAMP-stimulated water flow showed a similar dependence on serosal calcium (Taylor 1981; Taylor *et al.* 1987).[2]

Similar types of observations have been made in experiments involving manipulation of the sodium gradient across the peritubular surface of mammalian renal tubules. Exposure of isolated perfused rabbit proximal or cortical collecting tubules to low peritubular sodium resulted in a decrease in isotopic sodium efflux and net sodium absorption which was shown to be dependent on peritubular calcium, being attenuated at low calcium concentrations (Friedman *et al.* 1981; Frindt and Windhager 1983). Furthermore, in isolated collecting tubules the water permeability response to exogenous cAMP was reduced in a calcium-dependent manner on exposure to low peritubular sodium (Frindt, Windhager, and Taylor 1982).

[2] Hydrogen ions and a basolateral Na–H exchanger have recently been implicated in the negative feed-back control of apical membrane sodium permeability in frog skin (Harvey and Ehrenfeld 1988). It is noteworthy that the inhibitory effects of low serosal sodium on net sodium transport (A. S. Brem and A. Taylor, unpublished observations) and on cAMP-stimulated water flow (Taylor *et al.* 1987) in the toad bladder were reduced under conditions leading to cellular acidification, a situation in which Na–H exchange would be expected to be inhibited (Philipson 1985), whereas Na–H exchange is likely to be activated (Chaillet, Lopes, and Boron 1985). These findings argue against the possibility that the inhibitory effects of low serosal sodium observed in the toad bladder are due to an alteration in Na–H exchange and a decrease in cell pH.

These various results have been interpreted to indicate that a Na–Ca exchange system in the basolateral membrane plays a role in the regulation of cytosolic free calcium levels in the transporting epithelial cells, and that cytosolic calcium ions, in turn, can modulate the rate of transepithelial sodium transport and vasopressin-dependent water absorption (Taylor and Windhager 1985).

Other functional evidence consistent with the operation of a basolateral Na–Ca exchanger has been obtained in studies on active calcium transport by the kidney and gut; however, the role of the exchanger relative to that of ATP-driven calcium extrusion in active calcium absorption by these tissues is not yet established. In microperfusion experiments on proximal tubules of rat and hamster kidneys *in vivo*, Ullrich, Rumrich, and Kloss (1976) showed that active calcium re-absorption was markedly decreased when peritubular sodium was replaced by choline or on exposure to ouabain. In line with Blaustein's (1974) earlier proposal, these investigators concluded that the active extrusion of calcium across the peritubular cell membrane is mediated by Na–Ca countertransport.

Studies designed to evaluate the effects of changes in the transerosal sodium gradient on active calcium absorption by various segments of the gut have yielded conflicting results. While net calcium transport in rat caecum was found to be substantially inhibited by ouabain or removal of serosal sodium, indicative of the involvement of a Na–Ca exchange process (Nellans and Goldsmith 1981), similar studies in rat small intestine have produced variable results. Whereas Martin and DeLuca (1969) found that small intestinal calcium transport is sodium-dependent, other investigators have failed to demonstrate a sodium requirement (see Ghijsen, DeJong, and Van Os 1983; Favus 1985).

13.4.2 Na–Ca exchange in isolated basolateral membrane preparations

Direct evidence for the presence of a basolateral Na–Ca exchanger in both renal and intestinal epithelia has now been obtained in studies with isolated plasma membrane vesicles. Measurements of sodium-dependent calcium fluxes in vesicles prepared from basolateral membranes derived from toad bladder epithelium, rat and dog kidney cortex, and rat small intestine have been reported from a number of laboratories (Chase and Al-Awqati 1981; Gmaj, Murer and Kinne 1979; see reviews by Gmaj and Murer 1988; Van Os 1987). For example, ^{45}Ca efflux from preloaded basolateral membrane vesicles from dog kidney was stimulated by an inwardly-directed sodium gradient; this effect was inhibited by the sodium ionophore, monensin, and enhanced by inside-negative potentials, indicative of electrogenic Na–Ca exchange activity (Scoble, Mills, and Hruska 1985). A high affinity CaATPase (stimulated by calmodulin) has also been demonstrated in basolateral membranes of kidney cortex and small intestine (see Gmaj and Murer 1988; Van Os 1987). Indirect evidence for the operation of a Na–Ca exchanger in parallel with the CaATPase comes from the finding that the sodium gradient influences ATP-driven calcium accumulation by basolateral membrane vesicles (Gmaj, Murer, and Kinne 1979; Van Heeswijk, Geertsen, and Van Os 1984). Data obtained using both the direct and indirect approaches suggest that the Na–Ca exchanger derived from renal cortex is reversible and operates symmetrically *in vitro* (e.g. Van Heeswijk, Geertsen, and Van Os 1984).

In a recent study, the basolateral Na–Ca exchanger from beef kidney cortex was partially purified and incorporated into liposomes, allowing precise determination of the stoichiometry (Talor and Arruda 1985). The latter was ascertained using the thermodynamic approach of Reeves and Hale (1984), and was found to be 3.09 ± 0.22 and 2.89 ± 0.2 Na^{+}:$1Ca^{2+}$ in the presence of constant negative and positive membrane potentials, respectively. Thus, the stoichiometry of the renal basolateral Na–Ca exchanger is approximately 3:1, confirming previous estimates for the epithelial exchanger (e.g. Chase and Al-Awqati 1981) and in agreement with findings for the Na–Ca exchange system in cardiac and other plasma membranes (e.g. Rasgado-Flores and Blaustein 1987; Reeves and Hale 1984).

Table 13.1 shows the kinetic data obtained in studies on renal basolateral membrane preparations; data for the Na–Ca exchanger and CaATPase of heart sarcolemma are included for comparison. As discussed by Gmaj and Murer (1988), measurements of both sodium-dependent calcium fluxes and CaATPase activity in basolateral membrane vesicles are complicated by problems of variable orientation and leakiness of the vesicles; sodium-dependent calcium fluxes are very rapid and are linear for only a few seconds (Jayakumar *et al.* 1984; Scoble, Mills, and Hruska 1985). The kinetic constants for the renal Na–Ca exchanger in Table 13.1 are likely to represent underestimates because of the use of incubation times (5–60 s) that precluded accurate determination of initial rates (see Gmaj and Murer 1988; Jayakumar *et al.* 1984; Scoble, Mills, and Hruska 1985; Van Os 1987). The data of Jayakumar *et al.* (1984) and Scoble, Mills, and Hruska (1985) would appear to provide the best available approximation of real values, inasmuch as relatively short (5 s) incubation times were employed in their experiments; moreover their estimates of the affinity of the exchanger for calcium (K_m 8–10 μM) are similar to that obtained in reconstitution studies, in which uptake rates were linear for longer time periods (Talor and Arruda 1985).

Despite some variation and uncertainties regarding the absolute values, these *in-vitro* findings indicate that the properties of the epithelial Na–Ca exchanger are similar in several aspects to those reported for the exchanger derived from heart sarcolemma and other tissues (Baker and DiPolo 1984; Philipson 1985). The data suggest that the basolateral Na–Ca exchanger is a

Table 13.1 Kinetics of Na^+–Ca^{2+} exchange and ATP-dependent Ca^{2+} uptake

Source	K_{mCa} (μM)	$K_{0.5Na}$ (mM)	V_{max} (nmol Ca^{2+}/min. mg protein)	Reference
RENAL BASOLATERAL MEMBRANES				
Na^+–Ca^{2+} Exchange				
rat	8	15	5.4	Jayakumar *et al.* 1984
dog	10	11	4.0	Scoble *et al.* 1985
rat	0.2	—	3.2	Van Heeswijk *et al.* 1984
cow (reconstituted)	2.7	—	—	Talor and Arruda 1985
ATP-dependent Ca^{2+} Uptake				
rat	0.14	—	5.8	Gmaj *et al.* 1983
rat	0.07	—	7.4	Heeswijk *et al.* 1984
CARDIAC SARCOLEMMA				
Na^+–Ca^{2+} Exchange				
various	15–40	7–32	—	Philipson 1985
dog	1.5	—	900	Caroni *et al.* 1980
ATP-dependent Ca^{2+}-Uptake				
dog	0.3	—	31	Caroni and Carafoli 1981

low-affinity calcium transport system, in comparison with the high-affinity basolateral CaATPase, in accord with findings in excitable membranes. However, the maximum transport capacity of the epithelial exchange system appears to be relatively low in contrast to that in excitable cells.

While vesicle studies have provided conclusive evidence for the presence of a basolateral Na–Ca exchanger in certain epithelia, it must be recognized that the conditions employed in studies on isolated plasma membranes may not accurately reflect the situation that obtains in the intact cell. For example, the affinity and/or maximal transport capacity of the basolateral Na–Ca exchanger might be modified by calcium- and ATP-dependent phosphorylation reactions *in vivo*, as appears to be the case for the exchange system in excitable cells (Baker 1978; Caroni and Carafoli 1983; DiPolo and Beaugé 1983, 1987). Interestingly, the maximal activity (but not the affinity) of Na–Ca exchange in renal basolateral membrane vesicles varies with the parathyroid status of the animal (Jayakumar *et al* 1984; Scoble, Mills, and Hruska 1985); moreover parathyroid hormone and cAMP have been shown to stimulate sodium-dependent calcium efflux from isolated renal cortical cells (Hanai *et al.* 1986), suggesting that the basolateral Na–Ca exchanger can be activated by cAMP-dependent phosphorylation events *in vivo*.

13.5 Asymmetric operation of the Na–Ca exchange system *in vivo*: kinetic considerations

In order to understand how the Na–Ca exchange system operates *in vivo* both kinetic and thermodynamic considerations must be taken into account. Potentially, the Na–Ca exchanger can affect net movement of calcium either out of or into the cell, depending on the relative affinities for calcium and sodium on either side of the cell membrane and on the electrochemical driving forces acting on the system. Although vesicle and reconstitution studies have revealed that the Na–Ca exchanger can operate symmetrically *in vitro*, experiments on dialysed axons have shown that the exchange process functions in a highly asymmetric manner in intact cells: the 'forward' mode of operation of the exchanger (calcium efflux–sodium influx) and the 'reverse' mode (calcium influx–sodium efflux) behave as though they have a number of different properties (Baker 1986). This behaviour seems to be due, in part, to the asymmetric ionic conditions that prevail on either side of the cell membrane, but also may depend on the existence of specific binding sites for ligands at the internal and external faces of the carrier (DiPolo and Beaugé 1986; see Baker 1986). In giant excitable cells, sodium apparently competes with calcium for a common binding site (on either side of the carrier) and thus reduces the affinity of the system for calcium (see Baker and DiPolo 1984; Blaustein 1974; Reeves and Sutko 1983). This competitive effect of sodium has largely been ignored in consideration of the kinetics and/or the function of the epithelial exchanger. However, Chase (1984) has reported preliminary evidence indicating that sodium reduces the calcium affinity of the basolateral exchanger in toad bladder membrane vesicles; in these studies the apparent K_m was estimated to increase from 0.7 μM in the absence of sodium, to 1.4 μM at 10 mM sodium, and to 0.13 mM at 110 mM sodium.

As a result of the competitive effect of sodium, and because of the asymmetric ionic conditions that normally prevail on either side of the cell membrane, the exchanger will have a very much higher affinity for calcium inside the cell than outside. In theory, this will permit the transfer of calcium ions from a region of low to high calcium ion activity, and thus enable the 'forward' mode of operation of the carrier to function as a potential mechanism for calcium extrusion across the basolateral cell membrane (see Chase 1984). However, the available kinetic data suggest that the 'forward' mode of operation of the epithelial exchanger will be relatively ineffective as a calcium efflux mechanism under normal conditions of intracellular calcium (~100 nM) and sodium (10–20 mM) (see below). Any increase in cell sodium will result in a decrease in the affinity for calcium at the internal site on the carrier; thus, the exchange system will be rendered less effective as a calcium efflux pathway when cell sodium is elevated (e.g. by ouabain).

Under normal circumstances, in the presence of high extracellular sodium, the affinity of the exchanger for calcium at the external side of the membrane will be relatively very low ($K_{Ca} > 0.1$ mM). Nevertheless, as pointed out by Gmaj and Murer (1988), the 'reverse' mode of operation of the exchange system (calcium influx–sodium efflux) will not be so constrained by its low calcium affinity since extracellular ionized calcium is ~1 mM. Theoretically, therefore, the kinetic properties of the exchanger would appear to equip it to operate as a significant calcium influx pathway under normal conditions. Under experimental conditions in which extracellular sodium is markedly reduced, the affinity for calcium at the external site will be considerably increased. Thus kinetic (as well as thermodynamic) considerations would suggest that the calcium influx mode of operation of the carrier would be stimulated under such circumstances (cf. Grinstein and Erlij 1978).

It is of interest to note that in dialysed giant excitable cells the 'reverse' mode of operation of the exchange system is activated by an increase in either internal sodium or calcium (DiPolo and Beaugé 1986, 1987; Rasgado-Flores and Blaustein 1987), and is inhibited by calcium chelators (including EGTA and quin-2) (Allen and Baker 1985). The activation of calcium influx by internal calcium appears to represent a positive-feedback system (DiPolo and Beaugé 1986; Rasgado-Flores and Blaustein 1987). If similar mechanisms modulate the activity of the epithelial exchanger, this could have important implications for its operation *in vivo*.

13.6 Measurements of cytosolic free calcium ion levels in epithelial cells: resting cytosolic free calcium levels

Interpretation of the kinetic data and evaluation of the functional role of the basolateral Na–Ca exchange system obviously requires precise information regarding the levels of ionized calcium and sodium that normally prevail in the epithelial cell cytosol.

Lee, Taylor, and Windhager (1980) obtained the first measurements of cytosolic calcium ion activity in epithelial cells in experiments with Ca^{2+}-selective microelectrodes in proximal tubules of *Necturus* kidneys perfused *in situ*; in these studies, under normal conditions, cytosolic calcium ion activity averaged 116 ± 32 nM (n = 11). Lorenzen, Lee, and Windhager (1984) subsequently reported somewhat lower values (~70 nM) in a series of experiments on isolated *Necturus* proximal tubules perfused *in vitro*. Snowdowne and Borle (1984) obtained similar values (~60 nM) for cytosolic free calcium ion concentration in studies with aequorin in cultured isolated monkey kidney cells. The fluorescent calcium indicators, quin-2 and fura-2, have been employed to estimate ionized calcium levels in a variety of absorptive epithelial cells and have yielded comparable values for cytosolic free calcium concentration under resting conditions. For example, in two separate studies with quin-2 on isolated toad bladder epithelial cells, cytosolic calcium ion concentration averaged 97 ± 3 nM (n = 216) (Taylor, Pearl, and Crutch 1985) and 86 ± 9 nM (n = 22) (Jacobs and Mandel 1987). In a cultured kidney cell line (LLC-PK_1), cytosolic calcium ion concentration averaged 98 ± 5 nM (n = 81) when measured with quin-2, and 101 ± 23 nM (n = 7) when arsenazo II and a null-point method were employed (Bonventre and Cheung 1986).

The absolute values for intracellular free calcium estimated by these various methods are based upon a number of assumptions and are obviously subject to some uncertainty. Nevertheless, these various studies indicate that the level of cytosolic free calcium in sodium-absorbing epithelial cells is normally maintained at levels comparable to those found in excitable cells under resting conditions, i.e. at about 100 nM. If the affinity of the basolateral Na–Ca exchange system for calcium at the internal side of the membrane is of the order of 1–10 μM, as suggested by biochemical studies, only a few sites on the exchanger will be occupied by calcium when cytosolic calcium ion activity is 100 nM, assuming intracellular sodium is ~10 mM (see below). At such ionized calcium levels, it would seem that the exchanger could not transport very much calcium out of the cell, even if the electrochemical driving forces were favourable, and therefore could not function as a major calcium extrusion mechanism (cf. Gmaj and Murer 1988). Thus the kinetic data derived from isolated membrane preparations and the cytosolic calcium measurements, taken together, point to the conclusion that the ATP-driven calcium pump, rather than the Na–Ca exchange system, must be the major determinant of steady-state cytosolic calcium ion levels under physiological conditions. Nevertheless, even small variations in the rate and/or direction of Na–Ca exchange would be expected to exert some influence on the level of intracellular free calcium.

13.7 Changes in cytosolic calcium in response to a decrease in the transerosal sodium gradient

The assumption that an experimentally-induced decrease in the electrochemical gradient for sodium entry across the basolateral cell surface results in an increase in cytosolic calcium (Grinstein and Erlij 1978; Taylor and

Windhager 1979) has been tested directly in a series of experiments on *Necturus* proximal renal tubules (Lorenzen, Lee, and Windhager 1984; see Frindt *et al.* 1988). In these studies, intracellular microelectrodes were used to measure cytosolic calcium and sodium ion activities and basolateral membrane potential during manipulation of the sodium gradient across the peritubular cell surface. Due to the large size of the epithelial cells, and the ability to control the composition of both the luminal and contraluminal bathing solutions, the *Necturus* preparation is uniquely well-suited for such electrophysiological studies.

In experiments on isolated perfused tubules (Lorenzen, Lee, and Windhager 1984), a reduction in peritubular sodium from 100 mM to 10 mM (substituting choline) resulted in a rapid increase in cytosolic calcium ion activity from a control level of 73 ± 14 to 382 ± 69 nM ($p<0.001$, $n = 4$) (while the basolateral membrane potential depolarized from −50 ± 12 to −36 ± 7 mV), confirming earlier results on the intact *Necturus* preparation (Lee, Taylor, and Windhager 1980) (see Fig. 13.3); in separate tubules, intracellular sodium ion activity decreased from 12.8 ± 1.9 to 8.2 ± 1.8 mM ($p<0.001$, $n = 12$). In experiments in which isolated tubules were exposed to ouabain (10^{-4} M), cystosolic free calcium ion activity increased from 71 ± 9 to 546 ± 121 nM ($p<0.005$, $n = 9$) over a period of ~50 minutes, while basolateral membrane potential decreased from −45 ± 3 to −29 ± 3 mV ($p<0.003$); in different tubules, intracellular sodium ion activity rose from

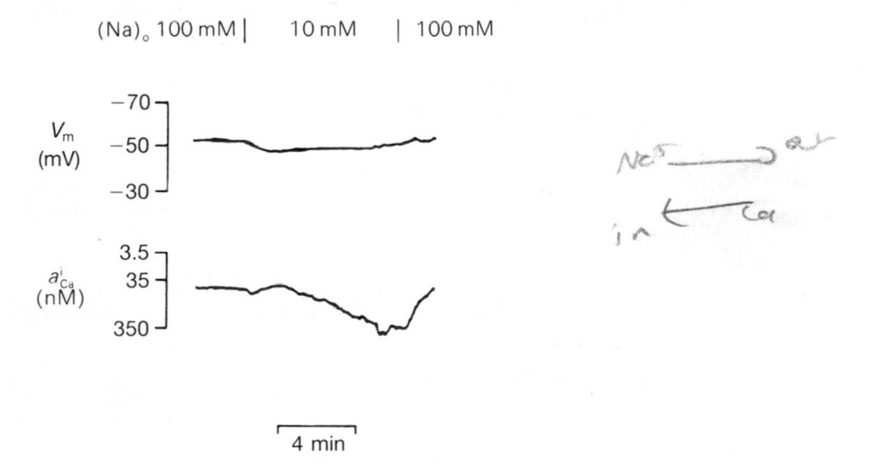

Fig. 13.3. Recordings of basolateral cell membrane potential (V_m) and cytosolic calcium ion activity (a^i_{Ca}) in epithelial cell of *Necturus* proximal renal tubule (perfused in situ) during reduction in peritubular sodium concentration (choline substitution). Exposure of tubule to peritubular low sodium resulted in an increase in a^i_{Ca}; this was reversed on restoration of normal extracellular sodium (from Lee *et al.* 1980, with permission).

15.1 ± 1.8 mM to 70.1 ± 6.3 mM over the same time period ($p<0.001$, $n = 9$) (see Fig. 13.4). These results are consistent with the existence of a Na–Ca exchanger in the basolateral cell membrane, and they indicate that an experimentally-induced decrease in the electro-chemical gradient for sodium entry across this cell surface can indeed lead to a substantial increase in cytosolic calcium.

These studies on isolated perfused *Necturus* tubules have recently been extended and measurements have been made under experimental conditions designed to change cell sodium through alterations in the rate of apical sodium entry, including lowering of luminal sodium concentration and luminal addition (or deletion) of organic solutes, i.e. conditions that might actually obtain physiologically *in vivo* (Yang, Lee, and Windhager 1988; reviewed in Frindt *et al.* 1988). In these experiments, alterations in intracellu-

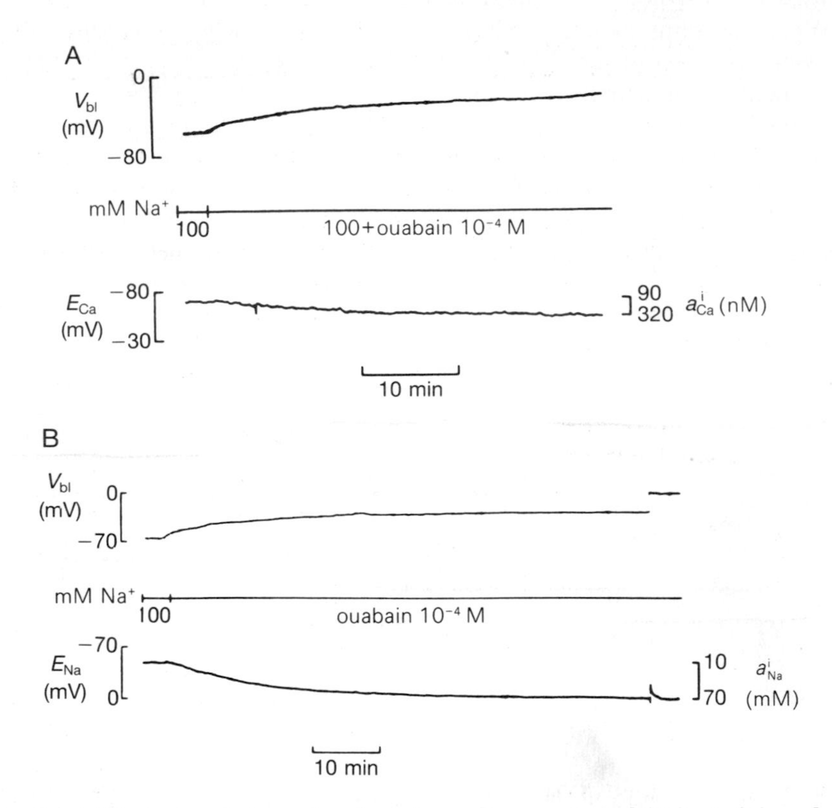

Fig. 13.4. Effect of ouabain on basolateral membrane potential (V_{bl}), a^i_{Ca} (A), and a^i_{Na} (B) in isolated perfused proximal tubules from *Necturus* kidneys. In (A) and (B), lower trace = potential recording obtained with Ca^{2+}– or Na^+– selective microelectrode (E_{Ca}, E_{Na}) corrected for V_{bl} by electronic subtraction. Exposure to ouabain resulted in increases in both a^i_{Na} and a^i_{Ca} (from Lorenzen *et al.* 1984, with permission).

lar sodium ion activity were accompanied by parallel changes in cytosolic calcium ion activity. For example, on removal of organic solutes from the luminal perfusate, cell sodium fell from 14.6 ± 0.6 to 8.3 ± 0.7 mM (n = 9, $p < 0.001$), and in different tubules cytosolic calcium decreased from 73 ± 11 to 61 ± 11 nM (n = 9, $p < 0.001$) (Yang, Lee, and Windhager 1988). These results are consistent with the notion that deletion of organic solutes reduces the cotransport of sodium into the cell across the apical plasma membrane, and the consequent decrease in cell sodium leads in turn to a reduction in cytosolic calcium, as predicted if a Na–Ca exchange process operates across the basolateral cell membrane (Yang, Lee, and Windhager 1988).

Taken together, the results of these various studies in *Necturus* tubules indicate that the level of ionized calcium in the epithelial cell cytosol is function of the electrochemical potential gradient for sodium across the basolateral cell membrane. The findings have provided support for the view that changes in the rate and/or direction of basolateral Na–Ca exchange may play a determining role in the regulation of cytosolic calcium ion activity under physiological conditions.

Some of the findings with Ca^{+}-selective microelectrodes have been confirmed by Snowdowne and Borle (1985) in experiments with aequorin in cultured monkey kidney cells (LLC-MK_2). These investigators found that a reduction in extracellular sodium (substituting choline, TMA, or lithium) resulted in a rapid rise in the level of intracellular ionized calcium. For example, when extracellular sodium was reduced from 144 to 24 mM (choline substitution) cytosolic free calcium rose from 63 ± 15 nM to a peak value of 298 ± 48 nM within 1–3 minutes, and then declined to a plateau at about twice the basal level. The peak ionized calcium level was inversely related to the extracellular sodium concentration and varied directly with the extracellular calcium concentration. Furthermore, the rise in free calcium in sodium-free media was enhanced at high pH and inhibited at low pH (consistent with the known properties of the Na–Ca exchange process (Philipson 1985)). These findings have provided further evidence that changes in the driving forces acting on Na–Ca exchange can result in substantial alterations in the level of cytosolic calcium.

Studies designed to assess the effects of a decrease in the sodium gradient on cytosolic free calcium levels using the fluorescent indicators, quin-2 and fura-2, have yielded somewhat different results. When isolated toad bladder epithelial cells were loaded with quin-2 and suspended in solutions in which sodium was partially replaced by choline or TMA, cytosolic free calcium concentration was only slightly (although significantly) increased (Crutch and Taylor 1983; Taylor, Pearl, and Crutch 1985). For example, in measurements on time-matched paired aliquots of cells, cytosolic calcium ion concentration averaged 83 ± 5 nM in controls and 108 ± 3 nM in cells suspended in a medium containing 2 mM sodium (n = 17, paired difference = 30 per cent,

$p<0.001$). Although only a small increase in ionized calcium was observed in response to a drastic decrease in external sodium, these results are consistent with the existence of a basolateral Na–Ca exchange system. However, in isolated cells suspended in a low sodium medium, cell sodium may be extremely low, possibly <1 mM (see Lewis and Wills 1981); thus the sodium gradient across the cell membrane may not be substantially altered under such conditions. When cells were preloaded with sodium by preincubating with ouabain in the presence of normal extracellular sodium, and were subsequently suspended in a medium containing 2 mM sodium, cytosolic calcium still only increased by of ~ 17 per cent relative to paired controls ($p<0.01$). Exposure to ouabain alone had no detectable effect on the free calcium ion concentration (B. Crutch *et al.*, unpublished observations). Essentially identical results have been obtained in studies with fura-2. To test whether Na–Ca countertransport plays a more prominent role in intracellular calcium homeostasis when cytosolic calcium levels are elevated, fura-2 -loaded toad bladder epithelial cells were exposed to varying doses of ionomycin prior to suspension in a potassium-free medium containing 55 mM sodium and 5 mM calcium. However, even under these conditions, cytosolic calcium ion concentration was not elevated relative to ionomycin-treated controls (Pinches and Taylor, unpublished observations).

Essentially comparable results have been obtained by other investigators using the fluorescent indicators to measure free calcium levels in toad bladder epithelial cells (Jacobs and Mandel 1987; Wong and Chase 1986), cultured renal (LLC-PK_1) cells (Bonventre and Cheung 1986), and rat cortical collecting tubules *in vitro* (Morel, Taniguchi, and Marchetti 1988). In each of these studies, only modest increases (if any) in intracellular ionized calcium were observed following a reduction in the extracellular sodium concentration or exposure to ouabain.

The results of these experiments with quin-2 and fura-2 would seem to indicate that the Na–Ca exchange system exerts relatively little influence on the level of cytosolic free calcium in these epithelial cell types, even under experimental conditions in which the sodium electrochemical gradient is drastically reduced. However, the discrepancy between the results obtained with the fluorescent indicators and those obtained with Ca^{2+}-selective microelectrodes or aequorin in other renal epithelial cells might be explained if the basolateral Na–Ca exchanger is operating in the calcium influx (rather than efflux) mode under the experimental conditions employed. Calcium chelators may selectively block the 'reverse' mode of operation of the exchange carrier, as demonstrated in squid axon (Allen and Baker 1985). If calcium influx via the exchanger were inhibited in quin-2- or fura-2-loaded epithelial cells, changes in cytosolic calcium in response to a decrease in the electrochemical gradient for sodium would be minimized (cf. Baker 1986). Isotope flux measurements may help to clarify the seemingly paradoxical results obtained in studies in which the fluorescent calcium indicators have been employed to monitor changes in free calcium in response to alterations in the sodium gradient.

13.8 Possible role of Na–Ca exchange as a calcium influx pathway

13.8.1 Flux studies

Although the Na–Ca exchanger has mostly been thought of as a calcium extrusion mechanism (e.g. Blaustein 1974; Chase 1984; Taylor and Windhager 1979; Ullrich, Rumrich, and Kloss 1976), several investigators have concluded that the epithelial exchanger may operate in the reverse (calcium influx) mode when extracellular sodium is decreased or intracellular sodium elevated experimentally (Grinstein and Erlij 1979; Jacobs and Mandel 1987; Taylor *et al.* 1987). Direct evidence that the basolateral exchange system operates as a calcium influx (rather than efflux) pathway when the driving force for sodium entry is reduced has recently been presented by Snowdowne and Borle (1985). These investigators measured ^{45}Ca and ^{22}Na fluxes, as well as cytosolic calcium, in isolated cultured kidney cells exposed to sodium-free media. The increase in ionized calcium (measured as aequorin luminescence) that occurred in response to a reduction in the sodium gradient was associated with an increase, rather than a decrease, in fractional ^{45}Ca efflux. Furthermore, there was a concomitant increase in both ^{45}Ca uptake and fractional ^{22}Na efflux. Significantly, the increases in ionized calcium and in isotopic calcium and sodium efflux were all abolished in calcium-free media. Snowdowne and Borle (1985) concluded that removal of extracellular sodium leads to an increase in cytosolic calcium by stimulating calcium influx via the Na–Ca exchanger, and not by reducing calcium efflux. They inferred that the increased rate of calcium influx via the exchanger is followed by enhanced calcium efflux via the high-affinity basolateral CaATPase. According to this view, calcium entering the cell via the exchanger is rapidly extruded by the ATP-driven calcium pump. These conclusions are consistent with an earlier report (Borle 1982) that total epithelial cell calcium content is unchanged following exposure to reduced extracellular sodium (a result confirmed by other investigators working with other epithelial cell preparations (Bonventre and Cheung 1986; Jacobs and Mandel 1987; Mandel and Murphy 1984; also E. Eich and A. Taylor, unpublished observations).

13.8.2 Energetic considerations

The direction in which the basolateral Na–Ca exchange system operates under any particular set of conditions will ultimately be determined by the balance of electrochemical driving forces acting on the sodium and calcium ions. Estimates of the driving forces, based on available measurements of intra- and extracellular sodium and extracellular calcium ion activities, and basolateral membrane potential, in mammalian renal tubular cells (assuming a value for cytosolic calcium ion activity of 100 nM and a stoichiometry of $3Na^{+}–1Ca^{2+}$), suggest that the exchange system is close to equilibrium under normal resting conditions (Snowdowne and Borle 1985). If this is the case, calcium efflux and influx will be more or less equal and the exchanger will in effect be 'idling'. However, if cell sodium is increased, or external sodium decreased, the balance of driving forces would be tipped in favour of sodium efflux, and the exchanger would operate in the reverse mode, bringing calcium into the cell and moving sodium out (Gmaj and Murer 1988; Snowdowne and Borle 1985).

Calculations based on directly measured values of basolateral membrane voltage and intracellular sodium and calcium ion activities obtained in isolated *Necturus* tubules, indicate that the net driving force for calcium entry exceeded that for sodium entry under control conditions as well as under a variety of conditions in which the sodium electrochemical gradient was reduced experimentally (see Fig. 12 in Yang, Lee, and Windhager 1988).[3] These calculations imply that the increase in cytosolic calcium ion activity actually observed in response to a reduction in the sodium gradient in *Necturus* proximal tubules is caused by stimulation of calcium influx, rather than by inhibition of calcium efflux, via the exchanger. The existence of a net inward driving force for calcium influx even under control conditions suggests that the reverse mode of the exchanger could be the normal and physiological mode of operation of the carrier—as proposed by Snowdowne and Borle (1985). Thermodynamic considerations predict that the exchanger could function in the forward mode only if intracellular sodium were lower, or basolateral membrane voltage higher, than recorded in isolated tubules perfused *in vitro*. It is, however, uncertain whether the values recorded in isolated tubules accurately reflect the situation that obtains *in vivo* (see Frindt *et al.* 1988) and further measurements in intact preparations will be required to clarify this issue.

If the basolateral Na–Ca exchange system normally operates in the calcium

[3] In isolated tubules from *Necturus* kidneys under control conditions (using measured values for basolateral membrane voltage = −49 mV; intra- and extracellular sodium ion activity = 12.8 and 74.6 mM, respectively; and intra- and extracellular calcium ion activity = 82 nM and 0.612 mM, respectively), the electrochemical potential gradient for sodium entry across the basolateral cell membrane is calculated to be 27 kJ/mol, and that for calcium 37 kJ/mol, assuming a stoichiometry of $3Na^{+}–Ca^{2+}$ (calculated from Yang *et al.* 1988).

influx (rather than efflux) mode, it follows that the exchanger could not mediate transepithelial active calcium absorption, as has been postulated by some investigators (see above). The inhibition of active calcium re-absorption observed when mammalian proximal renal tubules are exposed to ouabain or to low peritubular sodium (Ullrich, Rumrich, and Kloss 1976) might, however, be explained if increased cytosolic calcium down-regulates apical membrane permeability to calcium as well as to sodium, as proposed by Scoble, Mills, and Hruska (1985). The notion that cytosolic free calcium ions may exert parallel effects on apical membrane sodium and calcium permeability is appealing since changes in cytosolic calcium, mediated by alterations in Na–Ca exchange, might then account for the parallelism between sodium and calcium re-absorption that has been documented in proximal renal tubules (see Ullrich, Rumrich, and Kloss 1976).

13.9 Conclusions

Taking both kinetic and thermodynamic considerations into account, the currently available data point to the conclusion that the basolateral Na–Ca exchange system in sodium-transporting epithelial cells may operate predominantly in the reverse (calcium influx), rather than 'forward' (calcium efflux), mode. According to this view, the basolateral CaATPase would be primarily responsible for the maintenance of low intracellular free calcium levels, and the Na–Ca exchanger would function to modulate the level of ionized calcium, predominantly by mediating the entry of calcium into the epithelial cell cytosol. When operating in the calcium influx mode, the basolateral Na–Ca exchange carrier would be expected to be highly sensitive to changes in intracellular sodium ion activity—as well as basolateral membrane potential—resulting from alterations in the rate of apical entry and/or basolateral extrusion of sodium. As a calcium entry mechanism that is sensitive to cell sodium, the Na–Ca exchange system would seem well-suited to participate as a 'governor' in the negative-feedback control of apical membrane sodium permeability (Fig. 13.5). If calcium influx via the exchange carrier is activated by increases in both intracellular sodium and calcium (as appears to be the case in giant excitable cells), the feedback system might be expected to exhibit a threshold (cf. Jensen, Fisher, and Spring 1984; Wills and Lewis 1980) beyond which it would have a high gain; it could thus operate as an effective fail-safe mechanism serving to stabilize cell ion content and volume in the face of acute changes in the rate of sodium entry or exit. The relative importance of cytosolic calcium ions (versus, e.g., hydrogen ions (Garty, Asher, and Yeger 1987; Harvey and Ehrenfeld 1988) in co-ordinating the events that occur at the apical and basolateral membranes of sodium-transporting epithelial cells is unknown. Moreover, the

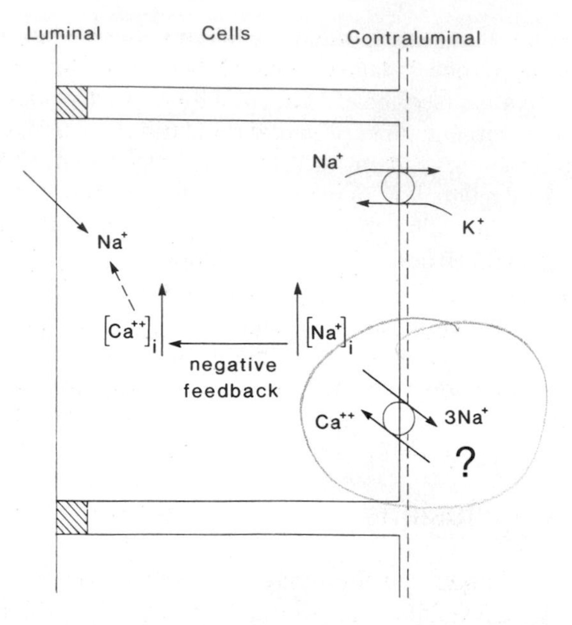

Fig. 13.5. Revised model for calcium-mediated negative feedback control of apical membrane sodium permeability in sodium-absorbing epithelial cells, incorporating concept of basolateral Na–Ca exchanger as calcium influx mechanism (see text).

extent to which such intrinsic feedback mechanisms operate under normal conditions, permitting controlled alterations in transepithelial sodium transport in response to physiological changes in the rate of sodium entry or exit, remains to be fully evaluated.

Acknowledgements

The author's investigations have been supported by funds from the United States Public Health Service (Grants AM–19022 and AM–28409) and from the Medical Research Council of Great Britain.

References

Allen, T. J. A. and Baker, P. F. (1985). Intracellular Ca indicator Quin-2 inhibits Ca^{2+} inflow via Na_i/Ca_o exchange in squid axon. *Nature* **315**, 755–6.

Arruda, J. A. L. and Sabatini S. (1980). Effect of quinidine on Na, H^+, and water transport by the turtle and toad bladders. *J. Membr. Biol.* **55**, 141–7.

Baker, P. F. (1978). The regulation of intracellular calcium in giant axons of *Loligo* and *Myxicola*. *Ann. NY Acad. Sci.* **307**, 250–68.

Baker, P. F. (1986). The sodium–calcium exchange system. In: *Calcium and the Cell* (Ciba Foundation Symposium 122) pp. 73–86, Wiley, Chichester.
Baker, P. F. and DiPolo, R. (1984). Axonal calcium and magnesium homeostasis. *Curr. Top. Membr. Trans.* **22**, 195–247.
Bentley, P. J. (1960). The effects of vasopressin on the short-circuit current across the wall of the isolated bladder of the toad, *Bufo marinus. J. Endocrinol.* **21**, 161–70.
Biber, T. U. L. (1971). Effect of changes in transepithelial transport on the uptake of sodium across the outer surface of the frog skin. *J. Gen. Physiol.* **53**, 131–44.
Blaustein, M. P. (1974). The interrelationship between sodium and calcium fluxes across cell membranes. *Rev. Physiol. Biochem. Pharmacol.* **70**, 33–82.
Bonventre, J. V. and Cheung, J. Y. (1986). Cytosolic free calcium concentration in cultured renal epithelial cells. *Am. J. Physiol.* **250**, F329–38.
Borle, A. B. (1982). Effect of sodium on cellular calcium transport in rat kidney. *J. Membr. Biol.* **66**, 183–91.
Caroni, P. and Carafoli, E. (1981). The Ca^{2+}–pumping ATPase of heart sarcolemma. *J. Biol. Chem.* **256**, 3263–70.
Caroni, P. and Carofoli, E. (1983). The regulation of the Na–Ca exchanger of heart sarcolemma. *Eur. J. Biochem.* **132**, 451–60.
Caroni, P., Reinlib, L., and Carafoli, E. (1980). Charge movements during the Na^{+}–Ca^{2+} exchange in heart sarcolemma vesicles. *Proc. Natl. Acad. Sci. USA* **77**, 6354–8.
Chaillet, J. R., Lopes, A. G., and Boron, W. F. (1985). Basolateral Na–H exchange in the rabbit cortical collecting tubule. *J. Gen. Physiol.* **86**, 795–812.
Chase, H. S. (1984). Does calcium couple the apical and basolateral membrane permeabilities in epithelia? *Am. J. Physiol.* **247**, F869–76.
Chase, H. S. and Al-Awqati, Q. (1981). Regulation of the sodium permeability of the luminal border of toad bladder by intracellular sodium and calcium. *J. Gen. Physiol.* **77**, 693–712.
Chase, H. S. and Al-Awqati, Q. (1983). Calcium reduces the sodium permeability of luminal membrane vesicles from toad bladder. Studies using a fast reaction apparatus. *J. Gen. Physiol.* **81**, 643–66.
Crutch, B. and Taylor, A. (1983). Measurement of cytosolic free Ca^{2+} concentration in epithelial cells of toad urinary bladder. *J. Physiol.* **345**, 109–27.
Davis, C. W. and Finn, A. L. (1987). Interactions of sodium transport, cell volume, and calcium in frog urinary bladder. *J. Gen. Physiol.* **89**, 687–702.
DiPolo, R. and Beaugé, L. (1983). The calcium pump and sodium–calcium exchange in squid axons. *Ann. Rev. Physiol.* **45**, 313–24.
DiPolo, R. and Beaugé, L. (1986). Reverse Na–Ca exchange requires internal Ca and/or ATP in squid axons. *Biochim. Biophys. Acta* **854**, 298–306.
DiPolo, R. and Beaugé, L. (1987). Characterization of the reverse Na/Ca exchange in squid axons and its modulation by Ca_i and ATP. *J. Gen. Physiol.* **90**, 505–25.
Eaton, D. C. (1981). Intracellular sodium ion activity and sodium transport in rabbit urinary bladder. *J. Physiol.* **316**, 527–44.
Erlij, D. and Smith, M. W. (1973). Sodium uptake by frog skin and its modification by inhibitors of transepithelial sodium transport. *J. Physiol.* **228**, 221–39.
Erlij, D., Gersten, L., Sterba, G., and Schoen, H. F. (1986). Role of prostaglandin release in the response of tight epithelia to Ca^{2+} ionophores. *Am. J. Physiol.* **250**, C629–36.

Essig, A. and Leaf, A. (1963). The role of potassium in active transport of sodium by the toad bladder. *J. Gen. Physiol.* **46**, 505–15.

Favus, M. J. (1985). Factors that influence absorption and secretion of calcium in the small intestine and colon. *Am. J. Physiol.* **248**, G147–57.

Finn, A. L. (1975). Action of ouabain on sodium transport in toad urinary bladder. Evidence for two pathways for sodium entry. *J. Gen. Physiol.* **65**, 503–14.

Friedman, P. A., Figueiredo, J. S., Maack, T., and Windhager, E. E. (1981). Sodium–calcium interactions in the renal proximal convoluted tubule of the rabbit. *Am. J. Physiol.* **240**, F558–68.

Frindt, G., Lee, C. O., Yang, J. M., and Windhager, E. E. (1988). Potential role of cytoplasmic calcium ions in the regulation of sodium transport in renal tubules. *Min. Electrol. Metab.* **14**, 40–7.

Frindt, G. and Windhager, E. E. (1983). Effect of quinidine, low peritubular [Na] or [Ca] on Na transport in isolated perfused cortical collecting tubules. *Fed. Proc.* **42**, 305.

Frindt, G. and Windhager, E. E. (1986). Effect of ionomycin on Na fluxes in perfused cortical collecting tubules of rabbit kidneys. *Fed. Proc.* **45**, 540.

Frindt. G., Windhager, E. E., and Taylor, A. (1982). Hydrosmotic response of collecting tubules to ADH or cAMP at reduced peritubular sodium. *Am. J. Physiol.* **243**, F503–13.

Fuchs, W., Larson, E. H., and Lindemann, B. (1977). Current voltage curve of sodium channels and concentration dependence of sodium permeability in frog skin. *J. Physiol.* **267**, 137–66.

Garty, H. and Asher, C. (1985). Ca^{2+}–dependent, temperature-sensitive regulation of Na^{+} channels in tight epithelia. *J. Biol. Chem.* **260**, 8330–35.

Garty, H., Asher, C., and Yeger, O. (1987). Direct inhibition of epithelial Na^{+} channels by a pH-dependent interaction with calcium and by other divalent ions. *J. Membr. Biol.* **95**, 151–62.

Garty, H. and Lindemann, B. (1984). Feedback inhibition of sodium uptake in K^{+}-depolarized toad urinary bladders. *Biochim. Biophys. Acta* **771**, 89–98.

Ghijsen, W. E. J. M., DeJong, M. D., and Van Os, C. H. (1983). Kinetic properties of Na^{+}/Ca^{2+} exchange in basolateral membranes of rat small intestine. *Biochim. Biophys. Acta* **730**, 85–94.

Gmaj, P. and Murer, H. (1988). Calcium transport mechanisms in epithelial cell membranes. *Min. Electrol. Metab.* **14**, 22–30.

Gmaj, P., Murer, H., and Kinne, R. (1979). Calcium ion transport across plasma membranes isolated from rat kidney cortex. *Biochem. J.* **178**, 549–57.

Gmaj, P., Zurini, M., Murer, H., and Carafoli, E. (1983). A high affinity calmodulin Ca pump in the baso-lateral plasma membranes of kidney cortex. *Eur. J. Biochem.* **136**, 71–76.

Grinstein, S. and Erlij, D. (1978) Intracellular calcium and the regulation of sodium transport in the frog skin. *Proc. Roy Soc. Lond. B* **202**, 353–60.

Hanai, H., Ishida, M., Liang, T., and Sacktor, B. (1986). Parathyroid hormone increases sodium/calcium exchange activity in renal cells and the blunting of the response in aging. *J. Biol. Chem.* **261**, 5419–25.

Harvey, B. J. and Ehrenfeld, J. Role of Na^{+}/H^{+} exchange in the control of intracellular pH and cell membrane conductances in frog skin epithelium. *J. Gen. Physiol.* (in press).

Jacobs, W. R. and Mandel, L. J. (1987). Fluorescent measurements of intracellular free calcium in isolated toad urinary bladder epithelial cells. *J. Membr. Biol.* **97**, 53–62.

Jayakumar, A., Cheng, L., Liang, C. T., and Sacktor, B. (1984). Sodium gradient-dependent calcium uptake in renal basolateral membrane vesicles. *J. Biol. Chem.* **259**, 10827–33.

Jensen, P. K., Fisher, R. S., and Spring, K. R. (1984). Feedback inhibition of the NaCl entry in Necturus gallbladder epithelial cells. *J. Membr. Biol.* **82**, 95–104.

Koefoed-Johnsen, V. and Ussing, H. H. (1958). The nature of the frog skin potential. *Acta Physiol. Scand.* **42**, 298–308.

Lee, C. O., Taylor A., and Windhager, E. E. (1980). Cytosolic calcium ion activity in epithelial cells of *Necturus* kidney. *Nature* **287**, 859–61.

Lewis, S. A., Eaton, D. C., and Diamond, J. M. (1976). The mechanism of Na^+ transport by rabbit urinary bladder. *J. Membr. Biol.*, **28**, 41–70.

Lewis, S. A. and Wills, N. K. (1981). Interaction between apical and basolateral membranes during sodium transport across tight epithelia. In: *Ion Transport by epithelia* (ed. S. G. Schultz) pp. 93–107. Raven Press, New York.

Lorenzen, M., Lee, C. O., and Windhager, E. E. (1984). Cytosolic Ca^{2+} and Na^+ activities in perfused proximal tubules of *Necturus* kidney. *Am. J. Physiol.* **247**, F93–102.

Ludens, J. H. (1978). Studies on the inhibition of Na^+ transport in toad bladder by the ionophore A-23187. *J. Pharmacol. Exp. Ther.* **206**, 414–22.

Mandel, L. J. and Curran, P. F. (1973). Response of the frog skin to steady state voltage-clamping. II. The active pathway. *J. Gen. Physiol.* **62**, 1–24.

Mandel, L. J. and Murphy, E. (1984). Regulation of cytosolic free calcium in rabbit proximal renal tubules. *J. Biol. Chem.* **259**, 11188–96.

Martin, D. L. and DeLuca, H. F. (1969). Influence of sodium on calcium transport by the rat small intestine. *Am. J. Physiol.* **216**, 1351–9.

Morel, F., Taniguchi, S., and Marchetti, J. (1988). Cell permeability to calcium measured with fura 2 in single rat cortical collecting tubules (CCT). *Kidney Int.* **33**, 166–83.

Nellans, H. N. and Goldsmith, R. C. (1981). Transepithelial calcium transport by rat cecum: high-efficiency absorptive site. *Am. J. Physiol.* **240**, G424–31.

Palmer, L. G. (1985). Modulation of apical Na permeability of the toad urinary bladder by intracellular Na, Ca and H. *J. Membr. Biol.* **83**, 57–69.

Palmer, L. G. and Frindt, G. (1987). Effects of cell Ca and pH on Na channels from rat cortical collecting tubule. *Am. J. Physiol.* **253**, F333–9.

Philipson, K. D. (1985). Sodium–calcium exchange in plasma membrane vesicles. *Ann. Rev. Physiol.* **47**, 561–71.

Rasgado-Flores, H. and Blaustein, M. P. (1987). Na/Ca exchange in barnacle muscle has a stoichiometry of $3Na^+/1Ca^{2+}$. *Am. J. Physiol.* **252**, C499–C504.

Reeves, J. P. and Hale, C. C. (1984). The stoichiometry of the cardiac sodium–calcium exchange system. *J. Biol. Chem.* **259**, 7733–9.

Reeves, J. P. and Sutko, J. L. (1983). Competitive interactions of sodium and calcium with the sodium–calcium exchange system of cardiac sarcolemmal vesicles. *J. Biol. Chem.* **258**, 3178–82.

Schultz, S. G. (1981). Homocellular regulatory mechanisms in sodium-transporting epithelia: avoidance of extinction by 'flush-through'. *Am. J. Physiol.* **241**, F579–90.

Scoble, J. E., Mills, S., and Hruska K. A. (1985). Calcium transport in canine renal basolateral membrane vesicles. *J. Clin. Invest.* **75**, 1096–105.

Snowdowne, K. W. and Borle, A. B. (1984). Measurement of cytosolic free calcium in mammalian cells with aequorin. *Am. J. Physiol.* **247**, C396–C408.

Snowdowne, K. W. and Borle, A. B. (1985). Effects of low extracellular sodium on cytosolic ionized calcium. Na^+–Ca^{2+} exchange as a major calcium influx pathway in kidney cells. *J. Biol. Chem.* **260**, 14998–5007.

Spring, K. R. and Giebisch, G. (1977). Kinetics of Na^+ transport in Necturus proximal tubule. *J. Gen. Physiol.* **70**, 307–28.

Talor, Z. and Arruda, J. A. L. (1985). Partial purification and reconstitution of renal basolateral Na^+–Ca^{2+} exchanger into liposomes. *J. Biol. Chem.* **260**, 15473–6.

Taylor A. (1975). Effect of quinidine on the action of vasopressin. *Fed. Proc.* **34**, 285.

Taylor, A. (1981). Role of cytosolic calcium and sodium–calcium exchange in regulation of transepithelial sodium and water absorption. In: *Ion transport by epithelia* (ed. S. G. Schultz) pp. 233–59. Raven Press, New York.

Taylor, A. and Eich, E. (1978). Evidence for Na–Ca exchange in toad urinary bladder. *Proc. VIIth Internl. Congress of Nephrology, Montreal.* C–8.

Taylor, A., Eich, E., Pearl, M., Brem, A. S., and Peeper, E. Q. (1987). Cytosolic calcium and the action of vasopressin in toad urinary bladder. *Am. J. Physiol.* **252**, F1028–41.

Taylor, A. and Palmer, L. G. (1982). Hormonal regulation of sodium chloride and water transport in epithelia. In: *Biological regulation and development*, Vol. III (eds. R. F. Goldberger, P. Berg, R. T. Schimke, K. Moldave, P. Leder, L. E. Hood, and K. Yamamoto) pp. 252–99. Plenum Press, New York.

Taylor, A., Pearl, M., and Crutch, B. (1985). Role of cytosolic free calcium and its measurement in a vasopressin-sensitive epithelium. *Mol. Physiol.* **8** 43–58.

Taylor, A. and Windhager, E. E. (1979). Possible role of cytosolic calcium and Na-Ca exchange in regulation of transepithelial sodium transport. *Am. J. Physiol.* **236**, F505–12.

Taylor, A. and Windhager, E. E. (1985). Cytosolic calcium and its role in the regulation of transepithelial ion and water transport. In: *Physiology and pathophysiology of electrolyte metabolism* (eds. D. Seldin and G. Giebisch) pp. 1297–379. Raven Press, New York.

Turnheim, K., Frizzell, R. A., and Schultz, S. G. (1978). Interaction between cell sodium and the amiloride-sensitive sodium entry step in rabbit colon. *J. Membr. Biol.* **39**, 233–56.

Ullrich, K. J., Rumrich, G., and Kloss, S. (1976). Active Ca^{2+} reabsorption in the proximal tubule of rat kidney. *Pflügers Arch.* **364**, 223–8.

Van Heeswijk, M. P. E., Geertsen, J. A. M., and Van Os, C. H. (1984). Kinetic properties of the ATP-dependent Ca^{2+} pump and the Na^+/Ca^{2+} exchange system in basolateral membranes from rat kidney cortex. *J. Membr. Biol.* **79**, 19–31.

Van Os. C. H. (1987). Transcellular calcium transport in intestinal and renal epithelial cells. *Biochim. Biophys. Acta* **906**, 195–222.

Wiesmann, W., Sinha, S., and Klahr, S. (1977). Effects of ionophore A23187 on baseline and vasopressin-stimulated sodium transport in the toad bladder. *J. Clin. Invest.* **59**, 418–25.

Wills, N. K. and Lewis, S. A. (1980). Intracellular Na^+ activity as a function of Na^+ transport rate across a tight epithelium. *Biophys. J.* **30**, 181–6.
Windhager, E., Frindt, G., Yang J. M., and Lee, C. O. (1986). Intracellular calcium ions as regulators of renal tubular sodium transport. *Klin. Wochenschr.* **64**, 847–52.
Windhager, E. E. and Taylor, A. (1983). Regulatory role of intracellular calcium ions in epithelial Na transport. *Ann. Rev. Physiol.* **45**, 519–32.
Wong, S. M. E. and Chase, H. S. (1986). Role of intracellular calcium in intracellular volume regulation. *Am. J. Physiol.* **250**, C841–52.
Yang, J. M., Lee, C. O., and Windhager, E. E. (1988). Regulation of cytosolic free calcium in isolated perfused proximal tubules of *Necturus*. *Am. J. Physiol.* **255**, F787–99.

Index

Ions are represented by chemical symbols, e.g. K^+, Na^+. The alphabetical arrangement is letter-by-letter.